AIR POLLUTION RESEARCH REPORT 36

MONITORING OF GASEOUS POLLUTANTS
BY TUNABLE DIODE LASERS

The Symposium was organized in collaboration with the Commission of the European Communities, DG XII/E, Environment Programme.

Commission of the European Communities

MONITORING OF GASEOUS POLLUTANTS BY TUNABLE DIODE LASERS

Proceedings of the International Symposium
held in Freiburg, Germany, 17–18 October 1991

organized by the Fraunhofer Institut
für Physikalische Messtechnik (Freiburg)
under the auspices of
The German Federal Minister of Research and Development
The Commission of the European Communities

Edited by

R. GRISAR, H. BÖTTNER, M. TACKE

Fraunhofer Institut für Physikalische Messtechnik, Freiburg, Germany

and

G. RESTELLI

*Commission of the European Communities,
Joint Research Centre - Environment Institute, Ispra, Italy*

KLUWER ACADEMIC PUBLISHERS
DORDRECHT / BOSTON / LONDON

Library of Congress Cataloging-in-Publication Data

Monitoring of gaseous pollutants by tunable diode lasers : proceedings
  of the international symposium held in Freiburg, Germany, 17-18
  October 1991 / organized by the Fraunhofer Institut für
  Physikalische Messtechnik (Freiburg) under the auspices of the
  German Federal Minister of Research and Development [and] The
  Commission of the European Communities ; edited by R. Grisar ... [et
  al.].
       p.   cm.
  Papers presented at the third International Symposium on
Monitoring of Gaseous Pollutants by Tunable Diode Lasers.
  Includes index.
  ISBN 0-7923-1826-9
  1. Combustion gases--Environmental aspects--Analysis--Congresses.
2. Semiconductor lasers--Congresses. 3. Laser spectroscopy-
-Congresses.   I. Grisar, R. II. Fraunhofer Institut für
Physikalische Messtechnik. III. International Symposium on
"Monitoring of Gaseous Pollutants by Tunable Diode Lasers" (3rd :
1991 : Freiburg im Breisgau, Germany)
  TD885.5.C66M66 1992
  628.5'3'0287--dc20   92-17520

ISBN 0-7923-1826-9

Publication arrangements by
Commission of the European Communities
Directorate-General Telecommunications, Information Industries and Innovation,
Dissemination Scientific and Technical Knowledge Unit, Luxembourg

EUR 14182 EN
© 1992 ECSC, EEC, EAEC, Brussels and Luxembourg

LEGAL NOTICE
Neither the Commission of the European Communities nor any person acting on behalf of the
Commission is responsible for the use which might be made of the following information.

Published by Kluwer Academic Publishers,
P.O. Box 17, 3300 AA Dordrecht, The Netherlands.

Kluwer Academic Publishers incorporates the publishing programmes of
D. Reidel, Martinus Nijhoff, Dr W. Junk and MTP Press.

Sold and distributed in the U.S.A. and Canada
by Kluwer Academic Publishers,
101 Philip Drive, Norwell, MA 02061, U.S.A.

In all other countries, sold and distributed
by Kluwer Academic Publishers Group,
P.O. Box 322, 3300 AH Dordrecht, The Netherlands.

Printed on acid-free paper

CONTENTS

INTRODUCTION   xiii

SECTION I: ATMOSPHERIC TRACE GAS DETECTION

High precision tunable diode laser absorption spectroscopy: application for
measuring long-lived atmospheric gases   3
    A. FRIED, B. HENRY, J. FOX
    The National Center for Atmospheric Research, Boulder, Colorado,
    U.S.A.
    JAMES R. DRUMMOND
    University of Toronto, Department of Physics, Toronto, Ontario,
    Canada.
    R. SAMS
    The National Institute of Standards and Technology, Gaithersburg,
    Maryland, U.S.A.

Formation and emission of atmospheric greenhouse gases in the combustion of
fossil fuels: monitoring by tunable diode laser spectroscopy   13
    K.H. BECKER, K.J. BROCKMANN, R. HÜDEPOHL,
    R. KURTENBACH, P. WIESEN
    Physikalische Chemie/Fachbereich 9, Bergische Universität -
    Gesamthochschule Wuppertal, Wuppertal, FRG.

Measurements of H_2O_2 in the Grand Canyon by tunable diode laser spectrometry during the Navajo Generating Station visibility study   21
    H.I. SCHIFF, G.I. MACKAY, D.R. KARECKI, J.T. PISANO,
    S.D. NADLER
    Unisearch Associates Inc., Concord, Ontario, Canada.

Measurement of atmospheric species by mid-infrared and near-infrared
tunable diode laser absorption   31
    A.C. STANTON, D.S. BOMSE, J.A. SILVER, D.C. HOVDE, D.B. OH
    Southwest Sciences, Inc., Santa Fe, New Mexico, U.S.A.

Pollution gas detection in free atmosphere beyond atmospheric windows 41
    B. SUMPF, S. DIETRICH, KA. HERRMANN, F. KÜHNEMANN,
    J. ORPHAL, V.V. PUSTOGOV
    Abteilung für Molekülphysik/Photobiophysik, Humboldt-Universität zu
    Berlin, Berlin, FRG.

A Portable bistatic monitor for atmospheric methane density with up to 240m
optical path   51
    RYUJI KOGA
    Department of Electric and Electronic Engineering, Faculty of
    Engineering, Okayama University, Tsushima, Okayama, Japan.

SECTION II: DIODE LASERS AND THEIR PROPERTIES

Embossed grating DFB-BH lead chalcogenide diode lasers   63
    A. FACH, H. BÖTTNER, M. TACKE
    Fraunhofer-Institut für Physikalische Messtechnik, Freiburg, FRG.

Epitaxial growth of laterally structured lead chalcogenide lasers   69
    A. LAMBRECHT, A. FACH, R. KURBEL, B. HALFORD,
    H. BÖTTNER, M. TACKE
    Fraunhofer-Institut für Physikalische Messtechnik, Freiburg, FRG.

Semiconductor lasers and photodiodes for gas analysis in the spectral range
1.8-2.5 μm   79
    A.N. BARANOV, A.N. IMENKOV, M.P. MIKHAILOVA,
    YU.P. YAKOVLEV
    A.F. Ioffe Physico-Technical Institute, St. Petersburg, Russia.

Linewidth and noise of lead chalcogenide diode lasers   85
    G. SPILKER, R. DADDATO, U. SCHIESSL, A. LAMBRECHT,
    M. TACKE
    Fraunhofer-Institut für Physikalische Messtechnik, Freiburg, FRG.

Measurements of IV-VI diode laser near- and farfield distributions   93
    M. AGNE, U. SCHIESSL, A. LAMBRECHT, M. TACKE
    Fraunhofer-Institut für Physikalische Messtechnik, Freiburg, FRG.

PROGRESS IN IV-VI PHYSICS AND OPTOELECTRONIC DEVICES
(Proceedings of the Freiburg IV-VI Colloquium: H. Böttner, A. Lambrecht, M. Tacke Editors)

Recent developments in MBE Grown $Pb_{1-x}Eu_xSe_yTe_{1-y}/Pb_{1-v}Sn_vTe$ diode
lasers for high resolution spectroscopy   105
    Z. FEIT, D. KOSTYK, R.J. WOODS, P. MAK
    Laser Photonics Analytics Division, Bedford, Ma, USA.

Contributions to the theory of Auger recombination in lead chalcogenides   111
    M. MOCKER, F. LEMKE
    Department of Physics, Humboldt-University Berlin, Berlin, FRG.

Thermodynamics of quasiternary systems containing IV-VI-semiconductors   115
    V. LEUTE, D. MENGE
    Institute of Physical Chemistry, University of Münster, Münster, FRG.

Development of lead-chalcogenide tunable diode lasers for 3 to 4 μm spectral
region at Lebedev Physical Institute   125
    A.P. SHOTOV
    Lebedev Physical Institute, Moscow, USSR

Optical properties of high energy gap lead salts   139
    K.H. HERRMANN, K.-P. MÖLLMANN, J.W. TOMM
    Humbold-Universität zu Berlin, Institut für Festkörperphysik, FB
    Physik, Berlin, FRG.
    H. BÖTTNER, A. LAMBRECHT, M. TACKE
    Fraunhofer-Institut für Physikalische Messtechnik, Freiburg, FRG.

Picosecond infrared spectroscopy of lead chalcogenide semiconductors   140
    R. KLANN, R. BUHLEIER, T. ELSAESSER
    Technische Universität München, Physik Department E11, Garching,
    FRG.
    A. LAMBRECHT
    Fraunhofer-Institut für Physikalische Messtechnik, Freiburg, FRG.

Optical Nonlinearities of free carriers in PbSe   141
    K. LEIDIG
    Universität Würzburg, Physikalisches Institut, Abt. EP IV/1, Würzburg,
    FRG.

Limitations of compound segregation in the vapour crystal growth of solid
solution crystals   142
    A. SZCZERBAKOW
    Vigo Ltd., Warszawa, Poland.

PbSnMnTe semimagnetic semiconductors: carrier concentration controlled
magnetic properties   143
    T. STORY
    Polish Academy of Sciences, Institute of Physics, Warsaw, Poland.

Structural and electrical properties of PbSe grown by the vertical Bridgman
method   144
    R. POST, M. MÜLBERG
    Humboldt Universität zu Berlin, Institut für Kristallographie und
    Materialforschung, Berlin, FRG.

Magnetically tuned $Pb_{1-x}Mn_xSe$ and $Pb_{1-x-y}Mn_xSn_ySe$ diode lasers   145
    L. KOWALCZYK
    Polish Academy of Sciences, Institute of Physics, Warsaw, Poland.

Se-diffused $PbS_{1-x}Se_x$-homolasers in the temperature range from 5 to 100 K   146
    D.A. IWASCHKINA, D. SICHE
    Humboldt-Universität zu Berlin, FB Physik, Abteilung Molekül- und
    Photobiophysik, Berlin, FRG.

Photovoltaic infrared sensor arrays in monolithic lead chalcogenides on silicon   147
    H. ZOGG, C. MAISSEN, J. MASEK, T. HOSHINO, S. BLUNIER
    Arbeitsgemeinschaft für Industrielle Forschung (AFIF), Swiss Federal
    Institute of Technology, Zürich, Switzerland.

Thermal mismatch strain relaxation of epitaxial IV-VI layers on Si(111)
substrates at cryogenic temperatures   148
    C. MAISSEN, A. SULTAN, S. TEODOROPOL, H. ZOGG
    Arbeitsgemeinschaft für Industrielle Forschung (AFIF), Swiss Federal
    Institute of Technology, Zürich, Switzerland.

MBE growth and characterization of PbSe layers on Si(111) using (Ba, Ca)F$_2$ as
buffer layers   149
    F. NGUYEN-VAN-DAU, D.G. CRÉTÉ, V. MATHET
    Thomson CSF/LCR, Laboratoire Central de Recherches, Orsay Cedex,
    France.

RHEED investigations of PbTe and Pb$_{1-x}$Eu$_x$Te surfaces during MBE growth   150
    G. SPRINGHOLZ, G. BAUER
    Johannes-Kepler-Universität Linz, Institut für Halbleiterphysik, Linz,
    Austria.

Scanning tunneling microscopy analysis of single crystal PbSe-surfaces and
epitaxial layers on BaF$_2$ 151
    W. STOCKER, S.N. MAGONOV, H.-I. CANTOW
    Freiburger Material-Forschungszentrum (FMF), Freiburg, FRG.
    H. BÖTTNER, S. SCHELB, M. TACKE
    Fraunhofer-Institut für Physikalische Messtechnik, Freiburg, FRG.

SECTION III: TECHNIQUES, SYSTEMS, COMPONENTS

State of the art of TDLS in the USSR   155
    A.I. NADEZHDINSKII
    General Physics Institute Ac. Sc. USSR, Moscow, USSR

Development of a prototype IR-FM Absorption Spectrometer: design criteria
and system performance   169
    P. WERLE, R. MÜCKE and F. SLEMR
    Fraunhofer-Institut für Atmosphärische Umweltforschung, IFU,
    Garmisch-Partenkirchen, FRG.

A TTFM spectrometer for detection of transient radical species: 2 ν_1 overtone
absorption lines of HO$_2$ at 1.5 μm   183
    T.J. JOHNSON, F.G. WIENHOLD, J.P. BURROWS, G.W. HARRIS
    Max Planck Institute for Chemistry, Atmospheric Chemistry Depart-
    ment, Mainz, FRG.
    H. BURKHARD
    DBP Telekom Research Institute, Darmstadt, FRG.

Line narrowing and frequency control of lead-salt diode lasers by optical
feedback   191
    M. MÜRTZ, M. SCHAEFER, M. SCHNEIDER, J.S. WELLS,
    W. URBAN
    Institut für Angewandte Physik der Universität Bonn, FRG.
    U. SCHIESSL, M. TACKE
    Fraunhofer-Institut für Physikalische Messtechnik, Freiburg, FRG.

Fiber optic accessories for molecular spectroscopy and gas analysis with tunable diode lasers in the middle infrared   203
E.V. STEPANOV, A.I. KUZNETSOV, K.L. MOSKALENKO,
A.I. NADEZHDINSKII
Institute of General Physics of the Academy of Science of the USSR, Moscow, USSR.

New developments in computer-controlled diode laser spectroscopy   217
M. PETRI, T. FINK, W. URBAN
Institut für Angewandte Physik der Universität Bonn, Bonn, FRG.

Measurement of pressure-effects with a stabilized ir-diode laser spectrometer   231
N. ANSELM, TH. GIESEN, M. HARTER, R. SCHIEDER,
G. WINNEWISSER
I. Physikalisches Institut, Universität zu Köln, Köln, FRG.

Pressure broadening parameter of NOx using TDLAS   241
V.V. PUSTOGOV, Ka. HERRMANN, F. KÜHNEMANN, J. ORPHAL,
B. SUMPF
Abteilung für Molekülphysik/Photobiophysik der Humboldt-Universität zu Berlin, Fachbereich Physik, Berlin.
J. BERGER, H.-D. KRONFELDT
Optisches Institut der Technischen Universität Berlin, Berlin, FRG.

The ν_3 band of SO_2-line parameters for atmospheric monitoring   249
F. KÜHNEMANN, Y. HEINER, B. SUMPF and Ka. HERRMANN
Abt. Molekülphysik/Photobiophysik, Fachbereich Physik der Humboldt-Universität zu Berlin, Berlin, FRG.
N.V. LEMECHOV, E.V. STEPANOV
Institute of General Physics, U.S.S.R. Academy of Sciences, Moscow, USSR.

Pure absorption spectroscopy of molecular oxygen using a cw AlGaAs Laser   257
M. DE ANGELIS
Istituto di Cibernetica, CNR, Arco Felice, Napoli, Italy.
F. MARIN
Scuola Normale Superiore, Pisa, Italy.
F.S. PAVONE
European Laboratory for Non-Linear Spectroscopy (LENS), Firenze, Italy.
G.M. TINO
Dipartimento di Scienze Fisiche dell'Università di Napoli, Italy.
M. INGUSCIO
Dipartimento di Fisica dell'Università di Firenze, Italy.

Spectroscopic measurements of the $\nu_2 + 2\nu_3$ band of CH_4 with a 1.3 μm InGaAsP diode laser   265
R.A. ROOTH
N.V. KEMA, Arnhem, The Netherlands.

ix

Sensitive detection of acetylene absorption in the visible using a stabilized AlGaAs diode laser 275
F.S. PAVONE, F. MARIN, M. INGUSCIO
European Laboratory for Nonlinear Spectroscopy (LENS), Firenze, Italy.
K. ERNST
Istituto Nazionale di Ottica (INO), Firenze, Italy.
G. DI LONARDO
Dipartimento di Chimica Fisica ed Inorganica, Università di Bologna, Bologna, Italy.

Fast scanning laser DOAS, an ultrasensitive technique for monitoring tropospheric trace gases 283
W. ARMERDING, A. HERBERT, M. SPIEKERMANN, J. WALTER, F.J. COMES
Johann Wolfgang Goethe-Universität Institut für Physikalische und Theoretische Chemie Niederurseler Hang, Frankfurt, FRG.

Enclosive flow cooling: concept of a new method for simplifying complex molecular spectra 291
S. BAUERECKER, H.K. CAMMENGA
Institut für Physikalische und Theoretische Chemie der TU Braunschweig, Braunschweig, FRG.
F. TAUCHER, C. WEITKAMP, W. MICHAELIS
Institut für Physik, GKSS-Forschungszentrum Geesthacht GmbH, Geesthacht, FRG.

SECTION IV: SPECIAL APPLICATIONS

TDLS analysis of water vapour traces in semiconductor process gas 303
R. KÄSTLE, R. GRISAR, M. TACKE
Fraunhofer-Institut für Physikalische Messtechnik, Freiburg, FRG.
D. DORNISCH, C. SCHOLZ
CS GmbH, Semiconductor- und Solar Technology, Munich, FRG.

Determination of unburned fuel in the wall boundary layer of methane/air flames 311
H. EBERIUS, TH. JUST, M. OVERHAMM
Deutsche Forschungsanstalt für Luft- und Raumfahrt Institut für Physikalische Chemie der Verbrennung, Stuttgart, FRG.

Analysis of trace components in automomotive exhaust gas 319
W.J. RIEDEL, R. GRISAR, U. KLOCKE, M. KNOTHE, H. WOLF
Fraunhofer-Institut für Physikalische Messtechnik, Freiburg, FRG.
P. SCHOTTKA, E. BESSEY, N. PELZ
Daimler-Benz AG, Abt. FVA/VE, Stuttgart, FRG.

Application of diode laser spectroscopy on the measurement of boundary layer - induced temperature changes in shock tubes 325
L.K. MOSER, F.J. HINDELANG
Universität der Bundeswehr München, Institut für Strömungsmechanik, Neubiberg, FRG.

A TDL- and FT-IR study of the reaction $NO_3+HO_2 \longrightarrow OH+NO_2+O_2$ 329
J. HJORTH, F. CAPPELLANI, G. RESTELLI
Commission of the European Communities, Joint Research Centre-
Environment Institute, Ispra, Italy.

Measurements of the $^{13}C/^{12}C$ ratio in methane using a tunable diode laser
absorption spectrometer   343
M. SCHUPP, P. BERGAMASCHI, G.W. HARRIS
Max Planck Institute for Chemistry, Air Chemistry Department,
ment, Mainz, FRG.

Application of tunable diode lasers for human expiration diagnostics   353
E.V. STEPANOV, I.I. ZASAVITSKII, K.L. MOSKALENKO,
A.I. NADEZHDINSKII
Institute of General Physics of the Academy of Science of the USSR,
Moscow, USSR.

Author Index   371

INTRODUCTION

The 3rd International Symposium on Monitoring of Gaseous Pollutants by Tunable Diode Lasers, held at the Fraunhofer Institute in Freiburg on October 17th and 18th, continued a tradition, established by the two preceeding meetings held at the same location, in 1986 and 1988.

The steadily increasing number of participants and contributions emphasizes the need for such a meeting, which is unique in Europe and appreciated abroad. This third symposium in particular, was specially marked by the presence of the former Eastern Block Countries, with a number of scientists, and contributions, from the USSR (still existing at the time of the meeting) and from East Berlin.

The large number of contributions proposed, originally, as oral presentations, could not be fitted in the schedule of the two-day limit which the participants to the previous meetings had recommended. To take account of these contributions, poster sessions were held, combined with extended discussion time. This intensified personal interaction was much appreciated by the participants.

This volume contains the papers of all oral and poster contributions presented at the Symposium; also included are the four invited papers and the abstracts of 13 further papers presented to a satellite meeting held at the same location on October 16th, the Freiburg IV-VI Colloquium. The Colloquium was in fact an extension of the special session on lead chalcogenide diode laser development, of the previous (1988) Symposium. As it turned out, combining both meetings, the Colloquium and the Symposium, resulted in a highly desirable strong interdisciplinary interaction.

The volume altogether contains 36 papers and 13 abstracts, grouped into four sections:
- Atmospheric Trace Gas Detection.
- Diode Lasers and Their Properties.
- Techniques, Systems, Components.
- Special Applications.
The second section, Diode Lasers and Their Properties, includes the papers of the Freiburg Colloquium, assembled in the sub-section "Progress in IV-VI Physics and Optoelectronic Devices" edited by H. Böttner, A. Lambrecht and M. Tacke.

Atmospheric trace gas detection was again of central interest. In this area the invited contribution of Fried et al., dealing with OCS measurements in immission and in emission in the atmosphere, clearly demonstrate the impressive capabilities of the diode laser technique.

Reports on IV-VI mid-infrared diode lasers stressed those properties of these devices which are essential for the application to ultra high sensitivity pollutant monitoring, namely radiation noise and emission linewidth. Attention to these characteristics has somewhat replaced that on high-temperature operation which was the main focus of the previous meeting.

Noteworthy are two reports on III-V lasers emitting at wavelengths shorter than 3 μm and exhibiting less severe cooling requirements with the possibility of more compact set-ups. These lasers can be used to detect pollutants *via* overtone absorption bands, as discussed in papers of the Section on Techniques, Systems and Components. The Colloquium on IV-VI Physics and Optoelectronic Devices specifically highlighted physical and technological aspects of mid-infrared diode lasers; it thus addressed a different group of scientists.

Laterally structured diode lasers were discussed in detail; one of the *new* results was the cw operation at maximum temperatures of 204 K, as shown by the paper of Feit et al. This very encouraging development will prove as important for pollution detection as for example the inclusion of periodic structures for monomode emission (Fach et al.).

The comparison of traditional amplitude-modulation and the more recent frequency modulation techniques was another important topic discussed at the meeting. The competition for the best system approach is still undecided since new developments of AM spectroscopy, such as the use of increased modulation frequencies as discussed by A. Stanton, appear capable of achieving detection sensitivities comparable with those presently obtained using the FM technique, which, however, has not yet reached its theoretical limit.

There were a number of reports on special applications which point to new directions. Some of them are connected to atmospheric trace gas analysis, such as the measurement of isotope ratio discussed by Schupp et al. This technique has indeed a high potential for application in medical diagnosis. Trace impurity detection was in fact reported for such different areas as human breath and process gases.

The symposium was combined with the semiannual meeting of the EUROTRAC subproject JETDLAG (Joint European Development of Tunable Diode Laser Absorption Spectroscopy for Measurement of Atmospheric Trace Gases, coordinator: F. Slemr). Scientists and organizations which consider to use techniques developed within JETDLAG were invited to attend the meeting; this invitation was followed by a large number of participants coming from environmental monitoring organizations and from industry. This showed-up during the Concluding Discussion Session. After having touched topics of diode laser development and special applications, a central question raised was how to get the maturing technique of diode laser gas analysis into widespread application. An outcome of the discussion was that as an important first step, the inclusion into national and international standards has to be sought.

At the Symposium we experienced again a fruitful and rewarding meeting with colleagues sharing the same field of research, which is becoming more and more important in applications such as environmental monitoring; we hope that this impression holds for all attendants. We also hope that this Symposium Volume will support the participants to further evaluate the technical results presented as well as to report the results to those who were unable to attend the meeting.

THE EDITORS

The Proceedings of the First and Second International Symposium were published respectively in 1987 and 1989 in the present series and with CEC sponsorship:

Air Pollution Research Report
Monitoring of Gaseous Pollutants by Tunable Diode Lasers
R. Grisar, H. Preier, G. Schmidtke, G. Restelli Eds.
D. Reidel Publ. 1987

Air Pollution Research Report
Monitoring of Gaseous Pollutants by Tunable Diode Lasers
R. Grisar, G. Schmidtke, M. Tacke, G. Restelli Eds.
Kluwer Academic Publ. 1989.

LIST OF PARTICIPANTS

Mr. M. Agne
Fraunhofer-Institut
für Physikalische Messtechnik
Heidenhofstrasse 8
W-7800 Freiburg, Germany

Dr. F. d'Amato
ENEA CRE Frascati
C.P Box 65
I-00044 Frascati RM, Italy

Mrs. M. de Angelis
Dipartimento di Scienze Fisiche
dell'Università di Napoli
Mostra d'Oltremare, PAD. 20
I-80125 Napoli, Italy

Mr. N. Anselm
1. Physikalisches Institut
der Universität zu Köln
Zülpicher Strasse 77
W-5000 Köln 41, Germany

Mr. D.W. Arlander
ICH3, Forschungszentrum Jülich
Postfach 1913
W-5170 Jülich, Germany

Dr. W. Armerding
Institut für Physikalische
und Theoretische Chemie
J.W. Goethe-Universität
W-6000 Frankfurt/M 50, Germany

Dr. V. Baev
Institut für Experimentalphysik
Universität Hamburg
Jungiusstrasse 9
W-2000 Hamburg 36, Germany

Dr. G. Baldacchini
ENEA CRE Frascati
C.P Box 65
I-00044 Frascati RM, Italy

Mrs. A. Ballangrud
Norsk Elektro Optikk
P.O. Box 17
N-2001 Lillestrøm, Norway

Mr. S. Bauerecker
Institut für Physik
GKSS Forschungszentrum Geesthacht
Max-Planck-Strasse 1
W-2054 Geesthacht, Germany

Mr. P. Bergamaschi
Max-Planck-Institut für Chemie
Saarstrasse 23
W-6500 Mainz, Germany

Dr. P. Bergner
Dr. Ing. h.c. Porsche AG
Postfach 1140
W-7251 Weissach, Germany

Dr. D. Bicanic
Dept. Physics and Meteorology
Acgricultural University
Duivendaal 1
NL-6701 AP Wageningen, The
Netherlands

Dr. K. Bobey
Inst. für Optik und Spektroskopie
Humboldt-Universität zu Berlin
Invalidenstrasse 110
O-1040 Berlin, Germany

Dr. H. Böttner
Fraunhofer-Institut
für Physikalische Messtechnik
Heidenhofstrasse 8
W-7800 Freiburg, Germany

Mr. W. Bremser
Fachbereich Physik
Humboldt-Universität zu Berlin
Invalidenstrasse 110
O-1040 Berlin, Germany

Dr. K. Brenner
Abt. Forschung u. Entwicklung
Messer Griesheim GmbH
Wörthstrasse 170
W-4100 Duisburg, Germany

Dr. M. Buck
Landesanstalt für Immissionsschutz
Wallneyerstrasse 6
W-4300 Essen, Germany

Mr. J. Büsing
SDME 2/32
Commission of the European
Communities
200, rue de la Loi
B-1049 Bruxelles, Belgium

Dr. F. Cappellani
Environment Institute
Joint Research Centre Ispra
I-21020 Ispra (VA), Italy

Mrs. I. Carrasco
Service d'Aeronomie du CNRS
Université Paris 6, Tour 15
4 place Jussieu
F-75230 Paris Cedex 05, France

Dr. D. Courtois
Laboratoire de Physique Moleculaire
Université de Reims
B.P. 347
F-51062 Reims Cedex, France

Mr. R. Daddato
Fraunhofer-Institut
für Physikalische Messtechnik
Heidenhofstrasse 8
W-7800 Freiburg, Germany

Dr. P.B. Davies
Department of Physical Chemistry
University of Cambridge
Lensfield Road
GB-Cambridge CB2 1EP, UK

Mr. W. Diehl
Battelle-Institut
Am Römerhof 35
W-6000 Frankfurt/M 90, Germany

Dr. H. Eberius
DLR - 1442
Pfaffenwaldring 38
W-7000 Stuttgart 80, Germany

Mr. V. Ebert
Physikalisch-Chemisches Institut
Universität Heidelberg
Im Neuenheimer Feld 253
W-6900 Heidelberg, Germany

Dr. G. Ehret
Forschungszentrum
Deutsche Forschungsanstalt für Luft-
und Raumfahrt
W-8031 Oberpfaffenhofen, Germany

Mr. A. Fach
Fraunhofer-Institut
für Physikalische Messtechnik
Heidenhofstrasse 8
W-7800 Freiburg, Germany

Dr. N. Feher
Institut für Physikalische Chemie
Universität Basel
Klingelbergstrasse 80
CH-4056 Basel, Switzerland

Dr. Z. Feit
Laser Photonics Inc.
Analytics Division
25 Wiggins Avenue
Bedford MA 01730, USA

Dr. H. Fischer
Hahn-Schickard-Institut
Roggenbachstrasse 8
W-7730 VS-Villingen, Germany

Mr. H. Frahm
Max-Planck-Institut für Chemie
Saarstrasse 23
W-6500 Mainz, Germany

Dr. A. Fried
National Center for Atmospheric
Research
P.O. Box 3000
Boulder, Colorado 80307, USA

Mrs. B. Galle
Swedish Environmental Research
Institute
P.O. Box 47086
S-40258 Gothenburg, Sweden

Mr. S. Gillespie
Thornton Research Centre
Shell Research Ltd.
P.O. Box 1
GB-Chester CH1 3SH, UK

Mr. J. Grieser
Institut für Physikalische
und Theoretische Chemie
J.W. Goethe-Universität
W-6000 Frankfurt/M 50, Germany

Dr. R. Grisar
Fraunhofer-Institut
für Physikalische Messtechnik
Heidenhofstrasse 8
W-7800 Freiburg, Germany

Prof. D. Haaks
Aerolaser GmbH
Hauptstrasse 44
W-8100 Garmisch-Partenkirchen,
Germany

Dr. H.G. Hänsel
Jenoptik Carl Zeiss Jena GmbH
Carl-Zeiss-Strasse 1
O-6900 Jena, Germany

Mr. B. Halford
Fraunhofer-Institut
für Physikalische Messtechnik
Heidenhofstrasse 8
W-7800 Freiburg, Germany

Mr. K.H. Haugholt
Norsk Elektro Optikk
N-2001 Lillestrøm, Norway

Dr. G.W. Harris
Max-Planck-Institut für Chemie
Saarstrasse 23
W-6500 Mainz, Germany

Dr. R. Hartig
Sick GmbH
Sebastian-Kneipp-Strasse 2
W-7808 Waldkirch, Germany

Dr. R. Harzer
Abteilung AV
Uranit GmbH
Postfach 1411
W-5170 Jülich, Germany

Dr. M. Heise
Institut für Spektrochemie
und Angewandte Spektroskopie
Bunsen-Kirchhoff-Strasse 11
W-4600 Dortmund, Germany

Prof. P. Hering
Institut für Lasermedizin
Moorenstrasse 5
W-4000 Düsseldorf, Germany

Prof. Karin Herrmann
Fachbereich Physik
Humboldt-Universität zu Berlin
Invalidenstrasse 110
O-1040 Berlin, Germany

Prof. F.J. Hindelang
Institut für Strömungsmechanik und
Aerodynamik
Hochschule der Bundeswehr München
Werner-Heisenberg-Weg 39
W-8014 Neubiberg, Germany

Dr. J. Hoffmann
Abteilung LAF1
Kernforschungszentrum Karlsruhe
Postfach 3640
W-7500 Karlsruhe, Germany

Mr. A. Honne
Senter for Industriforskning
P.O. Box 125 Blindern
N-0314 Oslo, Norway

Mr. F. Huard
Bertin et Cie
B.P. 22000
F-13791 Aix en Provence Cedex,
France

Mr. G. Huhn
Institut für Angewandte Physik
der Universität Bonn
Wegelerstrasse 8
W-23000 Bonn, Germany

Dr. T.J. Johnson
Max-Planck-Institut für Chemie
Saarstrasse 23
W-6500 Mainz, Germany

Dr. J. Kändler
Physikalische Technologien
VDI-Technologiezentrum
Graf-Recke-Strasse 84
W-4000 Düsseldorf, Germany

Prof. R. Kaufmann
Institut für Lasermedizin
Moorenstrasse 5
W-4000 Düsseldorf, Germany

Dr. D. Klemp
Institut für Chemie 2
Forschungszentrum Jülich
Postfach 1913
W-5170 Jülich, Germany

Mr. U. Klocke
Fraunhofer-Institut
für Physikalische Messtechnik
Heidenhofstrasse 8
W-7800 Freiburg, Germany

Mr. M. Knothe
Fraunhofer-Institut
für Physikalische Messtechnik
Heidenhofstrasse 8
W-7800 Freiburg, Germany

Dr. E. Koch
Grundlagenentwicklung Optik
Draegerwerk AG
Postfach 1339
W-2400 Lübeck, Germany

Prof. R. Koga
Department of Electronics
Okayama University
J-Tsushima, Okayama 700, Japan

Dr. F. Kühnemann
Fachbereich Physik
Humboldt-Universität zu Berlin
Invalidenstrasse 110
O-1040 Berlin, Germany

Mr. R. Kurtenbach
Physikalische Chemie - FB 9
Bergische Universität-GH Wuppertal
Gaußstrasse 20
W-5600 Wuppertal 1, Germany

Dr. A. Lambrecht
Fraunhofer-Institut
für Physikalische Messtechnik
Heidenhofstrasse 8
W-7800 Freiburg, Germany

Mr. A. Linge
Servotek AB
Villa Fortuna
S-23291 Arlöv, Sweden

Mr. H. Luf
Abt. 178610
Volkswagenwerk AG
Postfach
W-3180 Wolfsburg, Germany

Mrs. H. Mac Leod
Service d'Aeronomie du CNRS
Université Paris 6, Tour 15
4 place Jussieu
F-75230 Paris Cedex 05, France

Dr. N.A. Martin
National Physical Laboratory
DQM-Bldg 95
Queens Road
GB-Teddington, Middlessex, UK

Mr. P. Martin
Institut für Physikalische Chemie
Universität Basel
Klingelbergstrasse 80
CH-4056 Basel, Switzerland

Mr. G. Moreau
LPCE/CNRS
3A Av. de la Recherche Scientifique
F-45071 Orléans Cedex 2, France

Dr. L.K. Moser
Institut für Strömungsmechanik und
Aerodynamik
Universität der Bundeswehr München
Werner-Heisenberg-Weg 39
W-8014 Neubiberg, Germany

Dr. W. Moser
ECO Physics
Bubikonerstrasse 45
CH-8635 Dürnten, Switzerland

Mr. R. Mücke
Fraunhofer-Institut
für Atmosphärische Umweltforschung
Kreuzeckbahnstrasse 19
W-8100 Garmisch-Partenkirchen,
Germany

Mr. M. Mürtz
Institut für Angewandte Physik
der Universität Bonn
Wegelerstrasse 8
W-23000 Bonn, Germany

Prof. A. Nadezhdinskii
General Physics Institute
Academy of Science
Vavilov Street 38
117942 Moscow, Russia

Mr. R. Otjes
Netherlands Energy Research
Foundation (ECN)
P.O. Box 1
NL-1755 ZG Petten, The Netherlands

Mr. T. Papenbrock
Geschäftsbereich LaserChem
INNO-TEC GmbH & Co. KG
Universitätsstrasse 142
W-4630 Bochum, Germany

Dr. U.K. Parchatka
Max-Planck-Institut für Chemie
Saarstrasse 23
W-6500 Mainz, Germany

Dr. F. Pavone
LENS
Università di Firenze
Largo E. Fermi 2
I-50125 Firenze, Italy

Dr. M. Petri
Institut für Angewandte Physik
der Universität Bonn
Wegelerstrasse 8
W-23000 Bonn, Germany

Mr. V. Pustogov
Fachbereich Physik
Humboldt-Universität zu Berlin
Invalidenstrasse 110
O-1040 Berlin, Germany

Dr. G. Restelli
Environment Institute
Joint Research Centre Ispra
I-21020 Ispra (VA), Italy

Mr. F. Retailleau
LPCE/CNRS
3A Av. de la Recherche Scientifique
F-45071 Orléans Cedex 2, France

Mr. W.J. Riedel
Fraunhofer-Institut
für Physikalische Messtechnik
Heidenhofstrasse 8
W-7800 Freiburg, Germany

Dr. G. Rinke
Institut für Radiochemie
Kernforschungszentrum Karlsruhe
Postfach 3640
W-7500 Karlsruhe, Germany

Mr. H. Riris
Electro-Optics Systems Laboratory
SRI International
Menlo Park, CA 94025, USA

Mr. R.A. Rooth
N.V. Kema
P.O. Box 9035
NL-6800 ET Arnhem, The Netherlands

Mr. J. Roths
Max-Planck-Institut für Chemie
Saarstrasse 23
W-6500 Mainz, Germany

Mr. M. Schaefer
Institut für Angewandte Physik
der Universität Bonn
Wegelerstrasse 8
W-23000 Bonn, Germany

Dr. R. Schieder
1. Physikalisches Institut
der Universität zu Köln
Zülpicher Strasse 77
W-5000 Köln 41, Germany

Mr. U. Schiessl
Fraunhofer-Institut
für Physikalische Messtechnik
Heidenhofstrasse 8
W-7800 Freiburg, Germany

Prof. H. Schiff
Unisearch Associates Inc.
222 Snidercroft Road
CDN-Concord, Ontario L4K 1B5, Canada

Mr. U. Schneider
Physikalisch-Chemisches Institut
Universität Heidelberg
Im Neuenheimer Feld 253
W-6900 Heidelberg, Germany

Mr. M. Schupp
Max-Planck-Institut für Chemie
Saarstrasse 23
W-6500 Mainz, Germany

Prof. A. Shotov
General Physics Institute
Academy of Science
Vavilov Street 38
117942 Moscow, Russia

Mr. D. Siche
Abtlg. MBP, FB Physik
Humboldt-Universität zu Berlin
Invalidenstrasse 110
O-1040 Berlin, Germany

Mr. U. Simon
Institut für Angewandte Physik
der Universität Bonn
Wegelerstrasse 8
W-23000 Bonn, Germany

Dr. F. Slemr
Fraunhofer-Institut
für Atmosphärische Umweltforschung
Kreuzeckbahnstrasse 19
W-8100 Garmisch-Partenkirchen,
Germany

Mr. G. Spilker
Fraunhofer-Institut
für Physikalische Messtechnik
Heidenhofstrasse 8
W-7800 Freiburg, Germany

Dr. A.C. Stanton
Southwest Sciences Inc.
Suite E-11
1570 Pacheco Street
Santa Fe, New Mexico 87501, USA

Dr. E.V. Stepanov
General Physics Institute
Academy of Science
Vavilov Street 38
117942 Moscow, Russia

Dr. B. Sumpf
Fachbereich Physik
Humboldt-Universität zu Berlin
Invalidenstrasse 110
O-1040 Berlin, Germany

Dr. M. Tacke
Fraunhofer-Institut
für Physikalische Messtechnik
Heidenhofstrasse 8
W-7800 Freiburg, Germany

Mr. F. Taucher
Institut für Physik
GKSS Forschungszentrum Geesthacht
Max-Planck-Strasse 1
W-2054 Geesthacht, Germany

Mr. J. Terhürne
Physikalisch-Chemisches Institut
der JLU-Universität Giessen
Heinrich-Buff-Ring 58
W-6300 Giessen, Germany

Dr. C. Thiébaux
Laboratoire de Physique Moleculaire
Université de Reims
B.P. 347
F-51062 Reims Cedex, France

Mr. H. Tups
Zentrale Forschung und Entwicklung
Bayer AG
W-5090 Leverkusen Bayerwerk,
Germany

Mr. N. Van Dau
Thomson CSF/LCR
Domaine de Corbeville
F-91404 Orsay Cedex, France

Mrs. A. Vermoesen
Faculteit van de
Landbouwwetenschappen
Coupure links 653
B-9000 Gent, Belgium

Dr. J. Vogel
MÜTEK GmbH
Arzbergerstrasse 10
W-8036 Herrsching, Germany

Prof. E. Wagner
Fraunhofer-Institut
für Physikalische Messtechnik
Heidenhofstrasse 8
W-7800 Freiburg, Germany

Dr. P. Werle
Fraunhofer-Institut
für Atmosphärische Umweltforschung
Kreuzeckbahnstrasse 19
W-8100 Garmisch-Partenkirchen,
Germany

Mr. F. Wienhold
Max-Planck-Institut für Chemie
Saarstrasse 23
W-6500 Mainz, Germany

Mr. J. Wietzorrek
Fraunhofer-Institut
für Atmosphärische Umweltforschung
Kreuzeckbahnstrasse 19
W-8100 Garmisch-Partenkirchen,
Germany

Dr. P. Wiesen
Physikalische Chemie – FB 9
Bergische Universität-GH Wuppertal
Gaußstrasse 20
W-5600 Wuppertal 1, Germany

Mr. H. Wolf
Fraunhofer-Institut
für Physikalische Messtechnik
Heidenhofstrasse 8
W-7800 Freiburg, Germany

Dr. Yu.P. Yakovlev
Joffe Physical Technical Institute
Politechnicheskaya 26
194021 St. Petersburg, Russia

Dr. A. Zemel
Solid State Physics Dept.
Soreq Nuclear Research Center
IL-Yavne 70600, Israel

Mr. T. Zenker
Max-Planck-Institut für Chemie
Saarstrasse 23
W-6500 Mainz, Germany

Section I

ATMOSPHERIC TRACE GAS DETECTION

HIGH PRECISION TUNABLE DIODE LASER ABSORPTION SPECTROSCOPY: APPLICATION FOR MEASURING LONG-LIVED ATMOSPHERIC GASES

A. Fried, B. Henry, J. Fox, James R. Drummond[*], and R. Sams[**]

The National Center for Atmospheric Research
PO Box 3000
Boulder, Colorado 80307   U.S.A.

[*] University of Toronto, Department of Physics
60 St. George, Toronto, Ontario, Canada

[**] The National Institute of Standards and Technology
Gaithersburg, Maryland 20899   U.S.A.

SUMMARY

The ambient fluctuations of long-lived atmospheric gases contain important information about sources, sinks and potential secular trends. Since such fluctuations can be quite small, typically less than a few percent, high precision instruments are required. The present study describes a versatile tunable diode laser system to acquire such measurements. This system employs a number of novel features for increased system control and versatility, and these will be discussed. Specific applications of this system for measuring gases and processes related to the earth's aerosol sulfate layer will be given.

1. INTRODUCTION

The consequences of global warming and ozone depletion have focused attention on long-lived trace atmospheric gases such as carbon dioxide (CO_2), methane (CH_4), and carbonyl sulfide (OCS), to name a few. The ambient fluctuations of these gases contain important information about sources, sinks, and potential secular trends. Since such fluctuations can be quite small, typically less than a few percent in air not directly influenced by local sources, high precision measurements are required.

In addition to attributes of high sensitivity, high selectivity, versatility, and fast response, the technique of tunable diode laser absorption spectroscopy (TDLAS) has the potential to achieve such high measurement precision. However, with a few exceptions, this potential has not been exploited in most atmospheric studies. In the present paper, we describe a versatile TDLAS system which is capable of carrying out high precision measurements of long-lived atmospheric gases. This system, which has recently been described in detail by Fried et al.[1], addresses the major factors affecting tunable diode laser (TDL) precision.

We also discuss in this paper a number of applications of this system for studying gases and heterogeneous processes related to the stratospheric aerosol sulfate layer, the Junge layer. Crutzen[2] first raised the concern that anthropogenic emissions of the non-reactive sulfur gas, OCS, could be important in altering this layer during non-volcanic time periods. The resulting long term perturbation could thus significantly influence the earth's radiation budget and climate through increased solar scattering.[3] Recent studies[4] also suggest the potential for enhanced destruction of ozone on a global scale due to heterogeneous reactions. Long term and high precision ambient OCS measurements are therefore needed to determine whether a secular trend exists. In addition, measurements of OCS sources and sinks are also badly needed.

The studies described herein were carried out as part of a long term effort by the National Center for Atmospheric Research (NCAR) to understand these global effects and the potential role played by OCS. The present paper is divided into 4 sections. First, we present a discussion of our high precision TDLAS system, highlighting some of the new features for increased precision, system control and versatility. We will then briefly give three applications of this system for: (1) high precision ambient measurements of OCS, (2) automotive emission studies of OCS, and (3) laboratory heterogeneous studies involving stratospheric gases with sulfuric acid aerosols.

2. INSTRUMENTATION

Optical Layout

The optical layout is illustrated schematically in figure 1. Up to four high temperature operation (T ≥ 80 K) diode lasers are mounted on a temperature controlled stage housed in a specially fabricated liquid nitrogen dewar (Infrared Laboratories, Tucson, Az). The main dewar reservoir is connected to the diode stage through an externally adjustable thermal impedance which is designed to reduce cryogen boil-off and to minimize the heater current necessary for control. The latter is important for improved temperature control stability. The lasers are mounted side by side on a common axis parallel with the Y axis of figure 1. The dewar is mounted on slides to effect translation in both the X and Y directions of this figure. Once set, the dewar is locked in place.

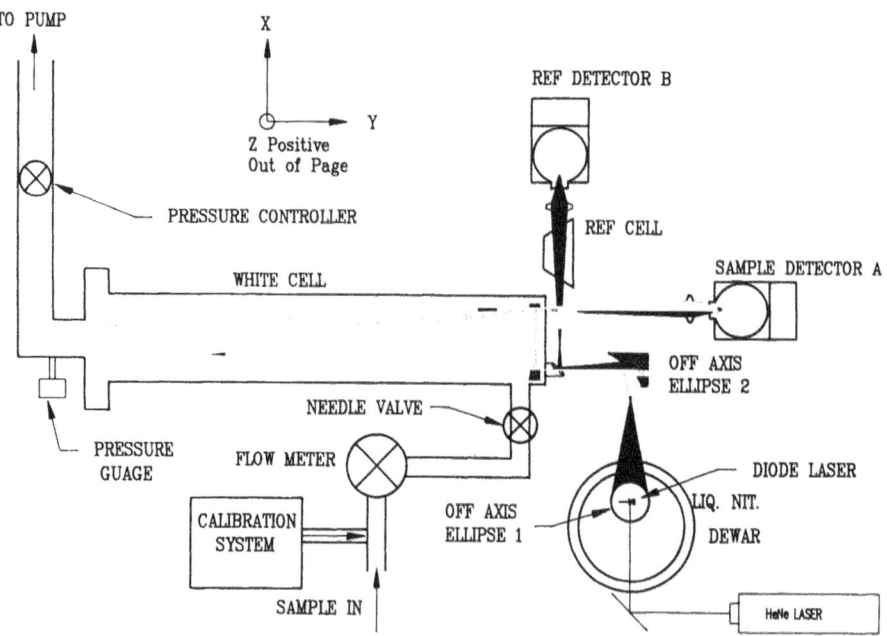

Figure 1:   Schematic of optical layout

The emerging IR radiation from the dewar, which is emitted downward, is collected and refocused by a 90 degree off-axis ellipsoidal (OAE) mirror placed beneath the dewar. A real image is formed at the conjugate focus of the OAE 20.31 cm from the mirror, and the subsequent diverging beam is then collected and re-imaged at the field mirror of an f/22 multipass White cell (Unisearch Incorp., Toronto, Canada) employing an identical OAE, as shown in figure 1. A small fraction of this beam is directed upward through a reference cell and onto a reference detector by the White cell entrance window. The main beam emerging from the White cell is collected and refocused onto an InSb sample detector. A HeNe alignment laser, which passes through a 1-mm diameter hole in the first OAE, establishes a reference axis for the entire optical system. The White cell has a 1.5-m base path and a corner cube reflector that produces 4 rows of spots.

Traditionally, off-axis parabolic (OAP) mirrors are used to collect and collimate the diverging IR laser radiation. These mirrors avoid some of the drawbacks associated with lenses, such as chromatic and spherical aberrations, optical feedback, and étalon effects. However, as Cooper and Carlisle [5] observe, OAPs tend to produce highly astigmatic beams which can cause performance-limiting optical noise. Employing an OAE presents several advantages over both lenses and OAPs. Like the OAP, the OAE avoids the drawbacks associated with lenses. However, like the lens, the OAE produces a real image at the conjugate focus, a feature we find particularly valuable for accurate and predictable system alignment and laser beam spatial characterization, as discussed in detail by Fried et al. [1] Both are important in minimizing sources of optical noise and hence in improving baseline precision. Our simple transfer optics system is yet a further improvement for high measurement precision. As can be seen in figure 1, our entire optical system is comprised of only two ellipsoidal mirrors at the White cell inlet and one lens detector assembly at the output. This simple system helps significantly in long term alignment stability, one of the dominant factors governing TDL precision.

Laser control & data handling system

The laser control and data handling are integrated into one computer-controlled system which consists of a signal generator and data acquisition (SGDA) unit and a 286-based computer connected by an interface card and cable, as shown in figure 2. The SGDA unit provides the control waveforms for the laser current controller and acquires the data from each detector employing the technique of second harmonic detection coupled with sweep integration. In this technique, the laser wavelength is rapidly scanned across absorption features of interest (20 Hz) while simultaneously

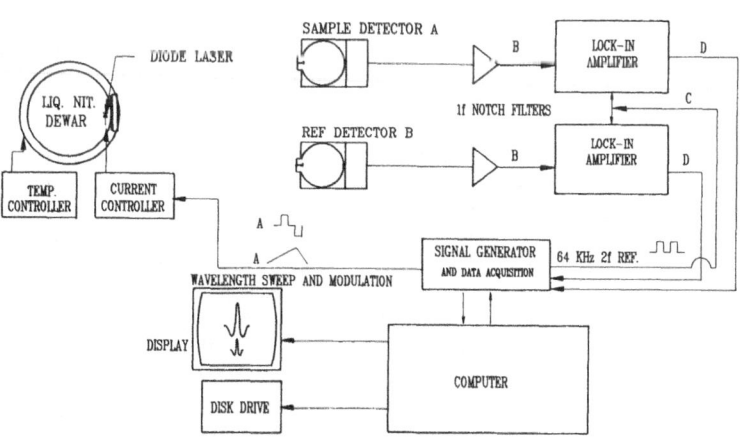

Figure 2:   Schematic of electronics showing input waveforms

wavelength-modulating the laser at much higher frequencies (32 KHz). The signal at twice the modulation frequency (2f) is recorded using a lock-in amplifier employing an appropriately small time constant. The diode laser is then repetitively scanned over the same spectral region and demodulated second harmonic data are co-averaged by the computer. Because of the small time constants required, the lock-in amplifier discrimination of unwanted signals at frequencies other than 2f is greatly reduced. To address this, our data acquisition system generates modulation and sweep waveforms that are synchronous with the computer sampling period, as will be further discussed.

The wavelength sweep and modulation waveforms are digitally generated from a master 4 MHz clock through the use of two 8192x16 memory systems and two digital-to-analog converters. The memories are cascaded using the contents of the "modulation" memory to advance the "sweep" memory address. The contents of the memories are generated by, and downloaded from, the computer. The great advantage of this approach is that the sweep and modulation waveforms and their interrelation are extremely flexible. This system is able to generate complex sweep and modulation waveforms simply by programming the functional form of the desired waveform into the computer. The waveforms chosen can be readily changed from a menu. In addition to various modulation waveforms such as triangular, sinusoidal, and quasi-square wave, one can select at the start of each measurement run standard linear scanning, non-linear scanning or jump scanning. As will be shown, the latter presents an advantage when attempting to simultaneously acquire widely-separated spectral lines. The overall and relative amplitudes of the two waveforms are independently controlled to allow simple flexible mixing and setting of the overall scan width. The waveform generators therefore permanently work at full-scale and the overall operation of the system is thereby simplified and the dynamic range expanded. All the attenuators are digital in form and the control values are downloaded from the computer. Once downloaded, the scan and modulation waveforms are continuously fed into the laser control module, an important aspect for laser control stability.

The data acquisition side of the SGDA is locked to the same master oscillator as the current modulation. The signal channels consist of a wideband preamplifier for each detector followed by an active twin-T notch filter, to effectively eliminate the 1f carrier and its potentially larger baseline structure, and a lock-in amplifier (Stanford Research Inc. SR510). The lock-in amplifiers are modified to allow the total elimination of the post-demodulation time constant. This permits the use of true integrators in the SGDA which have three advantages over simple RC time constants: First, the integrators are reset at the end of each integration period, eliminating any memory of previous events. This permits the maximum data acquisition rate whilst maintaining sample independence. Second, the integrator yields an exact average of the input signal over the integration time. Third, by making the integration time a multiple of the 1f modulation period, the 1f response is eliminated, as further shown by Fried et al.[1]. Thus, any baseline structure and associated changes due to the 1f signal are likewise eliminated.

Because of the requirement to shift the spectrum in the post-processing stage based on the output from the reference channel, it is critical that the sampling of the reference and signal channels be simultaneous. The integrators for each of the channels (either 2 or 4 channels can be selected) are synchronized and then the peak value of the integration is simultaneously sampled by a set of parallel sample and hold subsystems. The resulting values are then digitized in sequence by a 12-bit analog-to-digital converter. These last two functions are accomplished by a Data Translation DT5704 data acquisition module. Data from the module are then transferred to the computer using a direct memory access (DMA) system with a small FIFO buffer to smooth out the fluctuations in the DMA rate.

A simple menu-driven program allows the selection of a wide range of system control and data processing parameters which are then downloaded into the SGDA electronics box at the start of a measurement run. Scan parameters such as waveform type, frequency, amplitude, sampling rate, and integration time are selected at this point. The following additional data acquisition parameters, which are simply stored in program memory, can also be selected at this time: the total number of scans to be averaged, averaging mode (single block averaging, or cumulative linear or cumulative exponential averaging), and the acquisition mode (single acquisition sequence and stop or continuous time series measurements).

All of the software is written in TURBO PASCAL (Borland International Inc.) with assembler code used in some places for high-speed operations. With our current configuration (10 MHz 80286 processor), averaging speeds (sweep frequency x number of data points per scan) up to 30.5 kHz are attainable. Further details regarding data handling and timing are discussed by Fried et al.. [1]

Data analysis

The accumulated sweep-integrated spectrum is continuously displayed on the computer screen as a convenience for monitoring the progress of an experiment. Upon acquiring the desired number of averages, the program either returns to the main menu or automatically restarts a new averaging sequence for time series data acquisition. The cumulative sample and reference spectra can be stored on disk. This is done automatically in the continuous mode.

At the end of each averaging sequence, a rudimentary peak find routine finds the positive and both negative second harmonic lobes. A peak height is calculated based upon the difference of the former with the average value of the latter. All peak heights above an adjustable discriminator level can be output to the screen and to a printer along with the time and line center channel number. If the continuous mode is selected, these data are also output to a time series file stored on disk. The peak find routine operates on either the original data or on smoothed data employing a center-weighted running mean routine. The weighting function for this routine can be changed at any time. In either case, the original data is preserved and stored on disk for additional processing. A post-processing plotting routine can also be invoked to output data with any degree of smoothing to both the screen and to a plotter.

The peak find routine is used as an initial means in the determination of sample gas concentration. However, improved measurement precision is obtained using the entire second harmonic line profile employing an online linear-least-squares fitting routine. In this procedure, sample spectra are fit to a higher concentration reference spectrum, obtained by periodic addition of a reference standard to the White cell, using the following linear relationship:

$$Y_i = a + bX_i.$$

(1)

Here, Y_i and X_i represent the second harmonic signal amplitudes for the sample and reference spectra, respectively. The baseline offset (a) and the reference spectrum amplitude scaling factor (b) are simultaneously determined from the classical linear-least-squares relationships. The fitting is carried out over a window containing N points surrounding the peak. The window is selected by the user to encompass both negative lobes as well as some baseline on both sides. In order to account for a slow frequency drift in the interim between reference and sample spectra acquisition, the fitting procedure is carried out over a number of channels, usually 3 to 5, on both sides of the line center determined by the rudimentary peak find routine. The fit parameters a and b are thus determined by that frequency offset resulting in the minimum sum for the squared residuals.

The sample gas concentration is determined by the product of b and the reference concentration. The baseline in this fitting procedure is thus represented by a simple linear offset. A baseline slope and/or curvature does not present a problem unless it changes significantly with time. Such a changing profile, which was not evident in the fit residuals of the present study, would require frequent reference spectrum acquisition, or preferably, a more sophisticated baseline fit. This latter option will be incorporated in the future.

3. SYSTEM PERFORMANCE

Figure 3 illustrates the overall system performance achieved in carrying out OCS measurements employing a 0.600 part-per-billion by volume (ppbv) standard. The resulting signal response from individual 2-minute averages is displayed as a percentage deviation from the mean of all the data over a half-hour time period. Profile A shows the measurement precision employing our rudimentary peak find routine while B shows the corresponding precision employing the line fitting procedure. As can be seen, profile B is somewhat similar in structure to A but the major excursions from the mean are dramatically reduced. This demonstrates the improvements one can obtain using line fitting. The corresponding standard deviation improves from 0.82% (profile A)

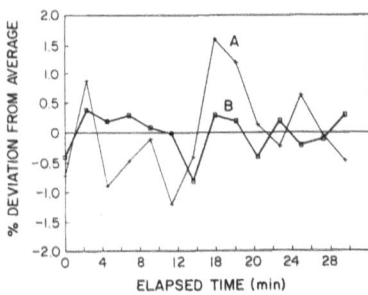

Figure 3: Signal stability as a function of time for a 0.600 ppbv OCS standard. Profile A shows the results using our peak find routine and profile B shows the improved precision obtained using line fitting.

to 0.35% (profile B). Figure 3 represents a realistic benchmark of overall instrument precision since the results not only fold in the imprecision from instrument gain and baseline stability but also from calibration output stability. Presently, the main cause of imprecision is due to optical noise, despite precautions to the contrary. This, however, is not a fundamental problem with our system but rather a direct result of poor spatial image quality of our present OCS laser. This laser produced a primary focal spot (at the conjugate focus of the first OAE) whose dimension in the Y-direction of Fig. 1 was 52 to 67% larger than all other lasers we tested.

Improved precision has been demonstrated by employing a laser with better spatial characteristics. In this case, a measurement precision of 0.34% was obtained over a 2-hour time period without line fitting. Employing line fitting only marginally improved the precision further to 0.30%.

4. APPLICATIONS

Ambient OCS Measurements

Figure 4 displays continuous ambient measurements of OCS and CO, acquired by a gas filter correlation spectrometer, taken on November 2, 1989 outside our laboratory at NCAR. Each measurement point is a 2-minute average. To accentuate differences, estimated background concentrations for each species were first subtracted off. Background concentrations of 0.538 ppbv and 94.4 ppbv were determined for OCS and CO, respectively, by taking the lowest recorded values for this measurement set. These values are not to be confused with the true continental background values which in all likelihood are somewhat lower, particularly in the case of OCS. We simply did not record enough data here for an accurate determination of the true OCS background unperturbed by local sources.

As can be seen, OCS and CO are highly correlated in this particular data set. A linear regression of the delta OCS concentrations as function of that for CO, results in a slope of 26.9 x 10⁻⁶, a Y-intercept of 0.006 ppbv, and an r^2 value of 0.68. The wind direction plays a big role in dictating the particular air mass being sampled in the present case. During the time period from 15:00 to 17:00 hours, the winds were from the east. Winds from this direction generally carry air influenced by anthropogenic sources from the Denver-Boulder, Colorado basin to our sampling site. Westerly winds, by contrast, generally carry much cleaner air to our sampling site. A change in wind direction from the west was observed at around 17:00 hours, at which time both OCS and CO concentrations dramatically dropped. This data set thus strongly indicates the influence of local anthropogenic sources of both gases, perhaps from fossil fuel combustion. Automobile emissions may, in part, be responsible for the observed profiles. Emission estimates of OCS from automobiles, however, are highly uncertain. The limited data available are based upon various estimates and assumptions and not on actual emission studies. To address this need, we subsequently employed our TDLAS system in automotive emission studies of OCS, as will now be described.

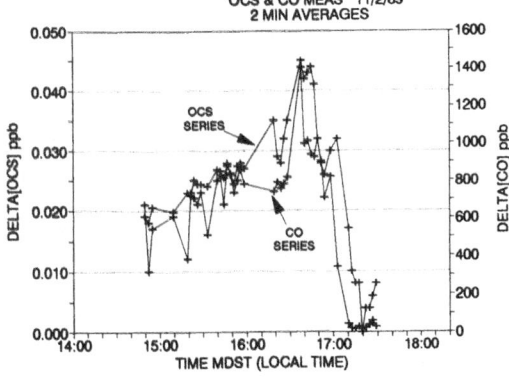

Figure 4: Continuous ambient 2-min. averages of OCS and CO above their respective background concentrations of 0.538 and 94.4 ppbv.

Automotive Emission Studies of OCS

The NCAR TDLAS system was transported to the Colorado Department of Health automotive test facility in Denver, Colorado. Both pseudo-continuous and integrated measurements of OCS were carried out on various vehicles employing the U.S. Federal Test Procedure (FTP).[6] A broad range of vehicles were studied including those without catalytic converters, those using older 2-way catalysts, and those employing newer 3-way catalysts. Integrated measurements of CO were also carried out as part of the FTP using a non-dispersive IR analyzer. Preliminary measurements were also carried out on 5 diesel-fuelled vehicles employing grab canisters. Measurements of both OCS and CO were acquired while four of the vehicles were idling. The fifth was tested at full throttle on a dynamometer. These diesel measurements, which have obvious shortcomings, are still useful for "order of magnitude" estimates. A detailed description of the entire study, including the experimental set-up, measurement approaches, estimates of error, and results are given by Fried et al..[7]

Carbonyl sulfide and CO concentrations measured using the FTP were all converted to an emission mass per unit mile from a knowledge of the particular gas density, the sample volume, and the total mileage driven during each test phase. Our complete data set, representing 44 measurements using two different sampling approaches, is displayed in Fig. 5. Here, OCS mass emission rates per unit mile are plotted as a function of that for CO. The resulting linear-least-squares regression yields a slope of $5.8 \times 10^{-6} \pm 2.6 \times 10^{-7}$, a Y-intercept of $-2 \times 10^{-5} \pm 6.3 \times 10^{-5}$, and an r^2 value of 0.92. These measurements span all three converter technologies mentioned above. As can be seen, OCS emissions are highly correlated with CO over a rather wide range of emission conditions. The CO emission range, in fact, spans some of the cleanest and some of the most polluting vehicles on the road today. This, together with the OCS-CO linearity ensures broad coverage with respect to OCS emissions.

The OCS-CO emission ratio of Fig. 5 when multiplied by the annual global estimate for automotive CO emissions (417.12 Tg, where 1 Tg = 10^{12} g)[8] yields a global estimate for automotive OCS emissions of 0.002 Tg yr[-1] from gasoline fuels. In a second approach, global OCS emissions are estimated using the data of Fig. 5 together with estimates for total gasoline consumed world-wide, the average fuel density, and the average fuel efficiency. Various OCS emission scenarios of Fig. 5 can be employed in this analysis. However, assuming the best and worst cases, we deduce a global emission range of 0.0004 to 0.006 Tg yr^{-1} for gasoline-fuelled vehicles. Given the uncertainties involved, the agreement between both determinations is considered good. Employing a similar procedure on our crude "order of magnitude" diesel-fuel measurements, we arrive at global emission range of 0.0002 to 0.002 Tg yr^{-1}. Thus, total emissions of OCS from gasoline and diesel-fuelled vehicles range between 0.0006 to 0.008 Tg yr^{-1} globally. It is significant to note that our highest estimate is a factor of 100 to 600 less than the sum of all global OCS sources. Automotive OCS emissions are thus unimportant on a global scale. Even with our crude diesel-fuel

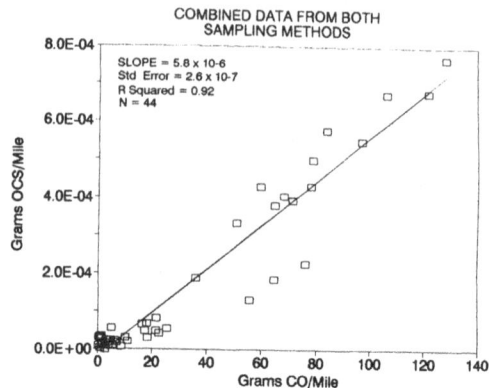

Figure 5: OCS and CO mass emission rates.

measurements, which should be repeated using a dedicated test facility, it is difficult to imagine any scenario that could change this conclusion. On a local scale, however, automotive emissions of OCS may be very important, particularly when one attempts to measure background concentrations and associated small secular trends. In this regard, it is interesting to note that the data of Fig. 4 yields a factor of 10 larger OCS-CO slope (26.9 x 10^{-6}) than our gasoline measurements of Fig. 5 when corrected to volumetric ratios using the appropriate ratio of molecular weights. This implies that either another source of OCS is prevalent or that approximately 27% of the automotive emissions being sampled originates from diesel fuel vehicles.

Laboratory Heterogeneous Studies

Recent modelling studies[4] suggest that heterogeneous reactions between N_2O_5 and sulfuric acid aerosols may be important in depleting stratospheric ozone on a global scale. Given the rather dramatic recent perturbations to the stratospheric aerosol layer from the volcanic eruptions of Pinatubo and Hudson, this reaction takes on even greater significance. We are presently employing our TDLAS system in a series of measurements using a low temperature flow reactor to study this heterogeneous reaction and its dependence with temperature and aerosol composition. Sulfuric acid aerosols are generated in these studies in the size and composition range representative of those in the mid-latitude stratosphere. The concentration of N_2O_5, as measured by changes in the concentration of added NO using a titration procedure, is measured as a function of interaction time with the aerosols. The NO concentrations using this procedure must be measured with high precision since the changes are small. In addition, H_2O must also be measured simultaneously. The concentrations of the latter, however, produce second harmonic signals whose amplitudes may be significantly different than that of NO.

To accomplish this, we have expanded our line fitting capability. Two independent fitting windows can be selected for simultaneous fitting of two spectral features acquired in the same scan. Each feature is fit to its own reference standard. For this purpose, the wavelength reference detector is disconnected and the sample detector is connected to both lock-in amplifiers. Each lock-in is set to a different input sensitivity to optimize the signal amplitude for the respective input channel. As shown in Fig. 6, two different spectral features, H_2O and an NO doublet in this case, can thus be quantified even though the former signal is 10 times the latter. The H_2O peak is fit using the 20 mv scale and the much smaller NO peak is fit using the 2 mv scale giving rise to nearly identical signal amplitudes. The fitting windows are indicated by the vertical lines. One can immediately see the advantage of this approach. Employing only a single input sensitivity, as is conventional, the H_2O peak would be too large for the 2 mv scale. As can be seen, this peak is clipped and highly non-linear in comparison to that on the 20 mv scale. The NO peak on the 20 mv scale, on the other hand, is too small to be digitized optimally by the input A/D converter, and hence measurement precision would suffer. Reducing the H_2O concentration had no effect on the NO peak. This confirms that the large H_2O peak of Fig. 6 does not affect the quantification of the smaller NO peak. Additional measurements confirm that the NO peak can be accurately quantified in the presence of a H_2O peak at least 26 times larger in the present set up. Larger H_2O signal amplitudes overload the 2 mv scale. However, even under these conditions the NO peak does not appear to be affected. The precise point where such an overload does present a problem has not been determined in this study. However, the factor of 26 can readily be extended, to perhaps a factor of 100 or so, by gating out the H_2O signal from the sample lock-in.

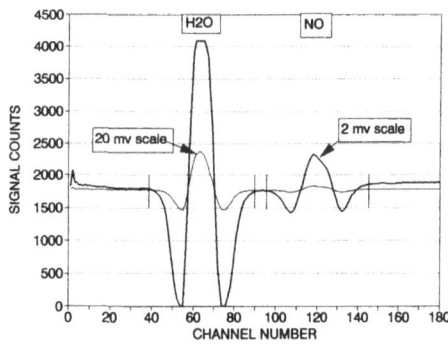

Figure 6: Second harmonic features for H_2O & NO recorded on two different sensitivities. The vertical lines indicate the two fitting windows.

Even without this modification, a larger signal amplitude ratio can be achieved if the line pairs are further apart. The present peaks are separated by 0.041 cm^{-1}. However, there is a practical limit to the maximum separation that can be achieved without incurring lost precision employing a conventional linear scanning profile. For peaks whose separation is greater than approximately 0.15 cm^{-1}, an increased scan rate must be employed in this case, resulting in a consequent reduction in the fraction of time spent on each feature and more time spent in transit. Our data acquisition system can also address this situation easily, as shown in Fig. 7. Here, we display two NO features separated by 0.424 cm^{-1}

employing a jump scanning profile instead of the usual linear sawtooth profile. Both features are recorded employing our normal scanning rate. After acquiring the first feature, the tuning current is jumped to the second spectral region, as shown by the step in the scanning profile. This optimizes the amount of time spent recording each line as well as the number of data points acquired across each line. As a result, the spectrum is recorded with higher signal-to-noise and higher resolution. The magnitude of the jump can be varied in the program menu. This jump capability thus greatly extends the number of spectral pairs that can be studied simultaneously.

5. SUMMARY

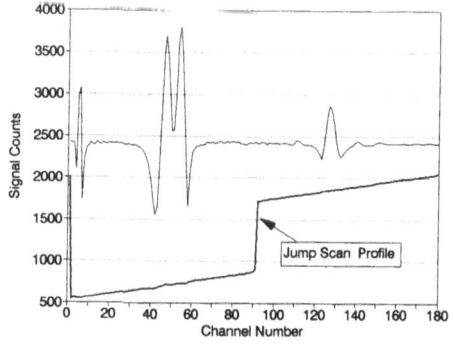

Figure 7: Second harmonic signals for two NO peaks separated by 0.424 cm^{-1} using a jump scanning profile.

We describe in the present paper a TDL system for high precision measurements of long-lived atmospheric gases and have implemented a number of new features for increased system control and versatility. Such versatility has allowed us to implement two novel data acquisition approaches for high precision ratio measurements. We have demonstrated the capability for quantifying absorption features that differ in amplitude by at least a factor of 26 and have demonstrated the utility of jump scanning for features that are separated by at least 0.424 cm^{-1}. Such new capabilities have many important atmospheric applications. We have successfully employed this system in acquiring ambient measurements of OCS with high precision, in automotive emission studies of OCS, and in laboratory heterogeneous studies.

REFERENCES

(1)   Fried, A., Drummond, J.R., Henry, B.E., and Fox, J. " Versatile Integrated Tunable Diode System for High Precision: Application for Ambient Measurements of OCS ", Appl. Opt., **30**, 1916 - 1932 (1991).

(2)   Crutzen, P.J., " The Possible Importance of CSO for the Sulfate Layer of the Stratosphere ", J. Geophys. Res. Lett., **3**, 73 - 76 (1976).

(3)   Hofmann, D.J., " Increase in the Stratospheric Background Sulfuric Acid Aerosol Mass in the Past 10 Years ", Science, **248**, 996 - 1000 (1990).

(4)   Brasseur, G.P., Granier, C., and Walters, S., " Future Changes in Stratospheric Ozone and the Role of Heterogeneous Chemistry ", Nature, **348**, 626 -628 (1990).

(5)   Cooper, D.E., and Carlisle, C.B., " High-Sensitivity FM Spectroscopy With a Lead-Salt Diode Laser ", Opt. Lett., **13**, 719 - 721 (1988).

(6)   U.S. Federal Test Procedure, published by the Office of the Federal Register, Code of Federal Regulations Number 40, " Protection of the Environment ", Part 86 (1986).

(7)   Fried, A., Henry, B.E., Ragazzi, R.A., Merrick, M., Stokes, J., Pyzdrowski, T., and Sams, R., " Measurements of Carbonyl Sulfide in Automotive Emissions and an Assessment of its Importance to the Global Sulfur Cycle ", in preparation for submission to J. Geophys. Res. (1991).

(8)   Cullis, C.F., and Hirschler, M.M., " Man's Emissions of Carbon Monoxide and Hydrocarbons into the Atmosphere ", Atm. Envir., **23**, 1195 - 1203 (1989).

ACKNOWLEDGMENTS

We wish to acknowledge the Colorado Department of Health in Denver, Colorado for allowing us to use their automotive test facility, David Parrish of the NOAA Aeronomy Lab. for ambient CO measurements, and Mike Mozurkewich of York University and Jack Calvert of NCAR for contributions in the heterogeneous studies. The National Center for Atmospheric Research is sponsored by the National Science Foundation.

FORMATION AND EMISSION OF ATMOSPHERIC GREENHOUSE GASES IN THE COMBUSTION OF FOSSIL FUELS: MONITORING BY TUNABLE DIODE LASER SPECTROSCOPY

K.H. BECKER, K.J. BROCKMANN, R. HÜDEPOHL, R. KURTENBACH and P. WIESEN

Physikalische Chemie / Fachbereich 9
Bergische Universität - Gesamthochschule Wuppertal
Postfach 10 01 27
D-5600 Wuppertal-1
Germany

SUMMARY

The N_2O and CH_4 emissions from different small sized furnaces, which were operated with charcoal, gas, oil and wood as fuel were determined by means of longpath infrared diode laser absorption spectroscopy. N_2O emissions were found to be comparable with those from conventional power plants. CH_4 emissions were found to be strongly dependent on the combustion conditions. Measurements from a full scale power plant with circulating atmospheric fluidized bed as NO_x reduction technique exhibit much higher N_2O concentrations. In addition, for the NCO + NO reaction a branching ratio of (0.25 ± 0.10) for N_2O formation was determined by photolysing ClNCO at 254 nm and following the N_2O formation by tunable diode laser spectroscopy.

1. INTRODUCTION

Human activities have caused a steady increase of atmospheric trace gas concentrations during the last 100 years. Two important consequences of this increase are a) a possible change of the earth´s climate since these gases modify the radiation energy balance of the atmosphere (greenhouse effect) and b) the precipitous loss of ozone in the Antarctic stratosphere every September during the past few years (ozone hole). More recently, substantial ozone losses have also appeared, although not as spectacularly as over Antarctica, in the Northern Hemispere´s temperate and polar regions. It has been estimated that carbon dioxide accounts for about half of the future temperature increase, while methane, nitrous oxide, ozone and chlorofluorocarbons (CFCs) will be responsible for the rest (see Fig. 1).

Among these gases, methane and nitrous oxide are of particular interest. As a result of the influence of CFCs on global ozone an international protocol to control the emissions of CFCs to the atmosphere was agreed upon in Montreal in 1987 and it is planned to stop CFC production and release by the year 2000. However, for nitrous oxide and methane no international agreements to reduce these emissions have yet been discussed.

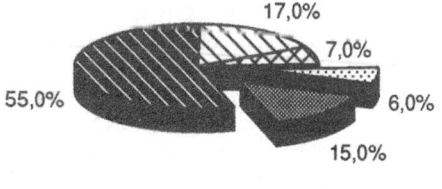

17,0%
7,0%
55,0%
6,0%
15,0%

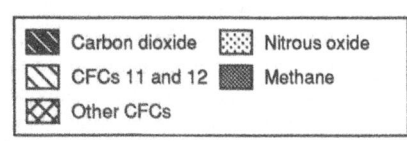

	Carbon dioxide		Nitrous oxide
	CFCs 11 and 12		Methane
	Other CFCs		

Fig. 1: Contribution of man-made greenhouse gases to global warming[1] (The contribution from ozone may also be significant, but cannot be quantified at present).

Methane is a key chemical species in the chemistry of both the troposphere and the stratosphere. In the troposphere it plays an important role in determining the hydroxyl radical concentration. As a contributor to greenhouse warming methane has a global warming potential (GWP) which is 21 times larger than that of carbon dioxide[1].

Nitrous oxide has an atmospheric lifetime of approximately 150 years and a global warming potential of 206. The transport of tropospheric N_2O to the stratosphere and its subsequent reaction with singlet oxygen atoms provides the major source of stratospheric NO^2.

Both gases, N_2O and CH_4, are produced by industrial and biological processes. However, sources and sinks for both gases are not well established[3-5]. Concentrations of atmospheric N_2O and CH_4 have been observed to increase at a rate of about 0.2 - 0.3% and 1.0% per year[6,7], respectively. For CH_4 it has been proposed that the increase is caused by natural gas releases, fossil fuel combustion, biomass burning, rice paddy cultivation and ruminants. The increase in N_2O is probably caused by the use of fertilizers, combustion of fossil fuels, biomass burning and changes in land use in the tropics. From the estimated loss of N_2O in the stratosphere by photochemical reactions together with the observed increase in N_2O a total source strength of about 12 -14 10^{12} g N_2O/N is to be expected at the earth´s surface.

During the last years the contribution of fossil fuel combustion to total N_2O emissions has been overestimated because the sampling method was inadequate[8]. It was shown that N_2O was formed in the sample containers by reactions involving NO_x, SO_2 and water[9]. Recent data, where this artificial N_2O formation has been controlled, show that the N_2O emission from fossil fuel combustion is much lower than previously reported[10]. However, it was shown that N_2O emissions from fluidized bed combustion and combustors with NO_x control systems are significantly higher[11]. Very recently, Nylon production was reported as a previously unrecognized source of atmospheric nitrous oxide that may account for about 10% of the observed increase[12]. However, about 30% of the N_2O sources remain unidentified.

In this work we have measured by means of longpath infrared diode laser absorption spectroscopy N_2O and CH_4 concentrations in gas samples which were drawn from various combustion sources. The influence of these sources on the global N_2O and CH_4 budgets is discussed. In addition, measurements on the elementary kinetics of the N_2O formation in combustion processes are reported.

2. EXPERIMENTAL METHOD

Gas samples were collected by a method which is based on the EPA recommended operating procedure for the analysis of nitrous oxide from combustion sources[13]. The equipment employed is shown schematically in Fig. 2. Exhaust gases were sampled through a stainless steel probe. The temperature of the exhaust gases was monitored by means of a thermocouple placed in the exhaust flow. At the end of the probe, the exhaust was drawn through a filter in order to remove particles. Both the filter and the probe were heated in order to prevent sample losses. The sample gas was dried by passing it through a condenser whose temperature was always kept below 3^0C. The remaining water vapour was removed by a $Mg(ClO_4)_2$ packed column. Special care was taken to remove SO_2 from the samples in order to avoid artifical formation of N_2O by the reaction of sulfur dioxde with NO during storage of the cylinders. This was achieved by passing the sample gas through a KOH solution. However, this led also to the removal of CO_2 and consequently to an increase in the apparent concentrations of the other components. Therefore, in order to obtain correct concentrations two gas samples were always drawn, one with passing the flue gas through the KOH solution and the second without KOH washing. Tests performed prior to the measurements showed that N_2O and CH_4 losses in the sampling system are negligible. Also no additional formation of N_2O was observed in the sample containers.

The N_2O and CH_4 concentrations in the gas samples were monitored by longpath infrared diode laser absorption spectroscopy by expanding the exhaust gas from the sample container into a White cell which had a base length of 2 m. All measurements were carried out with an absorption path length of 80 m at 10 Torr total pressure. The infrared absorption of N_2O and CH_4 was monitored by means of a tunable diode laser (Laser Analytics LS3) using vibration-rotation lines in the

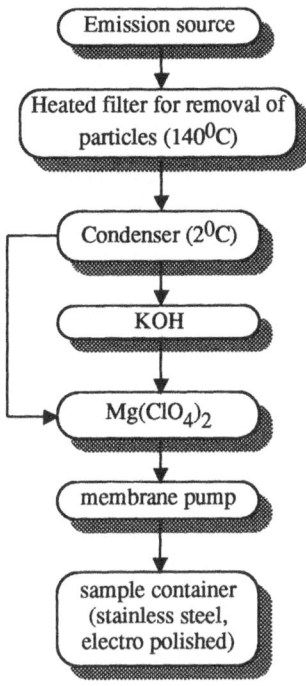

Fig. 2: Schematic diagram of the gas sampling system.

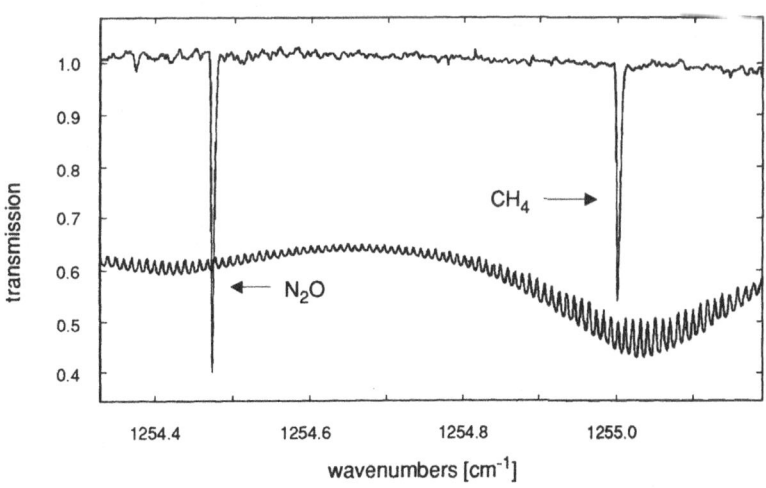

Fig. 3: Typical infrared absorption spectrum of a gas sample in the range 1254 -1255 cm^{-1}.

range 1254 - 1255 cm^{-1}. Fig. 3 shows a typical absorption spectrum of a gas sample. The laser beam from a Mesa-stripe diode was passed through a chopper for amplitude modulation at 400 Hz and a monochromator for single mode selection. The single mode beam was then collimated and split into three separate components which were directed into a reference cell containing N_2O for absolute wavelength calibration, a Fabry-Perot etalon for relative wavelength calibration and the longpath infrared absorption cell. The three beams were then focussed on separate Cu/Ge detectors, the signals of which were amplified and detected at the chopper frequency by means of three lock-in amplifiers. The wavelength range of interest was scanned by ramping the diode injection current. Scanning was controlled by a mini-computer which was also used for data collection, manipulation and storage. In order to better characterize the combustion conditions, in addition to CH_4 and N_2O, concentrations of NO, NO_2, CO, CO_2 and SO_2 were measured by using a FTIR spectrometer or in the case of O_2 and N_2 by using standard gas chromatography.

Concentrations were determined either by using calibration gas mixtures, tabulated absorption coefficients or air as a standard reference.

3. RESULTS AND DISCUSSION

Emission measurements

In this work we investigated the emission of N_2O, CH_4, NO, NO_2, SO_2, CO_2 and CO from different small sized furnaces which were operated with charcoal, gas, oil and wood as fuel. In addition, emissions from a full scale fluidized bed power plant were monitored. The results are summarized in tables 1 and 2. All given concentrations are based on a residual O_2 concentration of 7%.

N_2O and CH_4 emissions from the small sized furnaces are relatively low. With two exceptions, the N_2O concentrations are smaller than 10 ppm and are, therefore, comparable with

furnace fuel	N$_2$O [ppm]	CH$_4$ [ppm]	CO$_2$ [%]	CO [‰]	N$_2$ [%]	NO$_2$ [ppm]	NO [ppm]	SO$_2$ [ppm]
natural gas	11,7	39,6	27,8	1,5	64,6	<78,8	<13,9	<70,7
natural gas	1,3	7,2	9,7	<11 ppm	84,2	38,0	<25,5	<34,0
natural gas	2,7	7,8	9,5	<22 ppm	83,9	46,7	<50,0	<67,0
natural gas	1,6	42,2	0,4	<30 ppm	92,9	<43,4	<82,5	<109,0
oil	1,2	0,6	12,3	0,3	78,8	<8,2	<2,3	<10,6
oil	0,4	0,1	10,3	3,3 ppm	82,4	45,0	<9,0	<12,2
wood	10,3	1917,0	14,1	6,4	79,7	46,9	<31,0	<4,5
coke	3,8	0,9	15,1	0,3	78,7	78,0	<41,5	295,4
coke	4,3	3,3	14,9	1,6	79,9	88,6	<32,0	39,1
pit coal	6,9	8,1	13,9	2,1	79,6	16,1	<28,0	36,4
pit coal	8,9	3,2	13,9	0,9	81,2	97,3	<26,2	334,4
pit coal	6,5	13,6	9,7	0,6	82,4	28,9	<16,7	<23,0
pit coal	8,2	990,0	12,0	10,8	79,0	44,0	<64,9	<44,0
brown coal	3,4	47,2	17,0	15,0	75,6	44,0	<120,0	<92,7
brown coal	2,3	4,8	13,9	0,7	80,9	46,5	<54,9	329,6
gas (geyser)	0,4	4,9	9,5	1,4	83,8	40,4	<12,0	<16,0
gas (geyser)	0,5	3,9	8,3	0,7	84,4	46,3	<15,0	<20,0

Table 1: Emission data from different small sized furnaces.

sample point	N₂O [ppm]	CH₄ [ppm]	CO₂ [%]	CO [‰]	N₂ [%]	NO₂ [ppm]	NO [ppm]	SO₂ [ppm]
WBK/HBK 02	123,0	6,4	16,3	0,8	82,2	115,6	<8,4	<12,0
SZ 31	73,4	0,2	10,9	0,1	80,4	11,7	<9,5	<13,0
WBK/HBK 02	186,9	5,2	10,1	0,8	80,7	106,6	<15,0	<9,4
NBK/HBK 03	75,0	1,5	11,4	<2 ppm	80,1	30,6	<22,0	<16,6
SZ 31	87,0	0,4	11,0	0,2	80,5	11,0	15,0	<6,8

Table 2: Emission data from a coal fired power plant with circulating atmospheric fluidized bed (280 MW thermal power, fuel: pit coal, WBK = turbulent combustion chamber, HBK = main combustion chamber, NBK = secondary combustion chamber, SZ = induced draught)

emission data from conventional power plants whose emissions lie in the range 1 - 5 ppm N_2O[11]. However, two furnaces exhibit N_2O concentrations > 10 ppm, which is probably caused by a relatively low temperature in the furnace, which is known to favour N_2O formation. In addition, the measurements exhibit lower N_2O concentrations for natural gas and oil than for brown and pit coal, which is probably a result of additional N_2O formation on coal particles. Methane concentrations are very variable for the different furnaces. CH_4 concentrations of more than 1000 ppm were monitored for burning wood and pit coal with reduced air supply. Under these conditions methane is probably formed by pyrolysis. This is also confirmed by relatively high CO concentrations.

In contrast to the small sized furnaces the data from the coal fired power plant with a circulating atmospheric fluidized bed exhibit much higher N_2O concentrations. However, CH_4, NO_x and SO_2 emissions are small compared with conventional power plants. The high N_2O concentrations are probably a result of the relatively low temperature (~1100 K) and the high concentration of particles in the combustor. However, it is not clear to what extent homogeneous and heterogeneous reactions contribute to the N_2O formation.

It should be pointed out that also other NO_x reduction techniques tend to increase N_2O emissions, in particular, non catalytic selective NO reduction by injection of ammonia or urea and selective catalytic reduction of NO. Recently, it was also shown that N_2O emissions from cars with catalytic converter are significantly higher than those from uncontrolled cars[11]. It was also observed that N_2O emissions from aged catalysts appear to be much higher than those obtained from new catalysts.

The amount of fossil fuels which is burned in small sized furnaces represents a very small fraction of world's annual fuel consumption[14]. Therefore, if the results from the present study are taken into account, it is very likely that burning fossil fuels in small sized furnaces is only of secondary importance for the global N_2O budget. From the available data for CH_4, where the emission rates seem to be strongly dependent on the combustion conditions, more experimental data are needed in order to estimate the contribution from small sized furnaces to the global CH_4 budget.

Kinetics of N₂O formation

In addition to our emission measurements, we investigated the chemical mechanism leading to N_2O formation during combustion. N_2O can be formed both heterogeneously through different gas-solid phase interactions and homogeneously through gas phase reactions. However, the relative importance of the different mechanisms is not yet clear. Kramlich et al.[10] suggested that N_2O appears in pulverized coal flames as a result of HCN oxidation beyond the flame front, namely by the homogeneous gas phase reactions of NH, NH_2 and NCO radicals with NO. Probably the reaction NCO + NO is the dominant homogeneous source for N_2O formation[15]. The major uncertainty in the kinetic mechanism is related to this reaction since there are insufficient data available concerning the products. According to Perry[16] the most likely reaction channel is the one leading to nitrous oxide. However, other channels

$$NCO + NO \longrightarrow N_2 + CO_2$$
$$NCO + NO \longrightarrow N_2 + CO + O$$

are also thermodynamically accessible and no branching ratio for the various channels has been published up to now.

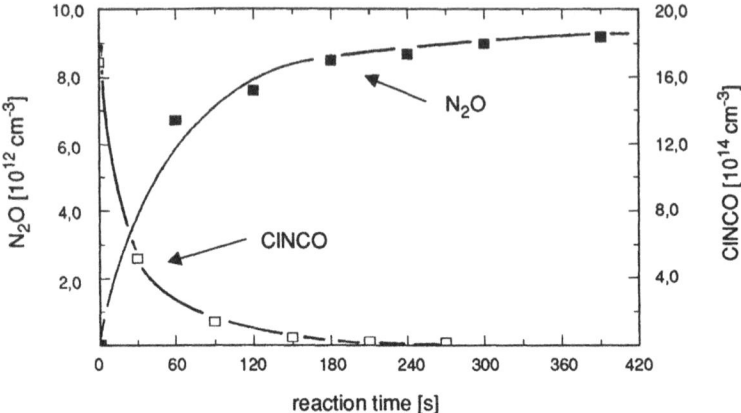

Fig. 4: Formation of N_2O after the 254 nm photolysis of ClNCO diluted in argon. Both species were monitored by tunable diode laser spectroscopy.

We investigated the NCO + NO reaction in a quartz reactor which was surrounded by photolysis lamps (λ=254 nm). The quartz cell was equipped with a White mirror system. By photolysing ClNCO diluted in argon and following the formation of N_2O, CO and CO_2 either by FTIR or diode laser spectroscopy the branching ratio of the reaction was determined.

$$
\begin{array}{c}
\text{254 nm} \qquad\qquad\quad +\,\text{NO} \\
\text{ClNCO} \longrightarrow
\begin{array}{l}
\text{NCO} \longrightarrow
\begin{array}{l}
N_2O + CO \\
N_2 + CO_2
\end{array} \\
\text{NO} \\
\text{Products}
\end{array}
\end{array}
$$

The photolysis of ClNCO, which is very fast, leads to the formation of NO, NCO and other products[17].

Under the experimental conditions applied in this study, from the absolute concentrations of N_2O and CO_2 formed after quasi-infinite reaction time and by using a rate constant of 3.2×10^{-11} cm^{3}s^{-1} which was determined by Perry[16] for the reaction NCO + NO $\longrightarrow$ products, a branching ratio of (0.25 ± 0.10) for N_2O formation was obtained at 300 K. Further experiments are in progress in order to investigate the possible temperature dependence of the branching ratio.

ACKNOWLEDGMENT

Financial support by the "Gesamtverband des Deutschen Steinkohlebergbaus" and the "Ministerium für Wissenschaft und Forschung des Landes Nordrhein-Westfalen" is gratefully acknowledged.

REFERENCES

(1) HOUGHTON, J.T., JENKINS, G.J. and EPHRAUMS, J.J., Eds. (1991). Climate Change: The IPCC Scientific Assessment, Cambridge University Press, Cambridge.

(2) McELROY, M.B. and McCONNELL, J.G. (1971). Nitrous oxide: a natural source of stratospheric ozone. J. Atmos. Sci. Vol. 28, 1095 - 1098.

(3) FUNG, I., JOHN, J., LERNER, J., MATTHEWS, E., PRATHER, M., STEELE, L.P. and FRASER, P.J. (1991). Three-dimensional model synthesis of the global methane cycle. J. Geophys. Res. Vol. 96, 13033 - 13065.

(4) DELMAS, R.A., MARENCO, A., TATHY, J.P., CROS, B. and BAUDET, J.G.R. (1991). Sources and sinks of methane in the african savanna. CH_4 emissions from biomass burning. J. Geophys. Res. Vol. 96, 7287 - 7299.

(5) CRUTZEN, P.J. (1991). Atmospheric chemistry - Methane sinks and sources. Nature Vol. 350, 380.

(6) CICERONE, R.J. (1989). Analysis of sources and sinks of atmospheric nitrous oxide (N_2O). J. Geophys. Res. Vol. 94, 18265 - 18271

(7) PRINN, R., CUNNOLD, D., RASMUSSEN, R., SIMMONDS, P., ALYEA, F., CRAWFORD, A., FRASER, P. and ROSEN, R. (1990). Atmospheric emissions and trends of nitrous oxide deduced from 10 years of ALE-GAGE data. J. Geophys. Res. Vol. 95, 18369 - 18385.

(8) MUZIO, L.J. and KRAMLICH, J.C. (1988). An artifact in the measurements of N_2O from combustion sources. Geophys. Res. Lett. Vol. 15, 1369 -1372

(9) De SOETE, G. (1989). Parametric study of N_2O formation from sulphur oxides and nitric oxide during storage of flue gas samples. IFP-Report No. 36732, Rueil-Malmaison.

(10) KRAMLICH, J.C., COLE, J.A., McCARTHY, J.M., LANIER, W.S. and McSCORLEY, J.A. (1989). Mechanisms of nitrous oxide formation in coal flames. Combust. Flame Vol. 77, 375 - 384.

(11) De SOETE, G., Ed. (1990). European Workshop on the Emisson of nitrous oxide [Proc.]. Lisbon, 6. - 8. June 1990.

(12) THIEMENS, M.H. and TROGLER, W.C. (1991). Nylon production: an unknown source of atmospheric nitrous oxide. Science Vol. 251, 932 - 934.

(13) Recommended operating procedure no. 45: Analysis of nitrous oxide from combustion sources. EPA-600/8-90-053.

(14) DEUTSCHE BP AG, Ed. (1991). Zahlen aus der Mineralölwirtschaft '91.

(15) KILPINEN, P. and HUPA, M. (1991). Homogeneous N_2O chemistry at fluidized bed combustion conditions: A kinetic modelling study. Comb. Flame Vol. 85, 94 - 104.

(16) PERRY, R.A. (1985). Kinetics of the reactions of NCO radicals with H_2 and NO using laser photolyis-laser induced fluorescence. J. Chem. Phys. Vol. 82, 5485 - 5488.

(17) BELL, D.D. and COOMBE, R.D. (1984). Photodissociation of chlorine isocyanate. J. Chem. Phys. Vol.82, 1317 - 1322.

MEASUREMENTS OF H_2O_2 IN THE GRAND CANYON BY TUNABLE DIODE LASER SPECTROMETRY DURING THE NAVAJO GENERATING STATION VISIBILITY STUDY

H. I. Schiff, G. I. Mackay, D. R. Karecki, J. T. Pisano and S. D. Nadler
Unisearch Associates Inc. 222 Snidercroft Rd., Concord, Ontario, Canada

SUMMARY

As a part of a visibility study in the Grand Canyon, a tunable diode laser absorption spectrometer (TDLAS) system was employed for the real time measurement of gas-phase hydrogen peroxide (H_2O_2). Measurements were made every 3 minutes for a total of 74 days in the winter of 1990 and subsequently reduced to 1 hour average values. Detection limits for the hourly averages were better than 50 parts per trillion (pptv) during 75% of the measurement period and better than 100 pptv for the remainder.

H_2O_2 maxima varied from 32 pptv to greater than 1.2 ppbv. On average maximum mixing ratios occurred during the night and minimum values during the day from 08:00 through 14:00. Occasionally, a small maximum was observed in the afternoon which could be attributed to photochemistry. For the most part, transport seems to have determined the temporal behaviour of the H_2O_2.

The highest concentrations were measured during a warm period which gave average values of 650 pptv compared with an average value of 250 pptv for the remainder of the measuring period.

The data obtained on the 23 days when the plume from the Navajo Generating station was expected to impact on the measurement site were not substantially different from those obtained during the non-intensive periods.

No correlation was found between the H_2O_2 concentrations and the concentrations of SO_2, particulate sulphate, or humidity. Apparently the H_2O_2 levels are controlled by long range transport of precursors and NO_x.

INTRODUCTION

During the winter of 1990 measurements were made to address the relative importance of a single point source, the Navajo Generating Station on the atmospheric transparency in the Grand Canyon. A variety of instruments were employed to measure various particulate and gaseous atmospheric species including SO_2, sulphate, organic particles, suspended particulates, photochemical oxidants (NO_x, O_3, H_2O_2) and hydrocarbons. In addition, tracer studies involving aircraft measurements were employed to follow the emissions from the power plant into the receptor area. In this paper we present tunable diode laser absorption spectrometer (TDLAS) measurements of hydrogen peroxide made during this mission.

Recombination of HO_2 radicals is the principle source of hydrogen peroxide in the gas phase. Measurements of H_2O_2 can therefore provide information about the concentration of these chain carriers (Logan et al, 1980). Furthermore, H_2O_2, in cloud droplets, is believed to be the most important oxidizer of S(IV) to S(VI) in the atmosphere and, therefore, plays an important part in acid deposition (Boyce, 1984; Penkett, 1979; Middleton, 1980). The sulphate particles formed from the oxidation of SO_2 are suspected of being a significant contributor to winter-time haze in the Canyon.

The measurements were made during the period January 10, 1990 through March 31, 1990. The mobile laboratory containing the TDLAS system was situated at the south rim of the Grand Canyon some 25 miles east of Grand Canyon Village, Grand Canyon National Park, Arizona. Other instrument systems were deployed in neighboring locations. The Navajo Generating Station is approximately 100 miles north west of the measurement sites in Page, Arizona. All instruments operated every day although several periods were designated as Intensive Operating Periods (IOP). During these IOPs the plume from the Navajo Generating Station was expected to be impacting the Grand Canyon and aircraft measurements of the plume trajectory were made.

The Unisearch Associates Inc. TAMS-150, TDLAS system was used to make real time measurements of gas-phase hydrogen peroxide (H_2O_2). The detection limits of the UNISEARCH Associate's TAMS-150 system for hourly average values was better than 100 pptv. This was adequate for following the temporal behaviour throughout the field study although on a few occasions the H_2O_2 mixing ratio remained close to the detection limit throughout a 24 hour period.

EXPERIMENTAL

The TDLAS method takes advantage of the high monochromaticity and rapid tunability of Pb salt diode lasers to measure absorptions from single rotational-vibrational lines in the mid-infrared spectrum of a molecule. To facilitate the measurement of very low optical densities (<10-5) at line center and to reduce the chances of overlap between adjacent absorption lines, reduced pressures ($\sim$25 Torr) are used to minimize pressure broadening of the rotational lines. The atmospheric sample is pumped rapidly at the reduced pressure through a White cell which also provides the long optical path lengths (153 m) required to achieve the desired detection limits.

Since a full description of the TAMS-150 field instrument together with the method of data analysis may be found in the papers by Mackay et al. (1990) and Schiff et al. (1987) only a brief description of the system will be presented here.

A schematic of the optical system and the electronic sub-assemblies is shown in Figure 1. Although this system is designed for simultaneous measurements of two species only H_2O_2 was monitored during this study. A diode laser characterized for H_2O_2 ($\sim$1280 cm-1) is housed in laser source assembly LSA-A which maintains the temperature of the laser in the 20 to 60 K range. The laser is modulated at 25 kHz so that phase sensitive detection techniques may be employed (see for example Hastie 1983).

The laser radiation is collected and collimated by an f/2 off-axis parabolic mirror, OAP1, which produces a 15 mm OD parallel beam of light. Plane fold mirror M1 and flip mirror S (which, in dual mode, is switched every three seconds to permit the beam from each diode to enter the White cell) manoeuvre the beam onto a second off axis-parabolic mirror, OAP3, which focuses the beam into the White cell.

The absorption cell is a 150 cm base path multiple reflection cell (Horn and Pimentel, 1971) with a total path length of 153 metres. Sample air enters through a 6 mm OD Teflon tube passing through one end plate and terminating just above, and in front of, the in-focus mirror. The gas is exhausted through a 2.5 cm ID tube in the aluminum plate at the opposite end of the cell.

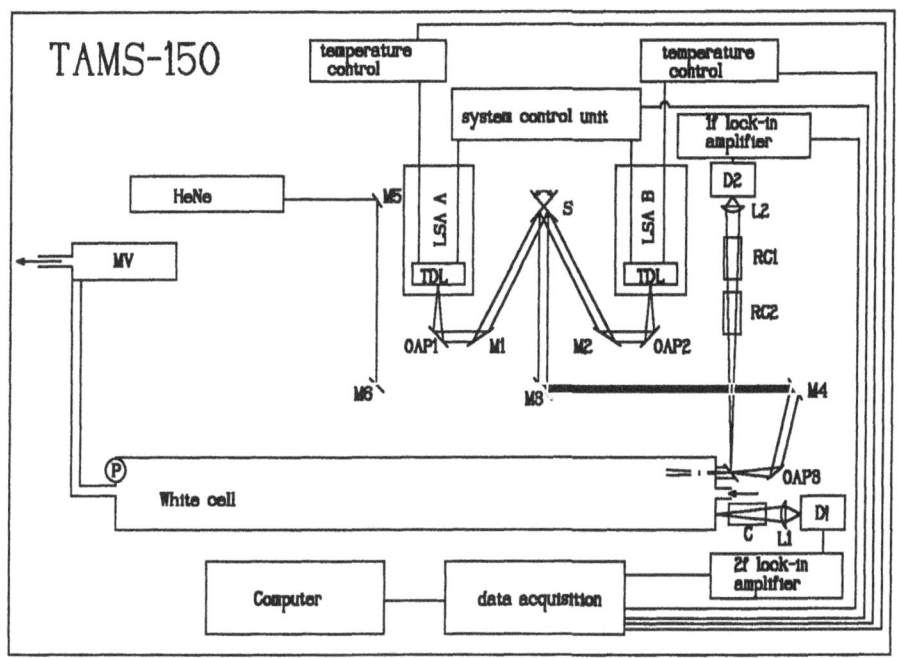

FIGURE 1. SCHEMATIC OF THE OPTICAL SYSTEM AND A BLOCK DIAGRAM OF THE CONTROL ELECTRONICS OF THE TAMS-150

Symbol	Description
MV	Pressure control valve
OAP1/2	f/2 off axis parabolic mirrors
OAP3	Off axis parabolic mirror with a focal length of 32 cm
M1-2	Beam directing mirrors
LSA A/B	Laser source assembly
D1/2	Infrared detectors
L1/2	BaF$_2$ lenses
S	Flip mirror
RC1/2	Reference cells for frequency-locking
HENE	HeNe alignment laser
TDL	Diode laser
M1-6	Beam directing mirrors
P	Baratron pressure gauge
C	Reference cell used in optical calibrations

The entrance window on the White cell is positioned at an angle of about 45° with respect to the laser beam and splits the beam into two parts. The majority of the laser radiation passes through the window and into the White cell. About 5% of the beam is reflected through cells, RC1 and RC2, which contain high concentrations of the monitored species and is then focussed by lens L2 onto detector D2. The output of this detector is passed to a specially designed lock-in amplifier which detects the center of the absorption line due to the trace gas being monitored and sends this information to the computer for frequency-locking.

The beam exiting the White cell is focussed by a BaF_2 f/1.5 lens, L1, onto a liquid-nitrogen cooled, mercury cadmium telluride, infrared detector, D1. The output of this detector is fed into an oscilloscope for visual representation and to a lock-in amplifier and data acquisition/computer control system for analysis and storage.

Sample air is drawn through a 6 mm OD.0.75 mm wall, FEP type Teflon tube approximately 8 m long. Particles are removed from the air by a 1.2 micron pore size Teflon filter located at the tubing entrance. A Teflon needle valve located 1 m downstream of the inlet maintains the flow into the White cell at 10 standard litres per minute (SLM). The section of the inlet line up to and including the needle valve was thermostated at 20°C. The low pressure portion of the line (8 m long) was at ambient temperatures. The sampling port was positioned 14 ft above the ground; 3 ft above the mobile laboratory roof and 3 ft away.

The air traverses the inlet line in a few tenths of a second while the residence time in the White cell is about 5 seconds. The flow of air entering the White cell is monitored with a calibrated mass flowmeter just prior to each calibration and is adjusted by the needle valve as necessary. A motorized valve, which is referenced to a MKS Baratron pressure gauge, controls the pressure in the White cell at 25 Torr.

The sensitivity of the TAMS-150 toward H_2O_2 is determined by introducing a 'spike' of known concentration of the target gas to the air stream where it is sampled into the system, immediately downstream of the particle filter. A detailed description of sampling integrity testing performed on H_2O_2 and the calibration procedure is given in Slemr et al, (1986).

The H_2O_2 standard used in this study was a permeation device of our own design. The permeation rate of the H_2O_2 calibration source was determined weekly during the field study by the colorimetric $TiCl_4$ method of Pilz and Johann (1974). Experience has shown that the permeation rate of our H_2O_2 device decreases slowly with time, approximately at the rate of 10% per month.

During the measurement periods the calibration gases flow continuously through the addition lines up to 3-way solenoid valves located close to the sampling gas inlet. When not required for calibration, these gas mixtures are exhausted far away from the inlet. During a calibration the solenoid valve closes the exhaust line and admits the calibration gas mixture to the sampled air stream. The addition lines are sufficiently short that negligible loss occurs and little delay is observed before stabilization of the measured calibration mixing ratio.

The TAMS-150 obtained a 1 minute average data point every 3 minutes. The raw data were stored on floppy diskettes for future analysis in the laboratory and the real time measurements were printed out to the computer monitor and printer at 3 minute intervals.

Calibration of H_2O_2 was performed once a day, usually between 06:00 and 08:00. The sensitivity of the instrument toward H_2O_2 was observed to have a day-to-day stability of better than 10%.

Hydrogen peroxide background spectra were obtained by inserting a scrubber in the sampling line that containing an amount of charcoal selected to efficiently remove the H_2O_2 without significantly altering the H_2O concentration. The objective was to retain the H_2O concentration close to ambient during the background measurements so that its contribution could be subtracted from the ambient measurements.

Although the chosen H_2O_2 absorption line was free of coincidental interferences, the pressure broadened tail of H_2O lines lying on either side of the H_2O_2 feature could cause changes in the background structure and hence degrade the detection limit. This would be particularly true during periods of high and variable humidity, e.g. at night and during rain events, when the scrubber might remove sufficient water vapour to modify the background.

Since the ambient H_2O_2 levels were close to the detection limit almost 50% of the time, changes in the background due to fluctuations in humidity could significantly degrade the detection limit. To minimize this source of noise the following ambient measurement sequence was used: 1) turn on scrubber and allow 30 seconds of stabilization, 2) measure background for 60 seconds, 3) turn off scrubber and allow 30 seconds of stabilization, 4) measure ambient for 60 seconds, 5) turn on scrubber and allow 30 seconds of stabilization, 6) measure background for 60 seconds, and so on.

For this measurement sequence, one ambient data point was obtained every 3 minutes. The results were subsequently reduced to 1 hour average values. Detection limits for the hourly averages were better than 100 pptv throughout the measurement period, better than 50 ppbv for 75% of the measurement period, and better than 25 pptv during 50% of the time.

The precision of the measurements over a 24 hour period is assessed to be + 5% while the overall accuracy is estimated to be + 20%. The major source of calibration error is the colorimetric calibration procedure which has an uncertainty assessed at + 10%. Additional uncertainties of + 2% are caused by temperature variations of the permeation device, and + 2% from the measurement of the carrier gas flow. The carrier gas flowmeter (Tylan, model FM361-20SLM) was checked against a bubble flowmeter prior to the study.

Measurements of H_2O_2 were made for a total of 74 days between 12:00 January 11 through 10:00 March 31 1990. Data coverage was better than 94% during the rest of the study. 110 hours of data out of a total of 1786 hours (6%) were lost due to calibration periods and to computer malfunction. Less than 1% data loss was recorded during the 23 designated Intensive Operating days.

RESULTS

Figure 2 shows the daily average H_2O_2 mixing ratios (obtained by fitting the daily average 2f signal to the calibration spectrum) measured during the study period. The values varied from between 32 pptv on March 12th to almost 800 pptv on March 25th. The highest values were measured during the period of unseasonably warm weather between March 20th and March 27th. The average values during this warm period was 650 ppbv, compared with 250 pptv measured during the rest of the study.

There was no obvious differences between measurements taken during the intensive operating periods when the plume from the generating station was expected to impact the site and on the non-intensive days when back trajectories indicated that the air mass over the site was more than 60 hours old with respect to any injection of gases from the generating station.

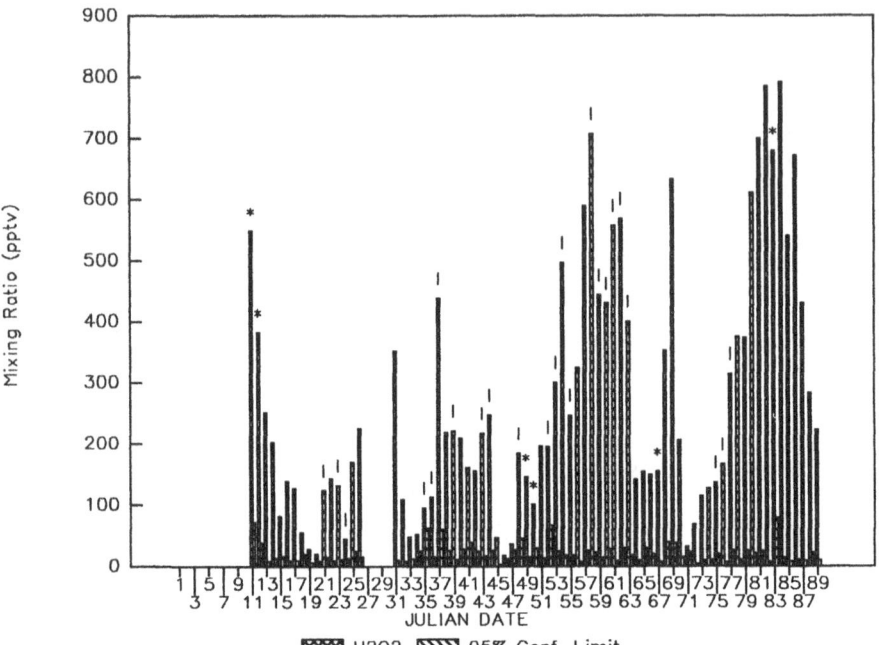

FIGURE 2.   Tunable diode laser absorption spectrometer measurements made at Desert View, Arizona during the Navajo Generating Station Visibility Study: Daily average H_2O_2 mixing ratios.
* indicates that data was acquired for less than 95% of the day but more than 50%; I indicates that this day was designated an Intensive Operating Period.

The H_2O_2 mixing ratios showed considerable variability, both in concentration and in diurnal behaviour. Daily maxima varied from below 100 pptv on January 24th to greater than 1.2 ppbv on March 10th. On average, the maximum value occurred during the night between 20:00 and 07:00 and minimum values during the daytime period between 08:00 and 14:00. Occasionally, an additional small maximum was observed between 12:00 and 16:00.

Figures 3a and 3b show respectively, the mean hourly H_2O_2 mixing ratios for the intervals designated Intensive Operating Periods and for the non-intensive periods. The data obtained on the 23 intensive days were not substantially different from those obtained during the non-intensive periods.

This diurnal behaviour is opposite to what we have observed previously, where maximum concentrations were found to occur during the afternoon and minima at night consistent with expected photochemical production during the day and removal by dry deposition at night. A possible explanation is provided by recognizing that our measurements were being conducted at the rim of the canyon which could be acting as a large lung. As the air subsides after sunset it brings higher concentration H_2O_2 from aloft past our sampling inlet. Conversely, in the morning, solar heating lifts canyon air which has been depleted in H_2O_2 as a result of deposition during the night, past our instrument. The secondary afternoon maximum seen occasionally is likely the result of the expected photochemical production normally seen at other locations.

The H_2O_2 measurements were examined for correlations with measurements of SO_2, fine and course sulphate particle concentrations and relative humidity. The sulphate measurements were made by a multiple impactor technique (Hering et al., 1991) and sulphur dioxide data were taken from the 4-hr filter pack data reported for Hopi Point (Richards et al., 1991). All data were averaged to the 8-hr time period for the sulphate aerosol data set. Day-to-day variations were examined by performing a running 24 hr average over each parameter.

On the basis of chemistry alone, one might have expected H_2O_2 concentrations to be correlated with sulphate particles and anti-correlated with SO_2, particularly under cloudy conditions or high humidity.

No statistically significant relationships were found. It is apparent that the H_2O_2 concentrations we measured were not under SO_2 control. This conclusion is also supported by the absence of detectable differences between H_2O_2 concentrations observed during the intensive periods when the measurement site was impacted by the plume from the generating station.

In addition, the H_2O_2 values are higher than model predictions for a clean site in winter. It is therefore likely that the H_2O_2 concentrations are governed by NO_x and precursors advected from source locations such as Phoenix or Los Angeles. This suggestion is similar to that reached by Kleinman (1991). Unfortunately, we were not able to check this hypothesis since the NO_x levels were generally below the detection limits of the instruments used during this study.

ACKNOWLEDGEMENTS

We gratefully acknowledge the financial support of Salt River Project provided under contract number VN-07992-CAS. We thank Dr. Dennis Fitz of AeroVironment Inc. for his efforts in obtaining the financial support for the TDLAS measurements and all the AeroVironment personnel involved for logistical support during the study. The efforts of Dr. Susanne Hering and Mr. Mark Stoltzenberg of Aerosol Dynamics who spent many hours cross correlating the suphate, SO_2, RH and H_2O_2 measurements are gratefully acknowledged. We also thank Mr. Michael Gordon of SRP for providing what limited NO_x data there was.

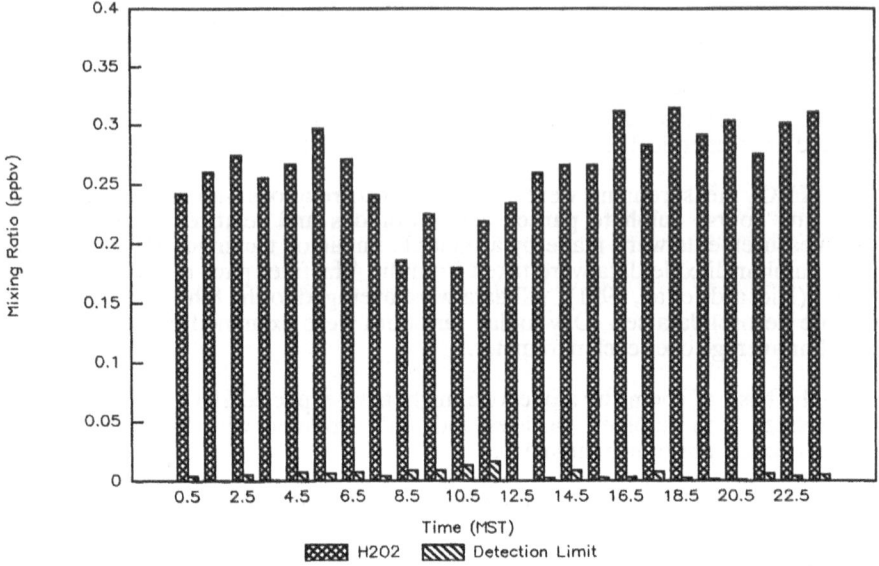

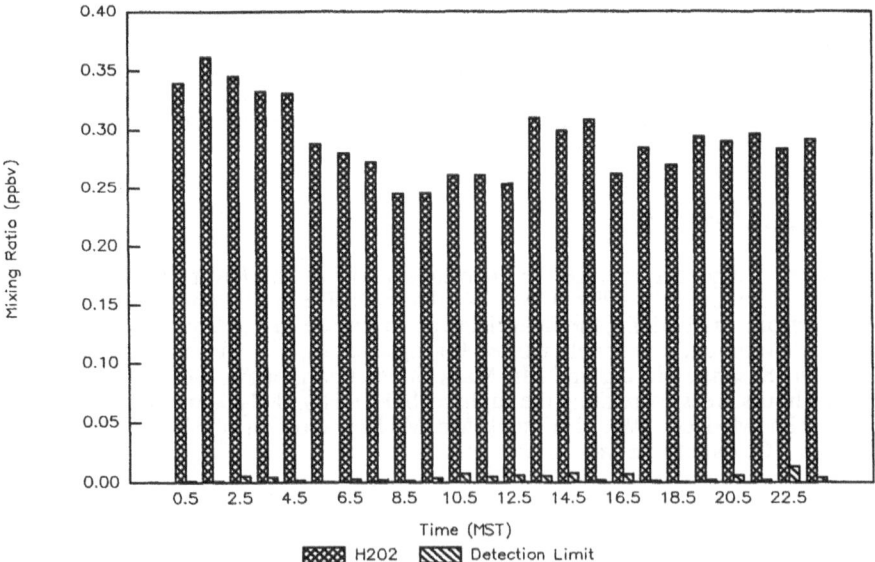

FIGURE 3. Mean hourly averaged H_2O_2 mixing ratios measured during all intensive operation periods (top) and non-intensive periods (bottom) between January 11 and March 19, 1990 by the tunable diode laser absorption spectrometer at the Desert View ground station, Grand Canyon Park, Arizona.

REFERENCES

Boyce, S.D., M.R. Hoffmann (1984) KINETICS AND MECHANISM OF THE FORMATION OF HYDROXYMETHANESULFONIC ACID AT LOW PH, J. Phys. Chem. 88, 4740-4748.

Hastie D.R., G.I. Mackay, T. Iguchi, B.A. Ridley, H.I. Schiff (1983) TUNABLE DIODE LASER SYSTEMS FOR MEASURING TRACE GASES IN TROPOSPHERIC AIR, Environ. Sci. Technol. 17, 352-364.

Hering, S.V., M.R. Stoltzenburg (1991) LOW PRESSURE IMPACTOR MEASUREMENTS OF SULFUR SIZE DISTRIBUTIONS FOR THE 1990 NAVAJO GENERATING STATION WINTER VISIBILITY STUDY, Final report to Salt River Project, Contract No. VN07990CAS

Horn, D., G.C. Pimentel (1971) 2.5-KM LOW-TEMPERATURE MULTIPLE REFLECTION CELL, Appl. Opt., 10, 1892-1898.

Kleinman, L.I. (1991) SEASONAL DEPENDENCE OF BOUNDARY LAYER PEROXIDE CONCENTRATION: LOW AND HIGH NO_x REGIMES, J. Geophys. Res., 96, 20,721-20,733.

Logan, J.A., M.J. Prather, S.C. Wofsy, M.B. McElroy (1980) TROPOSPHERIC CHEMISTRY: A GLOBAL PERSPECTIVE, J. Geophys. Res., 86, 7210-7254.

Mackay, G.I., L.K. Mayne, H.I. Schiff (1990) MEASUREMENTS OF H_2O_2 AND HCHO BY TUNABLE DIODE LASER ABSORPTION SPECTROSCOPY DURING THE 1986 CARBONACEOUS SPECIES METHODS COMPARISON STUDY IN GLENDORA, CALIFORNIA, Aerosol Sci. and Technol., 12, 000-000.

Middleton, P., C.S. Kiang, V.A. Mohnen (1980) THEORETICAL ESTIMATES OF THE RELATIVE IMPORTANCE OF VARIOUS URBAN SULFATE AEROSOL PRODUCTION MECHANISMS, Atmos. Environ., 14, 463-472.

Penkett, S.A., B.M.R. Jones, K.A. Bruce, A.E.G. Eggleton (1979) THE IMPORTANCE OF ATMOSPHERIC OZONE AND HYDROGEN PEROXIDE IN OXIDIZING SULPHUR DIOXIDE IN CLOUD AND RAIN-WATER, Atmos. Environ., 13, 123-137.

Pilz, W., I. Johann (1974) Die Bestimmung Lleinster Mengen von WASSERSSTUFFPEROXYD IN LUFT (DETERMINATION OF TRACE HYDROGEN PEROXIDE IN AIR), Int. J. Environ. Anal. Chem., 3, 257-270.

Schiff, H.I., G.W. Harris, G.I. Mackay, (1987) MEASUREMENT OF ATMOSPHERIC GASES BY LASER ABSORPTION SPECTROMETRY, ACS Symposium series 349, The Chemistry of Acid Rain: Sources and Atmospheric Processes, Ed. R.W. Johnson and G.E. Gordon, Chapter 24, 274-288.

Slemr, F., G.W. Harris, D.R. Hastie, G.I. Mackay, H.I. Schiff (1986) MEASUREMENT OF GAS PHASE HYDROGEN PEROXIDE IN AIR BY TUNABLE DIODE LASER ABSORPTION SPECTROSCOPY, J. Geophys. Res. 91, 5371-5378.

MEASUREMENT OF ATMOSPHERIC SPECIES BY MID-INFRARED AND NEAR-INFRARED TUNABLE DIODE LASER ABSORPTION

Alan C. Stanton, David S. Bomse, Joel A. Silver,
David C. Hovde, and Daniel B. Oh

Southwest Sciences, Inc.
1570 Pacheco St., Suite E-11
Santa Fe, New Mexico 87501
U. S. A.

SUMMARY

Recent progress with detection methods based on high frequency laser wavelength modulation has permitted the development of ultrasensitive instrumentation for measurement of gas phase species by diode laser absorption. The technique of high frequency wavelength modulation with harmonic detection is reviewed and its performance is compared with one-tone and two-tone FM spectroscopy. A new dual modulation method that provides an output signal proportional to absorbance while locking the diode laser wavelength to an absorption feature is presented. This method rejects drift in the signal baseline and is particularly well suited to measurement of weak absorption signals. The application of these techniques to the measurement of methane flux using near-infrared diode lasers is discussed.

1. INTRODUCTION

Absorption spectroscopy using tunable diode lasers is widely recognized as a valuable tool for sensitive measurement of gas species important in atmospheric chemistry and pollution research. In recent years, the sensitivity of diode laser instrumentation for detection of trace concentrations of atmospheric species has been enhanced by the development of new experimental techniques, such as frequency modulation (FM) spectroscopy, which permit the attainment of nearly shot noise limited performance in many instances.[1-5] When such techniques are used in systems employing multiple pass absorption cells to attain path lengths of 50 meters or greater, detection limits well below 1 part per billion (ppb) are possible for many species using lead salt diode lasers operating at mid-infrared wavelengths. Recently, substantial interest has also developed in the use of near-infrared diode lasers (InGaAsP or GaAlAs) for atmospheric species measurements.[6-9] Although these lasers operate in spectral regions where only overtone or combination absorption bands of most molecules can be accessed, sufficient sensitivity for measurements of some species in the sub part per million range can be achieved.

In this paper, we discuss recent work with diode lasers using high frequency wavelength modulation spectroscopy with harmonic detection. This technique operates in a frequency regime where most of the laser noise reduction achieved by FM spectroscopy is retained, yet significant operational advantages are found. We next describe a dual modulation technique that we have found useful for continuous diode laser measurement of weak absorption signals. This technique is a laser line-locking scheme, but it also achieves substantial suppression of low frequency drift in signal baselines while providing a continuous output signal proportional to absorbance. Finally, we provide an example of an application of these techniques to measurement of methane fluxes for greenhouse gas studies. In this work, methane absorption is measured in an open atmospheric path using InGaAsP diode lasers operating near 1.33 or 1.65 μm. The ambient methane concentration is measured at 10 Hz and combined with sonic anemometer measurements of the wind velocity components to determine methane flux by eddy correlation techniques.

2. HIGH FREQUENCY WAVELENGTH MODULATION SPECTROSCOPY

A variety of modulation schemes exists for applying diode laser absorption spectroscopy to trace species detection.[1-5,10-17] In all approaches, the diode laser frequency is modulated by adding an ac component to the laser injection current. Absorbance is quantified by phase-sensitive demodulation of the detector output at some multiple or combination of the modulation frequencies. Recently, much attention has been focused on frequency modulation spectroscopy (FMS), which employs modulation frequencies comparable to or greater than the absorption linewidth (100 MHz to several GHz). One-tone FMS detects the signal at the modulation frequency.[4] Two-tone FMS uses a pair of closely spaced frequencies to modulate the laser and quantifies absorbance by demodulating the detector output at the difference frequency.[11] Absorbance sensitivities in the low 10^{-7} regime have been demonstrated with these techniques using mid-IR and near-IR diode lasers.

"Wavelength modulation spectroscopy" (WMS) uses modulation frequencies which are much lower than the absorption linewidth. Typically, the modulation amplitude is adjusted so that the extent of the laser wavelength modulation is comparable to absorption linewidths. Detection is normally performed at a harmonic of the modulation frequency, with second harmonic detection being the most widely used. The literature contains hundreds of publications by diode laser spectroscopists who have used this technique, typically at modulation frequencies in the 1 to 10 kHz regime.

Recently, we examined the theories pertinent to FMS and WMS.[16] This work showed that WMS is just a subset of FMS, where the modulation frequencies are kept smaller than the linewidth, and that a single theory can be used to describe both FMS and WMS. Further, it was shown that high frequency WMS, with modulation frequencies of a few MHz, is usually as sensitive as one-tone or two-tone FMS and is substantially easier to implement. In particular, the impedance matching difficulties which arise in attempting diode laser current modulation at several hundred MHz are absent. It is not surprising that high frequency WMS should be comparable in sensitivity to two-tone FMS, because detection frequencies in the 5 to 20 MHz regime are typically used in both cases.

To confirm the results of these calculations, we performed an experimental comparison of wavelength modulation spectroscopy and one-tone and two-tone FM spectroscopy using a lead salt diode laser (11 μW output power).[17] Wavelength modulation spectroscopy was examined over a range of modulation frequencies spanning more than four orders of magnitude (500 Hz to 10 MHz). First and second harmonic detection and, in a few instances, fourth and sixth harmonic detection, were studied. Single-tone FM measurements at 150 MHz and two-tone FM at 10 MHz (with 345 and 355 MHz modulation frequencies) were also studied. Either a sine wave generator or a crystal oscillator was used as the modulation source. Demodulation at the detection frequency was accomplished using a lock-in amplifier at frequencies below 100 kHz or discrete components (mixer, phase shifter, etc.) at higher frequencies. These experiments used a low pressure cell containing N_2O as the sample absorbing gas. Our experimental results, which fall into the detector thermal noise-limited category, were compared with the theory and with other results reported for low power mid-IR lasers. A full description of the experimental setup and results is given in Reference 17.

Excellent agreement between predicted and experimentally observed signal to noise was obtained in these experiments. In particular, the prediction that high frequency WMS and FMS yield comparable sensitivities was confirmed. All of the high frequency modulation methods produced absorbance sensitivities between 2 and 7×10^{-7}. In separate studies, we have consistently shown that absorbance sensitivities better than 10^{-6} can be achieved with either lead salt mid-IR or InGaAsP or GaAlAs near-IR lasers using wavelength modulation spectroscopy with second or fourth harmonic detection, at detection frequencies as low as 100 kHz.

Because the method seems to provide the best combination of sensitivity and ease of implementation, we have adopted the use of high frequency wavelength modulation spectroscopy for most of our studies concerning trace species detection with diode lasers. Our typical systems use a modulation frequency of 5 MHz, with either 10 MHz or 20 MHz detection. In some instances with higher power lasers, it is likely that significant laser noise extends to higher frequencies. In that case, one-tone FM spectroscopy at frequencies of 100 MHz or greater may be useful.[3] In most practical cases, however, unwanted etalon fringes are likely to limit the sensitivity so that no further advantage is realized by moving to the higher detection frequencies.

3. A DUAL MODULATION METHOD FOR STABILIZING DIODE LASER WAVELENGTH

In applications where diode lasers are used to measure trace species concentrations, continuous monitoring of the absorbance signal is often desired. In most cases, active stabilization of the diode laser wavelength is necessary. In addition, low-frequency drift of the signal baseline in wavelength modulation spectroscopy can often limit the measurement of weak absorbance signals if only the peak absorbance is monitored. We have observed large fluctuations in the baseline in a variety of experiments with both mid-IR and near-IR diode laser devices. Low frequency baseline noise can be attributed to various effects such as feedback in optical fibers, beam steering from hot turbulent gases, fluctuations from mechanical vibrations, or thermal fluctuations in the laser.

We have recently developed a dual modulation method for stabilizing the diode laser wavelength that also nulls low frequency baseline noise in the sample absorbance. The method produces an output signal which is directly proportional to absorbance. This technique, which can be readily used with high frequency wavelength modulation, is described in detail in a recent note in *Applied Optics*.[18]

Briefly, the dual modulation method is implemented by using high frequency (e.g. $\Omega \sim 5$ MHz) wavelength modulation spectroscopy with $2f$ or $4f$ detection while simultaneously dithering the laser wavelength at a low frequency ω (say 500 Hz) between the troughs of the $2f$ or $4f$ signal. The modulation amplitude at frequency Ω is chosen to optimize the amplitude of the $2f$ or $4f$ waveform. The lower frequency modulation, using a sinusoidal or triangular waveform, provides the mechanism for nulling baseline noise and locking the laser wavelength.

If the sample absorbance is sufficiently large, a separate reference cell for line locking is not required. In this case, the detector output signal is split into two legs, one for the absorbance measurement and one for line locking. As indicated in Fig. 1(b), the output of the sample leg is demodulated at 2Ω to produce a second harmonic signal, as in conventional wavelength modulation spectroscopy. If this signal, denoted as $2f_\Omega$, is then demodulated at twice ω, an output that looks like a $4f$ spectrum is generated [Fig. 1(d)]. The technique nulls low frequency baseline drift because it effectively ac-couples the $2f_\Omega$ signal. That is, during every modulation period at frequency ω, the $2f_\Omega$ peak and troughs are each sampled twice at evenly spaced intervals. Demodulation at 2ω yields a signal proportional to the peak-to-trough difference independent of low frequency baseline fluctuations. We note that phase sensitive demodulation at 2ω is actually not required. Modulation of the $2f_\Omega$ signal at frequency ω generates a waveform with amplitude proportional to the $2f_\Omega$ peak-to-trough difference, independent of phase. Thus, processing of this signal with an rms detector may be sufficient.

Laser wavelength stabilization is also achieved by two sequential demodulations. First, as shown in Fig. 1(a), a $1f_\Omega$ signal is obtained by demodulating the detector reference leg output at frequency Ω. Then, a feedback voltage for line locking is produced by further demodulating the $1f_\Omega$ signal at 2ω. As shown in Fig. 1(c), this second demodulation produces a lineshape that looks like a $3f$ spectrum and, since it has a zero crossing at line center, is ideally suited for line locking.

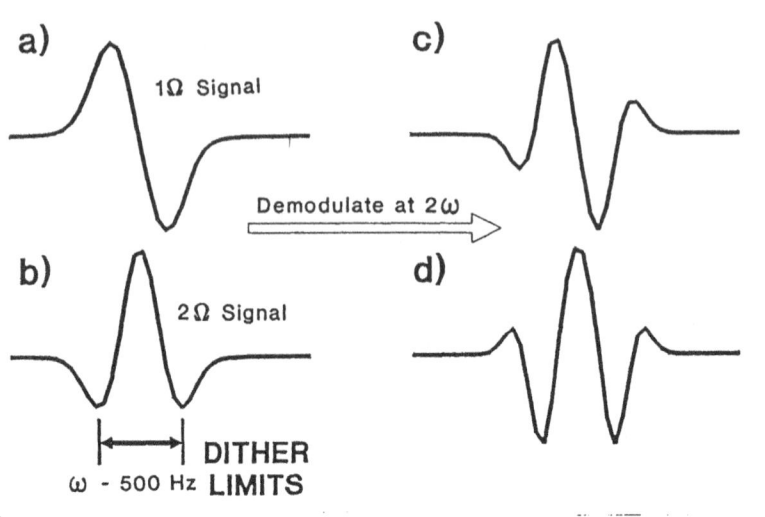

a) 1Ω Signal

b) 2Ω Signal

Demodulate at 2ω

c)

d)

$\mapsto$ DITHER
ω - 500 Hz LIMITS

Figure 1 - Example waveforms pertinent to dual modulation line locking scheme.

The ability of this method to stabilize laser wavelength and reject low frequency baseline fluctuations is shown by the experimental data in Fig. 2. These data were obtained with a lead salt laser tuned to the N_2O $\nu_3P(45)$ line at 1243.795 cm^{-1}. Wavelength modulation at 5 MHz and dither modulation with a 500 Hz triangle wave were used. One Torr of N_2O in an 11-cm long cell leads to an absorbance of 0.67. The signal traces labeled "standard line-locking" represent the amplitude of the $2f_\Omega$ absorbance signal when feedback is provided by the $1f_\Omega$ signal. This method for diode laser line locking has been employed in many previous studies. The traces labeled "dual modulation" show the amplitude of the $2f_\omega$ signal when our dual modulation scheme is implemented. The data on the left-hand side of the figure compare the two line locking schemes under conditions where the large absorbance is measured in the absence of significant baseline fluctuations. Both line locking methods work well under these conditions: the $2f_\Omega$ signal exhibits <0.5% rms fluctuation in a 1-Hz bandwidth. This amplitude stability corresponds to better than 0.1-ppm wavelength precision and can be maintained for at least 30 minutes. When the feedback control was removed, the $2f_\Omega$ signal dropped by half within 1 to 2 minutes.

The data on the right-hand side of Fig. 2 were taken when low-frequency baseline drift is simulated by using a differential amplifier to combine a 0.013-Hz sine wave with the $2f_\Omega$ signal. When these simulated baseline fluctuations were included at an amplitude equal to twice the $2f_\Omega$ signal, the fluctuations propagated unattenuated (upper right trace). In the lower right trace, where dual modulation line-locking is used, the fluctuations have disappeared. The observed $2f_\omega$ signal shows only a small increase in rms noise, from 0.33 to 0.39%, with the addition of low-frequency baseline drift at twice the $2f_\Omega$ signal amplitude.

We are using the dual modulation technique in several diode laser applications where continuous monitoring of an absorbance signal is required, including airborne measurements of H_2O and ground based measurements of methane flux. The application to methane flux measurements is described in the following section. We have found that a sample absorbance of ~0.002 or higher seems sufficient for line locking, eliminating the need for a separate line locking reference cell.

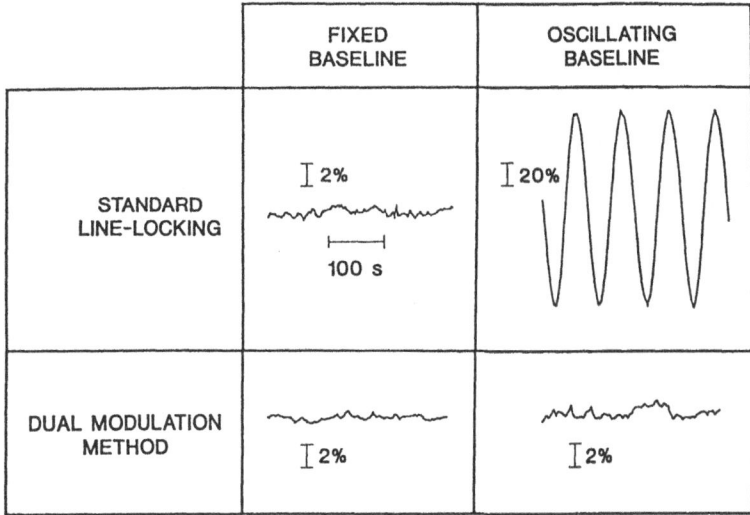

	FIXED BASELINE	OSCILLATING BASELINE
STANDARD LINE-LOCKING	I 2% ——— 100 s	I 20%
DUAL MODULATION METHOD	I 2%	I 2%

Figure 2 - Experimental results showing rejection of low-frequency baseline oscillations by the dual modulation technique.

4.   NEAR-INFRARED DIODE LASER MEASUREMENT OF METHANE FLUXES

Methane plays an important role in chemistry and radiative transport in the atmosphere. The diversity of global sources and sinks for this gas make it difficult to estimate the net annual flux of methane to the atmosphere. The traditional method for measuring methane flux from sources such as natural wetlands, rice paddies, landfills, etc. utilizes chamber techniques. The surface across which fluxes are to be measured is covered with a sealed chamber. Air samples are withdrawn from the chamber at intervals. The samples are later analyzed for methane using gas chromatography with flame ionization detection.

An alternative technique for flux measurements, the eddy correlation method, is preferred by many atmospheric researchers because perturbation of the measurement environment is minimal and gas concentrations are measured *in situ*, in real time. By measuring simultaneously the gas concentration and the three components of wind velocity, with time resolution of 0.1 second or faster, one can calculate the net flux of the gas perpendicular to the average wind streamline. The wind velocity measurement is easily achieved using commercially available sonic anemometers. Thus the key component for eddy correlation measurement of trace gas fluxes is a gas concentration sensor with adequate time response, sensitivity, and stability.

The possibility of using lead salt diode lasers for eddy correlation measurements of methane flux was demonstrated by Anderson and Zahniser.[19] An impediment to using this technique in remote field environments is the need for cryogenic cooling of the lead salt lasers, requiring either a supply of liquid nitrogen or substantial electrical power to operate a closed cycle refrigeration system. Near-infrared InGaAsP diode lasers, by contrast, operate at room temperature and are compatible with silica optical fibers. Although these lasers are not presently available commercially at any arbitrary near-IR wavelength, single frequency distributed feedback (DFB) lasers can be obtained near 1.33 μm, in the region of the $2\nu_3 + \nu_2$ combination band of methane. Very recently, DFB lasers have become available near 1.65 μm, in the region of the $2\nu_3$ overtone band of methane.

We have studied the sensitivity of methane detection using InGaAsP lasers operating near 1.33 μm. This laser wavelength is near the 1.31 μm wavelength preferred in many fiber optic communications applications, so DFB lasers are available from many manufacturers. Figure 3 shows experimental methane and water vapor absorption spectra obtained by scanning the temperature of a Toshiba DFB laser from 0 °C to 50 °C at fixed laser current. The second harmonic ($2f$) detection method was used in this scan, where only the positive-going portions of the $2f$ signals are shown to improve the clarity of the figure. Because of the compressed laser frequency scale in the figure, individual absorption lines appear as narrow spikes. For purposes of atmospheric methane detection, either the R(0) or R(1) line seems suitable, as both lines are relatively isolated within the CH_4 spectrum and therefore yield clean lineshapes which are readily analyzed. Also, each of these lines is free of water vapor interference, although nearby H_2O lines could possibly be exploited for near-simultaneous measurement of water vapor.

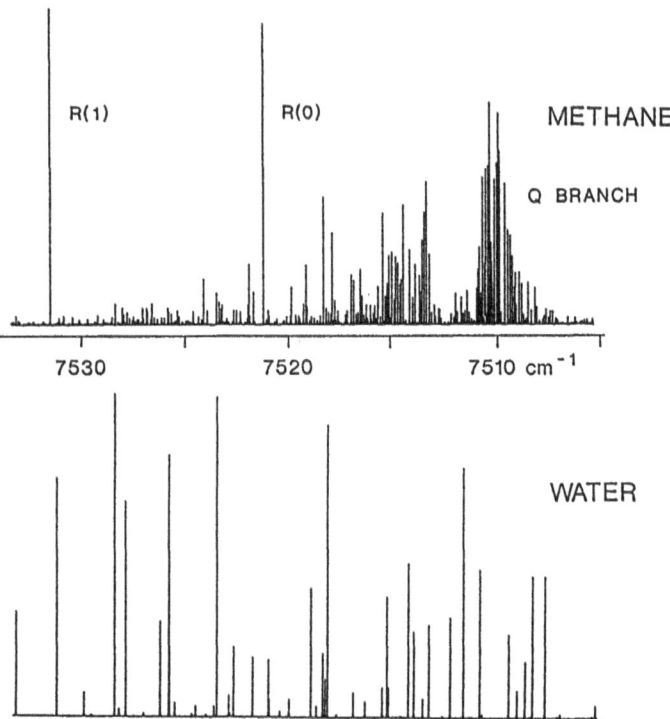

Figure 3 - InGaAsP DFB diode laser spectra of methane and water absorption near 1.33 μm.

The very limited prior studies of the methane $\nu_2 + 2\nu_3$ band do not provide quantitative information on absorption line strengths or pressure broadening parameters. To provide a more sound basis for evaluating the possibility of ambient methane detection in this band, we used the DFB diode laser to measure the strength and N_2 broadening coefficient for the R(1) line near 7531.4 cm^{-1}. These measurements were made using a multiple pass Herriott cell path length of 19.47 meters. The laser wavelength was current-tuned across the R(1) absorption feature and the direct absorption lineshape was recorded for known pressures of pure CH_4 or CH_4 broadened by known pressures of N_2. The diode laser transmission data were analyzed using a least-squares fitting routine to determine the best-fit Voigt profile. The integrated absorbance and linewidth are available from the fit, yielding the line strength, S, and pressure broadening coefficient (HWHM), γ. The results for the R(1) CH_4 line are S(296 K) = 6.62 × 10^{-23} cm^2 molecule^{-1} cm^{-1} and $\gamma(N_2$, 296 K) $\simeq$ 0.06 cm^{-1} atm^{-1}. The uncertainty in these measurements is estimated to be approximately 5%. An essentially identical result for the line strength was obtained from high resolution FTIR data.

These results may be compared with other recently reported data in the $\nu_2 + 2\nu_3$ band. Rooth, in another presentation in this symposium,[20] reports experimentally determined absorption coefficients for several methane lines in this band. Although the R(1) line was not studied by Rooth, he reports line center absorption cross sections for 10 Torr CH_4 which are on the order of 2 × 10^{-21} cm^2 for several lines from R(2) through R(6). The line center absorption cross section which we calculate from our R(1) line strength in the low pressure Doppler broadened limit is σ = 2.7 × 10^{-21} cm^2. Thus, the magnitude of this result seems to be quite consistent with Rooth's data. A recent publication by Carlisle and Cooper also discusses diode laser measurements in the same 1.33 μm methane band.[21] While the focus of the Carlisle and Cooper paper is not the measurement of ambient methane, they report a line strength of only 0.25 × 10^{-23} cm^2 molecule^{-1} cm^{-1} for a CH_4 line near 7563 cm^{-1}, a factor of approximately 25 lower than the R(1) line strength that we measured. From this result, they conclude that this methane band is too weak for ultrasensitive methane measurements. Our high resolution FTIR data indicate that lines near 7563 cm^{-1}, which are assigned to the R(4) transition, are comparable in strength to the R(1) line. Rooth's data show 10 Torr absorption cross sections for the R(4) lines ranging from 8 to 15 × 10^{-22} cm^2, within a factor of two of the R(1) cross section. We therefore conclude that Carlisle and Cooper's measurement is in error, or at a minimum represents a weak line which is not typical of the stronger lines in the $\nu_2 + 2\nu_3$ band.

Based on our line strength and broadening data, the line center absorption cross section for the R(1) line in air at atmospheric pressure is about 3.5 × 10^{-22} cm^2. For a 50 meter absorption path length, which is easily achieved with either White cell or Herriott cell designs, the ambient methane concentration of ~1.7 ppm in the troposphere corresponds to an absorbance of about 7.3 × 10^{-5}. This absorbance is comfortably above the detection limits that can be achieved with high frequency wavelength modulation or FM spectroscopy. If a detection limit of 10^{-6} is achieved, the minimum detectable methane concentration in an open path atmospheric pressure configuration using the 1.33 μm band would be in the range of 20 ppb. This level of sensitivity may make the technique suitable for use in eddy correlation flux measurements.

We have demonstrated the detection of ambient methane using a single mode fiber pigtailed DFB diode laser for measurement of the R(1) line in the 1.33 μm band. Figure 4 compares the 4f signal amplitudes obtained when a multiple pass absorption cell (~78 meter path length) is filled first with an analyzed 36.2 ppm methane-air mixture and then with ambient outside air. In both cases, a total pressure of 100 Torr was used. The dual modulation technique was used to process the 4f (20 MHz) signal, with a one-second time constant, but line locking was not used in this instance. By comparing the signal levels for the two cases, the outside air measurement implies an ambient methane concentration of ~1.8 ppm, reasonably consistent with the accepted value of 1.6 to 1.7 ppm for the mean tropospheric methane concentration. The signal to noise of the ambient methane measurement is about 10:1, implying a limiting sensitivity (S/N=1) of ~0.2 ppm with a one second time constant. The noise limitation, equivalent to about 7 × 10^{-6} absorbance, was apparently dominated by optical feedback to the laser from the cleaved end of the optical fiber. An optical isolator was not used in this experiment and could perhaps provide additional improvement in signal to noise.

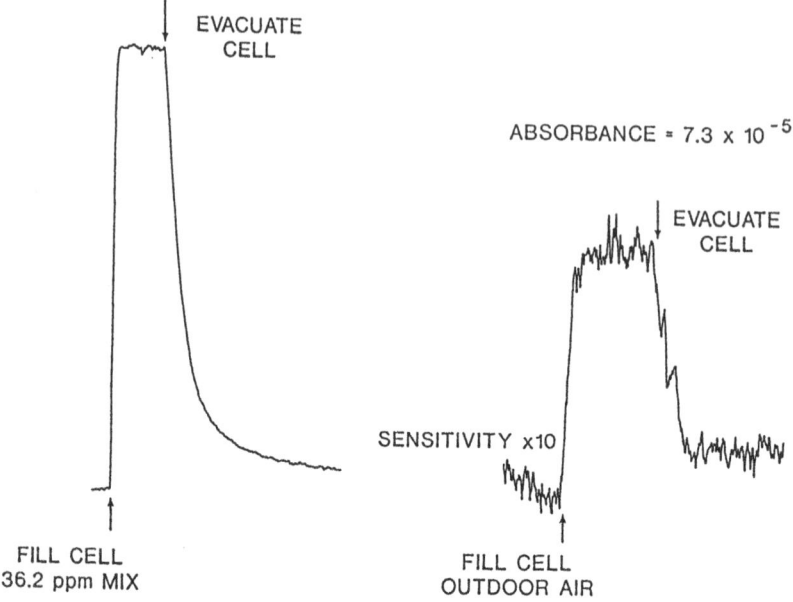

Figure 4 - Detection of ambient methane by DFB diode laser absorption in the $\nu_2 + 2\nu_3$ band.

While the above results imply that a 1.33 μm InGaAsP laser-based instrument could be used for methane flux measurements, recent developments in InGaAsP laser availability at other wavelengths have greatly improved the sensitivity of near-infrared diode laser detection of methane. Distributed feedback lasers are now commercially available at 1.65 μm from at least one manufacturer (Anritsu). This wavelength coincides with the much stronger $2\nu_3$ overtone band of CH_4. The strongest lines in this band have absorption cross sections approximately 40 times greater than the cross sections quoted above for the $\nu_2 + 2\nu_3$ band.[22] As a result, the absorbance due to 1.7 ppm methane in air for a 50 meter path length is about 3×10^{-3}. Absorbance signals at this level are very easily measured with high signal to noise using the high frequency detection techniques described in this paper, and the approach seems well suited to methane flux measurements.

We have developed an instrument for measurement of methane flux by eddy correlation, based on 1.65 μm diode laser absorption. The laser output is transmitted through an open path multiple pass Herriott cell to an InGaAs photodiode. The multiple pass mirrors (10 cm diameter, high reflectivity dielectric coating for 1.65 μm) are separated by approximately 50 cm. We have been able to achieve in excess of 200 optical passes in this cell, but we typically adjust the mirror spacing for about 110 passes for greater optical stability. The dual modulation detection and line locking approach described in this paper, with $\Omega = 5$ MHz and $\omega = 10$ kHz, is used with this system. The ambient methane signal from the measurement channel is used for line locking, so that no separate reference cell is required. A laptop computer (80386/80387 microprocessor) is used to control the diode laser instrument and acquire methane concentration data. The methane concentration is measured at 0.1 second intervals for correlation with the wind velocity data measured by a sonic anemometer.

We have successfully tested this instrument for measurement of methane flux from a capped landfill. These experiments were conducted in collaboration with researchers from the U. S. National Oceanic and Atmospheric Administration, Atmospheric Turbulence and Diffusion Division, Oak Ridge, Tennessee. A complete description of the instrumentation and the results of these initial experiments will be provided in forthcoming publications.

5. DISCUSSION

We have briefly reviewed techniques for achieving high sensitivity in detection of trace species by tunable diode laser absorption. The technique of high frequency wavelength modulation spectroscopy is found to be as sensitive as FM spectroscopy in most practical instances. This technique is relatively easy to implement. In particular, it does not require impedance matching of high frequency modulation to the diode laser. A dual modulation technique has been described that can be used in combination with high frequency wavelength modulation spectroscopy. The dual modulation method is an approach for laser wavelength stabilization that provides an output signal proportional to the sample absorbance. In addition, the method is effective in rejecting low frequency drift in the signal baseline. Finally, we have discussed the implementation of these techniques in instrumentation to measure methane flux for greenhouse gas studies, using near-IR InGaAsP diode lasers.

REFERENCES

1. CARLISLE, C. B., COOPER, D. E., and PRIER, H. (1989). Quantum Noise-Limited FM Spectroscopy with a Lead-Salt Diode Laser. *Appl. Opt.* **28**, 2567-2576.
2. WANG, L. G., TATE, D. A., RIRIS, H., and GALLAGHER, T. F. (1989). High-Sensitivity Frequency-Modulation Spectroscopy with a GaAlAs Diode Laser. *J. Opt. Soc. Am.* **B6**, 871-876.
3. WERLE, P., SLEMR, F., GEHRTZ, M., and BRAUCHLE, C. (1989). Quantum-Limited FM-Spectroscopy with a Lead-Salt Diode Laser. *Appl. Phys.* **B49**, 99-108.
4. LENTH, W. (1983). Optical Heterodyne Spectroscopy with Frequency- and Amplitude-Modulated Semiconductor Lasers. *Opt. Lett.* **11**, 575-577.
5. SILVER, J. A. and STANTON, A. C. (1988). Two-Tone Optical Heterodyne Spectroscopy Using Buried Double Heterostructure Lead-Salt Diode Lasers. *Appl. Opt.* **27**, 4438-4444.
6. JOHNSON, T. J., WIENHOLD, F. G., BURROWS, J. P., and HARRIS, G. W. (1991). Frequency Modulation Spectroscopy at 1.3 μm Using InGaAsP Lasers: a Prototype Field Instrument for Atmospheric Chemistry Research. *Appl. Opt.* **30**, 407-413.
7. STANTON, A. C. and SILVER, J. A. (1988). Measurements in the HCl 3 ← 0 Band Using a Near-IR InGaAsP Diode Laser. *Appl. Opt.* **27**, 5009-5015.
8. CASSIDY, D. T. (1988). Trace Gas Detection Using 1.3-μm InGaAsP Diode Laser Transmitter Modules. *Appl. Opt.* **27**, 610-614.
9. CARLISLE, C. B., and COOPER, D. E. (1990). Tunable Diode Laser Frequency Modulation Spectroscopy through an Optical Fiber: High-Sensitivity Detection of Water Vapor. *Appl. Phys. Lett.* **56**, 805-807.
10. COOPER, D. E. and GALLAGHER, T. F. (1985). Double Frequency Modulation Spectroscopy: High Modulation Frequency with Low-Bandwidth Detectors. *Appl. Opt.* **24**, 1327-1334.
11. JANIK, G. R., CARLISLE, C. B., and GALLAGHER, T. F. (1986). Two-Tone Frequency-Modulation Spectroscopy. *J. Opt. Soc. Am.* **B3**, 1070-1074.
12. JANIK, G., CARLISLE, C., and GALLAGHER, T. F. (1985). Frequency Modulation Spectroscopy with Second Harmonic Detection. *Appl. Opt.* **24**, 3318-3319.
13. REID, J. and LABRIE, D. (1981). Second-Harmonic Detection with Tunable Diode Lasers - Comparison of Experiment and Theory. *Appl. Phys.* **B26**, 203-210.

14. MOSES, E. I. and TANG, C. L. (1977). High Sensitivity Laser Wavelength-Modulation Spectroscopy. *Opt. Lett.* **1**, 115-117.
15. POKROWSKY, P., ZAPKA, W., CHU, F., and BJORKLUND, G. C. (1983). High Frequency Wavelength Modulation Spectroscopy with Diode Lasers. *Opt. Commun.* **44**, 175-179.
16. SILVER, J. A. (1992). Frequency Modulation Spectroscopy for Trace Species Detection: Theory and Comparison among Experimental Methods. *Appl. Opt.*, in press.
17. BOMSE, D. S., STANTON, A. C., and SILVER, J. A. (1992). Frequency Modulation and Wavelength Modulation Spectroscopies: Comparison of Experimental Methods Using a Lead-Salt Diode Laser. *Appl. Opt.*, in press.
18. BOMSE, D. S. (1991). Dual-Modulation Laser Line-Locking Scheme. *Appl. Opt.* **30**, 2922-2924.
19. ANDERSON, S. M. and ZAHNISER, M. S. (1991). Open-Path Tunable Diode Laser Absorption for Eddy Correlation Flux Measurements of Atmospheric Trace Gases. *Proceedings of the SPIE* **1433**, 167-178.
20. ROOTH, R. A. (1992). Spectroscopic Measurements of the $\nu_2 + 2\nu_3$ Band of CH_4 with a 1.3 μm InGaAsP Diode Laser. *Proceedings of the 3rd International Symposium on Monitoring of Gaseous Pollutants by Tunable Diode Lasers*.
21. CARLISLE, C. B. and COOPER, D. E. (1989). Tunable-Diode-Laser Frequency-Modulation Spectroscopy Using Balanced Homodyne Detection. *Opt. Lett.* **14**, 1306-1308.
22. ROTHMAN, L. S., *et al.* (1987). The HITRAN Database: 1986 Edition. *Appl. Opt.* **26**, 4058-4097.

POLLUTION GAS DETECTION IN FREE ATMOSPHERE
BEYOND ATMOSPHERIC WINDOWS

B. Sumpf, S. Dietrich, Ka. Herrmann,
F. Kühnemann, J. Orphal, V.V. Pustogov

Abteilung für Molekülphysik/Photobiophysik
der Humboldt-Universität zu Berlin
Invalidenstraße 110, Berlin, O-1040, Germany

SUMMARY

Under atmospheric conditions (normal pressure and humidity) the sensitivity of pollutant gas detection with Tunable Diode Laser Spectroscopy (TDLAS) is limited by the overlapping between the rovibrational bands of the trace gases and the CO_2 and H_2O bands. In the present paper the sensitivity limits are discussed for three types of gases using calculated spectra from known line parameters and own experiments:
- Gases with a line spacing greater than the linewidth at atmospheric pressure (CO, CO_2, NO); experiments in free atmosphere
- Gases with a line spacing smaller than the linewidth at atmospheric pressure (SO_2, NO_2);
  experiments under lab conditions at normal pressure
- Gases with a line spacing smaller than the Doppler linewidth (CF_2Cl_2);
  experiments under lab conditions.
It is shown, that even under atmospheric conditions the detection limits for these gases are of about 10 - 100 ppb using carefully selected wavenumber regions.

1. INTRODUCTION

Measuring trace gases at ppb-level under atmospheric conditions (normal pressure about 760 Torr and air humidity) the main spectroscopic problem for the detection of trace gases is the overlapping due to the rotational-vibrational bands of CO_2 and H_2O. The present paper presents:
- an estimation of sensitivity limits for the detection of NO, NO_2, SO_2, O_3 using the known line parameters and line by line calculations
- open path measurements for CO and NO and laboratory experiments for NO, NO_2 and SO_2 using a spectrometer with diode lasers working in pulsed regime
- experiments on freon CFC-12 (CF_2Cl_2)
- selected wavenumber regions for the detection of trace gases with a concentration limit of about 10 - 100 ppb

The features of the rotational-vibrational bands determine the sensitivity limits. One important parameter is the ratio of their line spacing compared to the line width. It is obvious that in the case of rather simple gases as CO or CO_2 the wavenumber selection is not that important because all the structures of the absorption spectra have the same shape and differ only in strength. For molecular spectra, where the line spacing is smaller than the line width (SO_2, NO_2, O_3, CF_2Cl_2) it is more important to select the best structure concerning strength and shape. This problem is illustrated comparing the spectra of CO and O_3 in the wavenumber region near 2100 cm^{-1} (see Figure 1). The value $S(\tilde{\nu})$ is the sum over all lines i (1..n) with a given strength S_i at the position $\tilde{\nu}_i$ multiplied with the Lorentz-shape :

$$S(\tilde{\nu}) = \sum_1^n S_i(\tilde{\nu}_i) \frac{1}{1 + \dfrac{(\tilde{\nu}-\tilde{\nu}_i)^2}{\Delta\tilde{\nu}^2}}$$

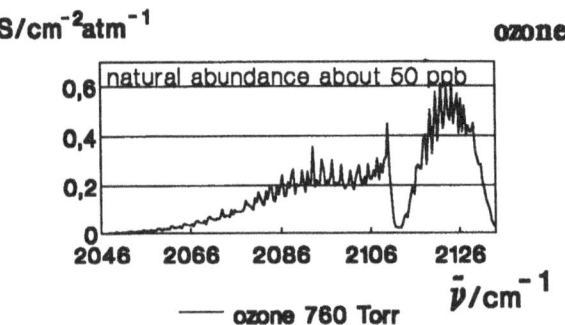

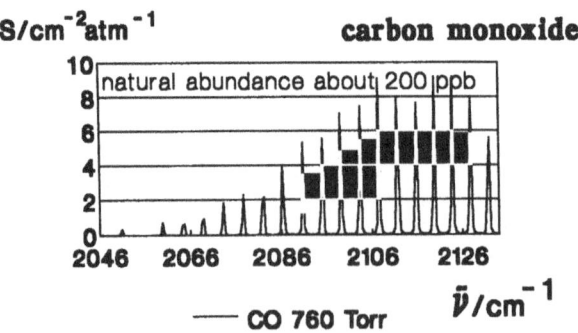

Fig. 1: Comparison of the ozone-spectrum and the CO fundamental near 2100 cm^{-1}

The second problem is the value of the absorption coefficient α of the trace gas compared to those of carbon dioxide and water vapour. Although the line strengths are in the same order of magnitude the natural concentration leads to an absorption coefficient, which is much greater than the α of the trace gases. Table 1 gives an summary for several gases:

Table 1:   Comparison of the absorption coefficient for several trace gases
            and carbon dioxide and water vapour

$\alpha_{gas} > \alpha_{water}$	$\alpha_{gas} = \alpha_{water}$	$\alpha_{gas} < \alpha_{water}$
0 - 1 CO	ν_1 SO_2	ν_3 SO_2
ν_3 CO_2		0 - 1 NO
		ν_3 NO_2
		$\nu_1 + \nu_3$ O_3
		CF_2Cl_2

Before we discuss the problem in more detail Table 2 gives a summary for the concentrations of ir-active components in troposphere and typical values under urban conditions.

Table 2: Concentration of ir-active components in troposphere

gas	ir-band	band centre $\tilde{\nu}/cm^{-1}$	background concentrat./ ppm	MIC/ppm -1-	typ. Berlin/ ppm -2-
CO_2	ν_3	2349.3	350	700	
H_2O	ν_2	1595.0	40000		
CO	0 - 1	2145.0	0,2	6,0	0,8-3,0
SO_2	ν_3	1361.0	0,002	0,1	0,01-0,03
NO	0 - 1	1880.0	0,02	0,15	0,01-0,05
NO_2	ν_3	1621.0	0,02	0,1	0,025
O_3	$\nu_1 + \nu_3$	2110.8	0,005	0,1	0,015

-1- maximum immission concentration (Ass. of German Engineers, VDI)
-2- mean values within a year /1/

2. FUNDAMENTAL BAND OF CARBON MONOXIDE
FUNDAMENTAL BAND OF NITROGEN MONOXIDE

In this paragraph results on the detection of carbon monoxide under atmospheric conditions and nitrogen oxide under laboratory condition will be discussed.

The first step in our research work in TDLAS was the measurement of background concentrations of CO with a sensitivity of about 100 ppb. Since the structure of absorption band is simple and the lines are very strong compared to the overlapping of water vapour the main problem was the design of the equipment. The result of our first routine experiment is shown in Figure 2. The experimental equipment was described elsewhere /2/. The peaks within the time dependence of the CO-concentration were caused by car exhaust on an unfrequented street near by the optical path.

Figure 2: CO-concentration measurement in a background region

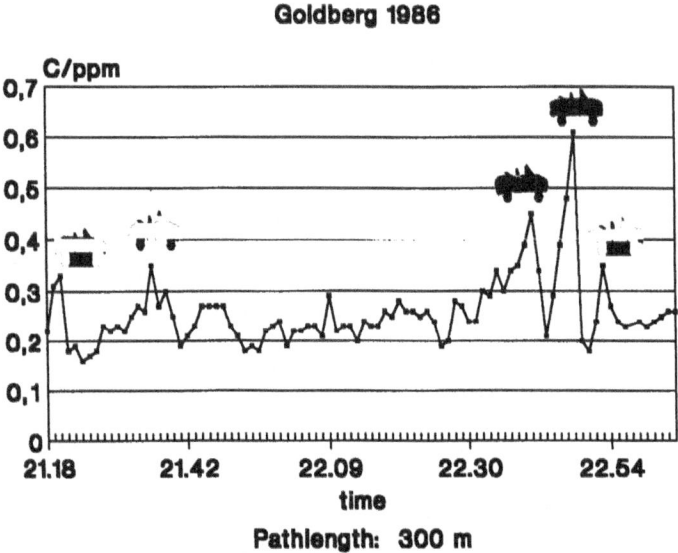

Goldberg 1986

Pathlength: 300 m
Sensitivity: 1.5 ppm•m

The equipment could be used under background conditions as well as in an urban situation.

Although NO has a spectrum similar to that of CO the detection is much more complicated due to the overlapping with water vapour. Calculations using the well known line parameters /3/ recommend only three wavenumber regions for the gas detection under atmospheric conditions. The most suitable line is near 1900 cm^{-1}, possible are those near 1912 cm^{-1} and 1935 cm^{-1}.

To estimate the influence of water vapour we started experiments under laboratory conditions. One example for these tests (line R(18.5)) is shown in Figure 3.

Figure 3: NO-experiment under lab conditions 1989

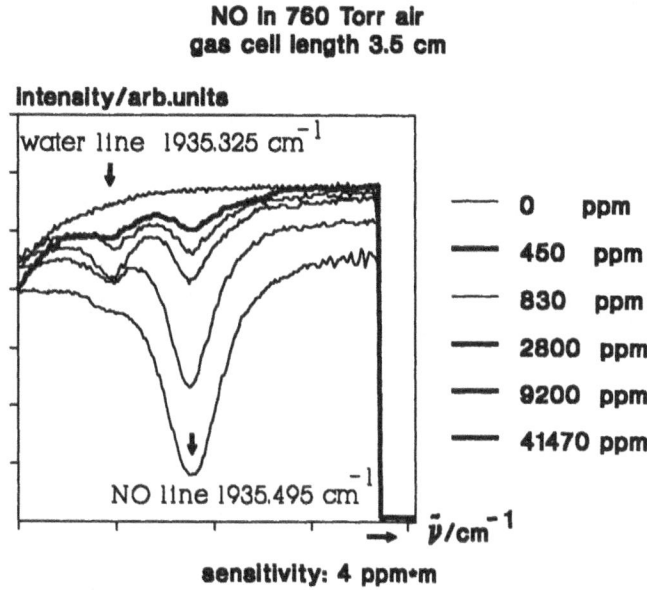

NO in 760 Torr air
gas cell length 3.5 cm

intensity/arb.units

water line 1935.325 cm^{-1}

—— 0 ppm
—— 450 ppm
—— 830 ppm
—— 2800 ppm
—— 9200 ppm
—— 41470 ppm

NO line 1935.495 cm^{-1}

$\bar{\nu}$/cm^{-1}

sensitivity: 4 ppm•m

Based on these experiments we select the best wavenumber region near 1900 cm^{-1} for the NO-detection under urban condition. The sensitivity with a signal-to-noise-ratio of 100:1 was estimated to be about 4 ppm*m. With the equipment developed by K. Bobey /4/ it is now possible to measure some 100 ppb over a Berlin downtown street continuously.

3. ν_3-BAND OF SULPHUR DIOXIDE
ν_3-BAND OF NITROGEN DIOXIDE

In the case of these gases additional to the problem of water vapour occurs the problem of their own band structure where at normal pressure the lines strongly overlap. Calculations based on the HITRAN database /3/ suggest for SO$_2$ one absorption structure near 1342.5 cm^{-1}. But all experiments show no significant structure in this region. Additional investigations of line positions, line strength and line broadening show big differences to the HITRAN database. In the case of line broadening parameters the differences were up to 30 %. This is discussed in the paper by F. Kühnemann /5,6/ in more details.

Here only the result of the "best" structure should be mentioned. Fig. 4 shows in the spectral region near 1382 cm^{-1} a SO_2 structure and the neighbouring H_2O line. The sensitivity limit could be estimated to 11 ppm*m.

Figure 4: SO_2 experiment under lab conditions 1988
a) gas cell without SO_2 b) gas cell with 28 ppm SO_2 c) absorption (magnified)
d) line positions and strength from the HITRAN-database /3/
   between 1381.722 cm^{-1} and 1381.956 cm^{-1} (10 lines)

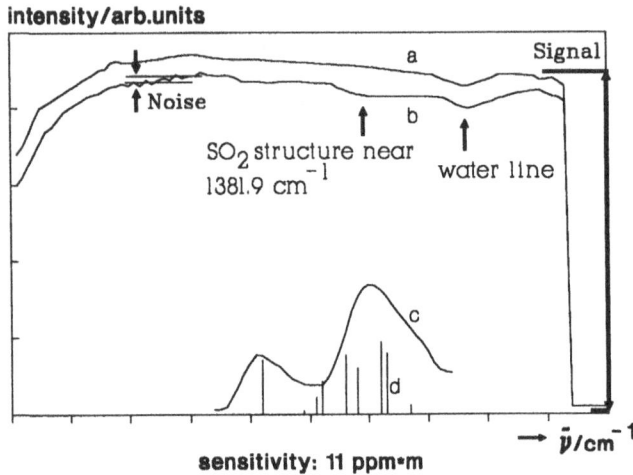

Sulphur dioxide in 760 Torr air
gas cell length 50 cm

sensitivity: 11 ppm·m

The sensitivity limit was estimated in all cases assuming a signal-to-noise-Ratio of 100:1, which is a typical value for our laboratory TDL-spectrometer measuring at normal pressure. A more specialized equipment reaches values up to 2000 : 1 at normal pressure, easily.

The structure of the ν_3-band of NO_2 is similar to that of the ν_3-band of SO_2, but more complicated because the band centre is near to the centre of the ν_2-band of water vapour . Even outside the centre of water lines the absorption coefficient of water is up to 3 orders of magnitude larger than for NO_2 at ppb - level. From calculations using 2094 lines in the ν_3-band of NO_2 between 1596 and 1654 cm^{-1} and 196 lines of water vapour a sensitivity for the NO_2 - detection of 2 ppm*m was determined at the best structure near 1600 cm^{-1}. For that the pathlength in free atmosphere could only be some 10 m. To measure the influence of water vapour even at shorter pathlengths we made some experiments with a Herriott-cell. Some of the results are shown in Figure 5. It is obvious that within the absorption line of water at 293 K and 40% air humidity the whole laser emission is absorbed. The lines of NO_2 could be measured as isolated lines at 10 Torr, but their line spacing is smaller than their broadening at normal pressure. Further experiments on NO_2 are discussed in the paper of V. V. Pustogov /7/.

Figure 5: NO₂ experiment using a Herriott-cell to estimate the influence of water vapour

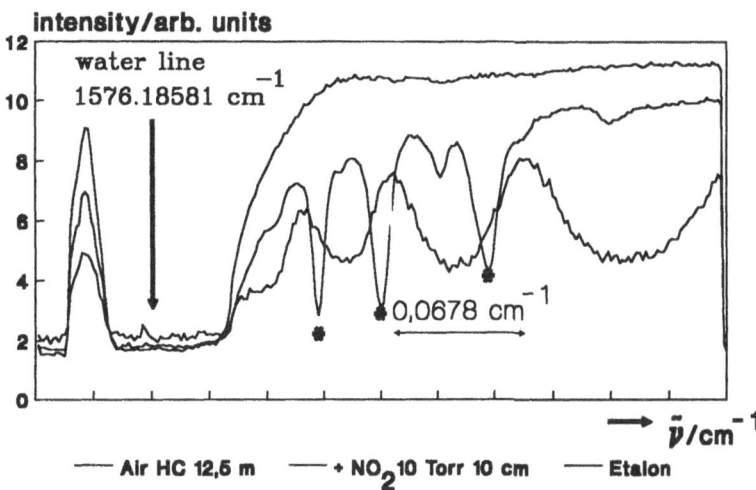

Measurement with Herriott-Cell (HC)
∗ - nitrogen dioxide line

intensity/arb. units

water line
1576.18581 cm^{-1}

0,0678 cm^{-1}

$\bar{\nu}$/cm^{-1}

—— Air HC 12,5 m   —— + NO₂ 10 Torr 10 cm   —— Etalon

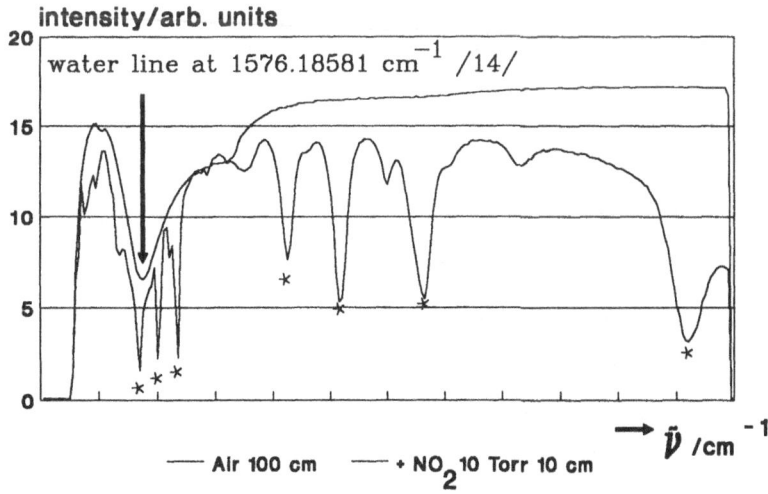

Measurement without Herriott-Cell
∗ - nitrogen dioxide lines

intensity/arb. units

water line at 1576.18581 cm^{-1} /14/

$\bar{\nu}$ /cm^{-1}

—— Air 100 cm   —— + NO₂ 10 Torr 10 cm

4. THE ν_8-BAND OF CF_2Cl_2

Here the high density of absorption lines causes a strong overlapping even at small pressure (Doppler-broadened lines). The spectrum exhibits no isolated lines as shown in Figure 6.

Figure 6: Transmission spectrum of CF_2Cl_2 between 1146.0 and 1147.2 cm^{-1}

P-branch of ν_8-band

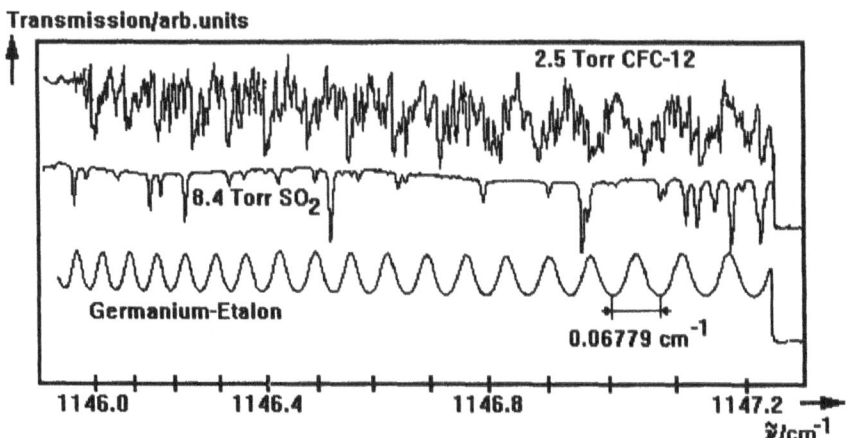

Figure 7: Q-branch maximum at 1161.07 cm^{-1} under 748 Torr

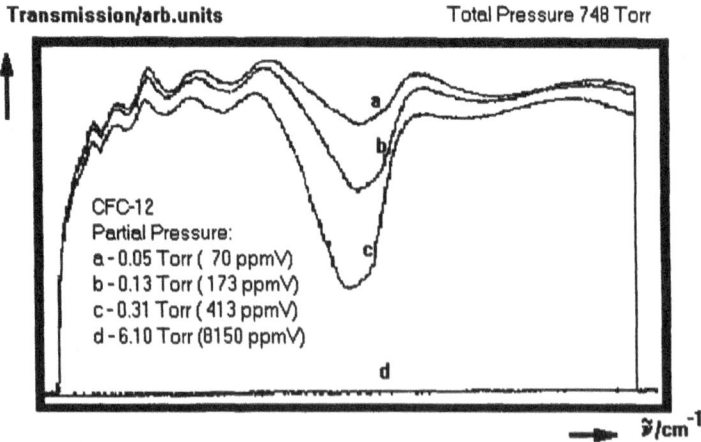

Only the ν_6- and the ν_1-band (band origins at 922 and 1101 cm^{-1}, respectively) have been investigated in the past with high resolution infrared spectroscopy /9-13/.

No suggestions for the determination of low atmospheric CF_2Cl_2 concentrations (average tropospheric volume mixing ratio 0.5 ppbV) with ir-spectroscopy could be made until today.

Using TDLAS we measured the high resolution ir-spectra of the P- and Q-branch of the ν_8-band, which is a c-type band with its origin at 1161 cm^{-1}. The lines of the ν_1-band of SO_2 served as calibration wavelength standards. Additional to this work in the field of basic research we made an estimation of minimum CF_2Cl_2 detection concentration with TDLAS under atmospheric pressures (Fig. 7 and 8) using the sharp Q-branch maximum at 1161.07 cm^{-1}.

Figure 8: Q-branch maximum at 1161.07 cm^{-1} under 100 Torr

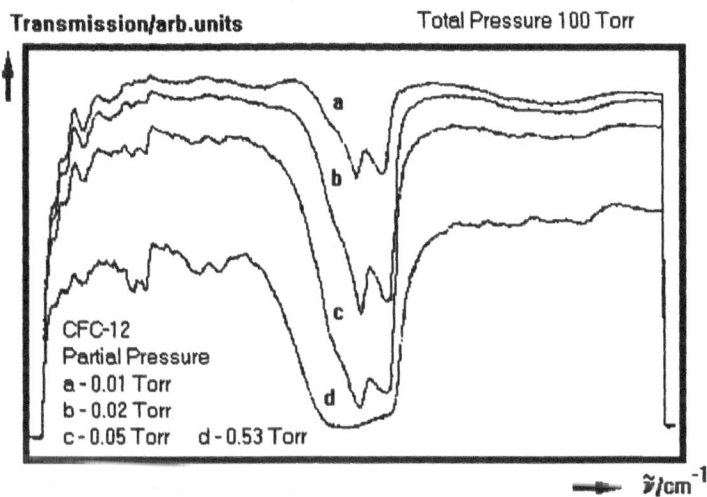

At smaller pressure the two peaks of the two Q-branches of each isotopic species of ^{35}Cl and ^{37}Cl are resolvable. For that reason it seams to be possible to measure the mixing ratio of these species using this wavenumber region. Table 3 summarizes the results for CFC-12 detection.

Table 3: Limits of CFC-12 detection

Signal-to-Noise-Ratio	Detection Sensitivity (ppmV*m) at P_{total}		Pathlength (m) for 0.5 ppbV at P_{total}	
	760 Torr	100 Torr	760 Torr	100 Torr
100 : 1	1000(20)	700(20)	2000(40)	1400(40)
1000 : 1	100(2)	70(2)	200(4)	140(4)

As conclusion we could say that CF_2Cl_2 concentrations should be detectable with a suitable TDLAS experimental setup using the second derivative and a multipass cell. Absorption lines of other atmospheric trace gases (O_3, NH_3, N_2O, SO_2) taken from catalogues /3/,/8/,/14/ seem to have no influence on the absorption structure at 1161.07cm^{-1}.

5. SUMMARY

Based on the known line parameters as well as on own experimental data it seems to be possible to detect the trace gases CO, NO, NO_2, SO_2, O_3 and CF_2Cl_2 in urban concentrations using spectroscopic techniques in free atmosphere, for example TDLAS with a typical pathlength of about 100 m and a signal-to-noise-ratio better 100 : 1. The main problem is the line overlapping due to water vapour or/and carbon dioxide. This effect limits the possible pathlength, in the case of SO_2 to about 200 m, of NO to about 100 m and for NO_2 to about 10 m. Table 4 summarizes all the results.

Table 4: Sensitivity limits

gas	MIC ------ ppb	$\tilde{\nu}$ -------- cm^{-1}	$C_{min}*L(th)$ ---------------- $ppm*m$	$C_{min}*L(exp)$ ----------------- $ppm*m$
CO	6000	2150		1,5
SO_2	100	1342	10	
		1382		11
NO	150	1900	2,5	4
NO_2	100	1600	2	
O_3	100	1055	7,5	
CF_2Cl_2		1161		1

6. LITERATURE

/1/ Das Luftgüte-Meßnetz (BLUME); Informationsreihe zur Luftreinhaltung in Berlin
Senatsverwaltung für Stadtentwicklung und Umweltschutz

/2/ B. Sumpf, D. Göring, R. Haseloff, Ka. Herrmann, J. W. Tomm
Collect. Czech. Chem. Commun. 54,284,1989

/3/ L. S. Rothman et. al   Appl. Opt. 26,4058,1987

/4/ K. Bobey, I. Dudeck, R. Strub   5. Kongreßmesse für ind. Meßtechnik, Wiesbaden Sept. 1991
Ein Diodenlasermeßsystem zur automatischen Fernmessung von Luftschadstoffen

/5/ F. Kühnemann, Y. Heiner, Ka. Herrmann, B. Sumpf, N. V. Lemichov, E. V. Stepanov
The ν_3-Band of SO_2: Line Parameters for Atmospheric Monitoring;   This book

/6/ F. Kühnemann, Y. Heiner, Ka. Herrmann, B. Sumpf   accepted in J. Mol. Spectr.

/7/ V. V. Pustogov, Ka. Herrmann, F. Kühnemann, J. Orphal, B. Sumpf, J. Berger, H.-D. Kronfeldt
Pressure Broadening Parameters of NO_x using TDLAS   This book

/8/ J.-M. Flaud, C. Camy-Peyret, C. P. Rinsland, M. A. H. Smith, V. M. Devi
Atlas of Ozone Spectral Parameters from Microwave to Medium Infrared
Academic Press, New York, 1990

/9/ R.J.Nordstrom, M. Morillon-Chapey, J.-C. Deroche, D.E. Jennings ,Phys. Letters 40,L37,1979

/10/ H.Jones, M. Morillon-Chapey   J. Mol. Spectr. 91,87,1982

/11/ G.Taubmann, H.Jones   J. Mol. Spectr. 117,283,1986

/12/ M.Snels, W.L.Meerts   Appl. Phys. B45,27,1988

/13/ S.Giorgianni, A.Gambi, A.Baldachini, A. DeLorenzi, S. Ghersetti;   J. Mol. Spectr. 144,230,1990

/14/ G. Guelachvili, K. N. Rao; Handbook of Infrared Standards; Academic Press, New York,1986

A PORTABLE BISTATIC MONITOR FOR ATMOSPHERIC METHANE DENSITY WITH UP TO 240m OPTICAL PATH

Ryuji KOGA

Department of Electric and Electronic Engineering
Faculty of Engineering, Okayama University
Tsushima, Okayama 700
JAPAN

SUMMARY

Results of field measurements of methane density through 24 hours are reported along with the description of the TDLAS device developed by the authors. The device is featured with its double-beam configuration, open and up to 240 m long optical path in the open atmosphere. An *in situ* and real-time measurements are possible with an rms fluctuation as small as 20 ppb·m. Subjects for further improvement are mentioned in relation with reducing the spectral-scanning period for a resistance power to atmospheric scintillation.

1. INTRODUCTION

As the possibility of methane's contribution to the global greenhouse effect is now clear, its mechanism of generation and extinctions has become important to be examined. Flux density of methane that is released across the ground or water surface to the atmosphere must be measured *in situ* under different weather. A medium scaled, not a point, measurement is required to know the flux density that means the total emission to the atmosphere per a unit time, an area of the ground surface. The scale should be 10 or 100 meters in length: an average value is desirable to discuss on the global effect. The authors have developed various apparatuses to measure density of atmospheric trace gas significant as air-pollutants as NO_2 or SO_2[1]. Since the development of PbSnTe diode lasers by Fujitsu Laboratories in Japan, a target gas has been methane because of its clear spectrum and the ease of system performance test. Meanwhile the density of methane has become of interest, and the authors are requested to join the group of scientists to help their measuring activities.

The main agricultural product in Japan and in southeastern Asia is rice, and rice-paddy fields are considered to be responsible for from 8% to 30% of anthropological methane yield.[2] The emission should be controlled through fertilization and feeding of tapping water. There is required a device that can monitor methane emission from a rice-paddy field which is *in situ*, real-time and gives a spatially averaged value without a sampling procedure.

The authors modified their system for that purpose extending their optical-path length. A double beam system with a reference leg and a measuring leg was constructed. The latter is folded back with a retroreflector. After its stability had been examined, field measurements were done in the experimental rice-paddy fields and the change of methane density through 24 hours was measured recording even a subtle change according to the change of sunshine conditions.

Principles of the tunable diode laser absorption spectrometry(TDLAS) is employed there, but a short optical path was employed to achieve an *in situ*, noncontact measurement: no White cell is employed. Sophisticated optical and electronic techniques have been developed to lower the noise level.[3] The dominant noise is the etalon fringe that arises between two surfaces of optical elements: among them the most inherent is that between the laser and the infrared detector, whose pitch is finer than a pressure broadened methane line. Employment of tailored modulation profile was effective to suppress it. Effect of drifts both in the base line and the diode temperature on the obtained methane density results are suppressed with an algorithm developed and named as "adjoint spectra" by the authors and is carried out on a microprocessor.[3,4]

2. APPARATUS

After some types had been build and tested, present system having an internal code M9 with A2 optics, was build. Schematic configuration of the system is shown in Fig.1. Two PC–type HgCdTe infrared detectors (IRD–R and IRD–X), are equipped each for the reference leg and the atmospheric leg, respectively. An electronic signal from an IRD is used twofold. One is to obtain the second harmonic component referring the clock signal for the laser frequency modulation with a lock–in amplifier system. The other is to know amplitude of the infrared power falling onto the IRD. The former result is normalized by the latter giving the true absorptance, which is independent of the throughput of a leg of the optical system as well as a responsivity of the IRD.

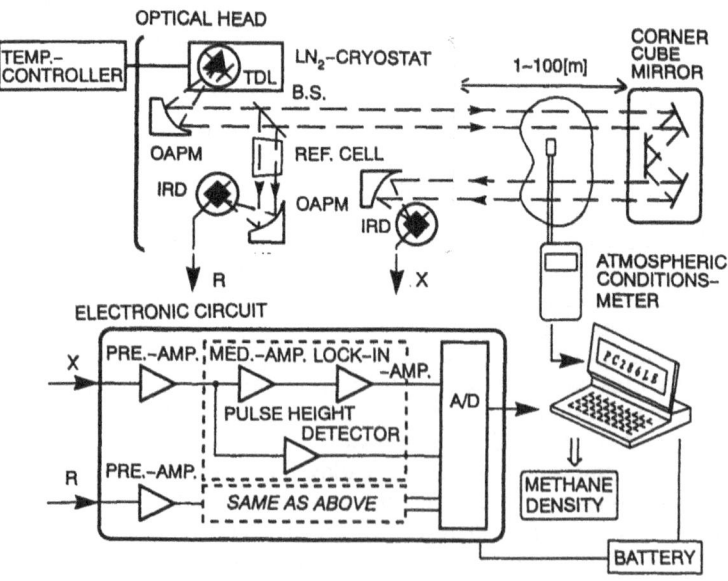

Fig.1 Schematic diagram of the total system. Internal code name
M13–A2 system.

The laser current is programmed as shown in Fig.2. The bias current is swept from its maximum to the minimum as the trace (a), which is for the sake of safe operation in an experiment where a mistake in a manual setting of the laser current program may destroy the diode laser element. The program was thus designed after a few tragedies. The bias current changes stepwise every 1/60 second as shown by the inset (b). This is synchronous to our 60Hz power mains to avoid the ham noise: a tedious shielding is not necessary for an experimental system. An application system that can be equipped with strict shielding will be free from the ham noise. On the bias current, a periodic signal of 7.68kHz(128×60Hz) is superimposed. Its profile is not a pure sinusoidal but a tailored one as shown by the inset (c) which has peaks at both extremes. The system response to etalon fringes with spectral pitches finer than FWHM of methane is suppressed by the tailoring of the modulation profile.[3]

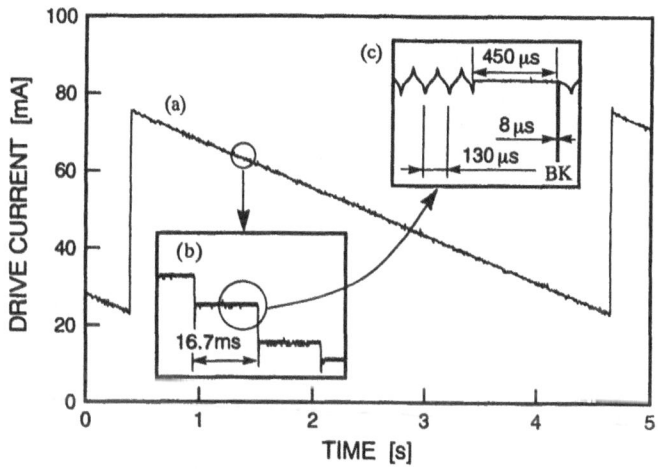

Fig.2 Temporal program of the laser current.

The total laser current is blanked out for 8 μs at the end of the time slot with 1/60 second duration. The impulse response that appears in the receiver preamplifier output is processed by the pulse–hight analyzser(PHD in Fig.1) triggered by a blanking signal, which is given by a central control system. Before the blanking, the modulation is suppressed off for 130 μs to avoid an undesirable fluctuation of the transmitted laser power that may be introduced by a jitter in the clock signal. The width 8 μs of the blanking was determined by trading the signal to noise ratio in the pulse–hight analyzer and undesirable diode temperature transient that is considered to be harmful for the health of the diode element due to thermal stress. The threshold current of our diode element made by Fujitsu Laboratories Ltd.[5] has not increased after 580 hours of the blanking mode operation.

Spectra of the atmosphere and reference spectra is correlated and gives an atmospheric methane density C_x, as

$$C_x = \frac{C_R L_R <S^*, S_X>}{L_X <S^*, S_R>}, \qquad (1)$$

where L_X is the optical path−length of the probing beam spanned in the open atmosphere and L_R is the length of the reference gas cell containing standard methane gas of known density C_R with known atmospheric pressure. The symbol $<x, y>$ stands for an inner product between two spectra's x and y both expressed as vectors composed of $n128$ data items each measured over a time slot of 1/60 second. The spectrum S^* is given by

$$S^* = S_R - \sum_{i=1}^{I} <S_R, U_i> U_i, \qquad (2)$$

where U_i's are orthonormalized interference spectra that are postulated beforehand. The interference may be an actual one if it is well described. If it cannot be described, virtual spectra can be employed retaining only their generous features: polynomials of the lowest a few orders of the laser frequency are possible.

Spectra actually measured by our system are shown in Fig.3: spectra of the atmosphere, of reference methane, and of water vapor. The atmospheric spectrum

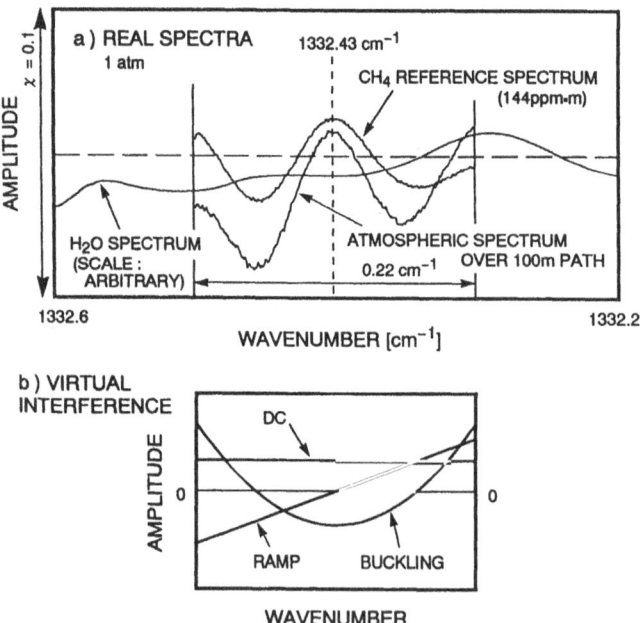

Fig.3 Spectra measured in the field experiments(a) and the virtual interference spectra employed in the numerical treatment to remove the interference of water vapor to the atmospheric methane spectra.

is found to be distorted by the water vapor spectrum and this overlap is the dominant spectral interference for atmospheric methane. The water vapor spectrum was substituted by the three virtual spectra of the dc, the ramp and the buckling as depicted by the inset (b) of Fig.3. The correlating operation is carried over the wavenumber region of 0.22 cm^{-1} width designated in the figure.

Actual optical configuration is shown in Fig.4, and Fig.5 is its photograph. Surface reflecting optics are employed to avoid etalon fringes. The PbSnTe buried heterostructure diode laser is mounted on a liquid nitrogen cooled cryostat that is temperature controlled in within 10 mK rms accuracy. Two mercury cadmium telluride infrared detectors are also mounted on LN$_2$ cooled Dewars. Cryostats and Dewar windows are AR-coated ZnS and wedged. The collimations are done with

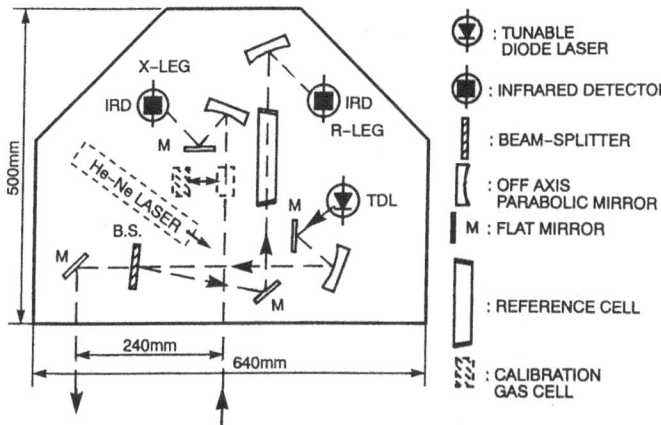

Fig.4 Arrangement of the optical system, which is internally called as A2 optical system.

Fig.5 Photograph of the A2 optical system.

two replicas, gold–plated 40 degree off–axis parabolic mirrors(OAPM) with focal length of 141 mm, and a hand shaped one of 45 degree and 80mm. Beam steering is achieved with small flat mirrors put between the OAPM and the laser or the IRD elements. The beam splitter is also wedged ZnS one side of it AR coated. The naked side is used as the partial reflecting surface. The incidence angle of the beam to the surface is chosen as small as 5 degrees to assure the beam splitting ratio should be the same between the p– and s– components of different polarization of the beam: the laser may operate in a transversely multiple mode. Transmitted beam diameter is about 30 mm. A distance between centers of the transmitted and re–ceived beams are designed to be 240 mm that is correspondent to the nominal distance between the two end–mirrors of the corner cube retroreflector. The base plate is of Al alloy with 40 mm thickness, which is planned its material should be removed to reduce weight that is now weighs about 50 kg: too heavy to carry. Whole optical system is accommodated in a housing of aluminum sheeting of 3 mm thick to prevent attacks of insects, leaves, gusts, and solar radiation that may disturb system temperature.

The retroreflector is a corner–cube type composed of 100x60 mm rectangular mirrors mounted on Al–alloy frame. A fine adjustment is possible using three micrometer–head screws and an accuracy less than 1 second is attained using an *ad hoc* jig. This corner–cube system is also accommodated in a housing as Fig.6.

Fig.6 Retroreflector accommodated in a housing to avoid debris and solar radiation.

Stability of the system was examined in our laboratory: a result is given in Fig.7. The optical system was encapsulated in a polyethylene film housing encompassing a folded optical path of 3m long inside it. Due to a residual instability of the laser cryostats, the beam sways for 5 minutes after liquid nitrogen is supplied to it. Otherwise the whole system work achieves 0.17 ppm(rms) stability for an average of 16.9ppm.

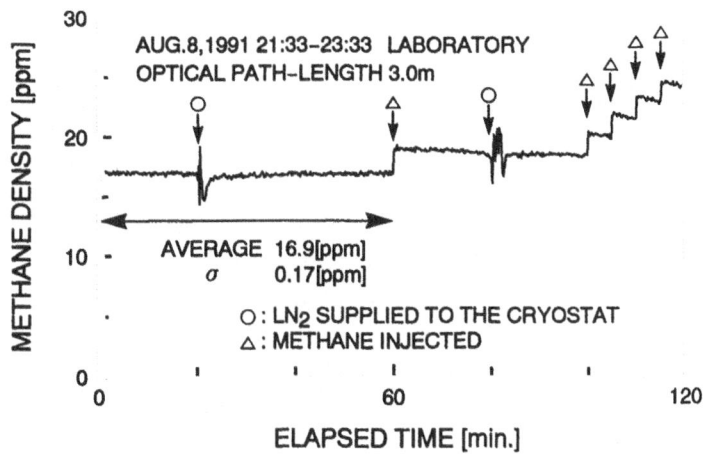

Fig.7 Stability of the system measured in the laboratory.

3. FIELD EXPERIMENTS

Field measurements were carried out in the experimental farm in the campus of Okayama University as Fig.8 since June to July 1991. It was amid the rainy season in Japan, and the emission of methane from rice-paddy fields is expected to be most active through an year[6].

The rainy season of this year was particularly stormy compared to ordinary years, and our experiments have been attacked by local storms so furious that electric poles on roadside were broken down. Our optical and electronic system could survive under these extremely bad conditions though measurements were interrupted.

In Fig.9, two results of the methane densities measured over 24 hours are shown. Optical path length was adjusted to be just 100m round trip that had been analyzed to give the best SNR. The upper trace was taken on July 2nd: it rained intermittently whole the day, and the optics was dewed at the midnight. Through-put for the laser beam was lowered and the return beam became so weak as it could not be detectable at all. It was recovered, however, by putting an electric lamp in the housing of the optics, warming its inside.

Measurements were repeated every one hour, and at the beginning calibration procedure were achieved for 10 minutes for the sake of confirmation. A couple of spectra, S_X and S_R was scanned over 4.2 seconds every 30 seconds and is recorded on a digital magnetic tape along with other parameters such as atmospheric and soil conditions, simultaneously.

The lower trace of Fig.9 was taken in September, when the rice paddies have grown up and its flowers are in full bloom. Weather had been fine for one week before this day and wind was calm. The methane density changes significantly as the soil temperature that is warmed up by sunshine.

Fig.8 Field measurement in an experimental rice paddy field of
Okayama University.

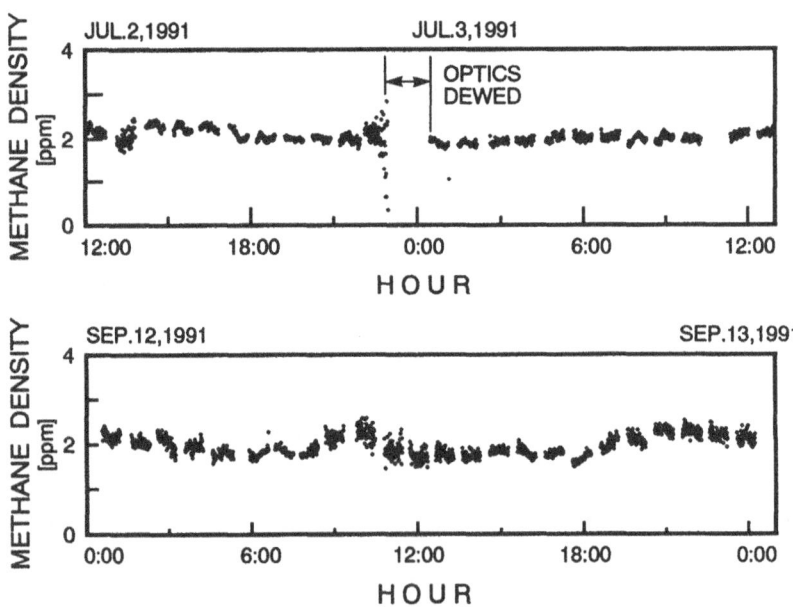

Fig.9 Record of measured methane densities above the rice paddy field
in Okayama University. Upper trace was taken amid the rainy season,
and the lower on a clear day when the effect of sunshine is
found on the methane density.

4. ATMOSPHERIC SCINTILLATION AT λ=7.6μm

Another experiment was made to test the longest optical path that our system can use. An optical path length of 240 m round trip was enough usable and the return beam was found only after 10 minutes of adjustments assisted by the overlaid HeNe laser beam even in the daytime.

For the longest optical path, atmospheric scintillation may limit the sensitivity deflecting the laser beam randomly. Theoretical and experimental examinations have been done by many researchers[7,8], but results for infrared region, especially in this 7 μm region, are scarcely found. Figure 10 is a result obtained using our system, though still smeared by trivial noise. Fluctuation for the 7.6 μm and 0.63 μm beams with the same optical path were measured. The data for TDL is found distinctly different from that forecasted from the results for HeNe laser beam using the theory by Tatarskii[9]. Contributions of water–vapor lines having dispersions around them are suspected, though many and precise examinations should be conducted until a qualitative result is obtained. This will lead to the future improvement of measuring techniques with the highest accuracy.

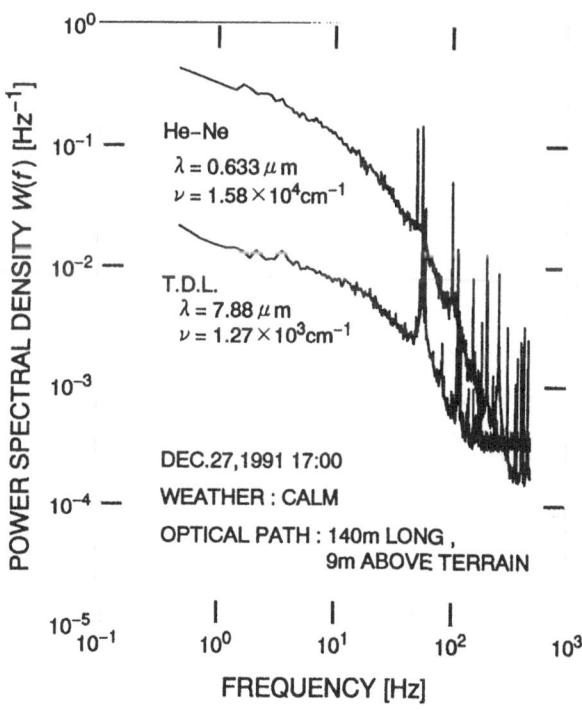

Fig.10 Fluctuation of the received laser beam power. Two spectral densities are shown for the HeNe of λ=0.63 μm and the PbSnTe TDL of λ=7.6 μm. Distinct difference is found between the two.

5. CONCLUSIONS

The temporal variation of atmospheric methane sensitivity was measured with a Pb–TDL based TDLAS system with open and up–to 240m round–trip long optical path. This configuration allowed a real–time and *in situ* measurements in the rice paddy–field where they may meet various disturbances such as debris, solar radiation, rainfall and even as furious storm. A robustness was proven with the experimental device with sophisticated optical configurations. A diurnal measurement was proven to be possible giving average value over the optical path traversing above a rice–paddy field.

The atmospheric scintillation was also measured and discrepancy was found for the beam of $\lambda=7\mu m$ from the results that are expected from the HeNe laser beam of 0.63 μm and the theory of Tatarskii. Water vapor is considered to be responsible to the discrepancy having dispersions around its lines. Further examinations are necessary to be made.

References
(1)   R.Koga,M.Kosaka, and K.Shinohara, High Sensitivity short–Path Monitoring of trace Gases Employing PbSnTe Tunable Diode Laser, Japan. J. Appl. Phys., **20**, 2145–2153(1981).
(2)   A.H.–Pschorn, W.Seiler, Methane Emission During a Cultivation Period From an Italian Rice Paddy, *J. Geophysical Res.*, **91**, 11803–11814(1986).
(3)   H.Sano, R.Koga and M.Kosaka, Analytical Description of Tunable Diode Laser Derivative Spectrometry, *Japan. J. Appl. Phys.* **22**, 1883–1888(1983).
(4)   M.Bouzidi, N.Kagawa, O.Wada and R.Koga, Adjoint Spectrum I: an Algorithm to Extract Target Spectra Under Spectral Interference for Use in On–Line Spectrometry, *submitted to Japan. J. Appl. Phys.* (1991).
(5)   K.Shinohara, Development of Lead–Chalcogenide lasers at *FUJITSU*, Proc. Intern. Symp. Monitoring of Gaseous Pollutants by Tunable Diode Lasers, held in Freiburg, F.R.G. 17–18, Oct.1988, pp.77–84(1989).
(6)   K.Yagi and K.Minami: Effect of organic matter application of methane emission from some Japanese paddy fields, *Soil Sci. Plant Nutr.*, **36**, 599–610(1990).
(7)   H.B.Janes, M.C.Thompson,Jr., D.Smith, and A.W.Kirkpatrick, Comparison of Simultaneous Line–of–Sight Signals at 9.6 and 34.5 GHz, *IEEE Trans.*, **AP–18**, 447–451(1970).
(8)   A.G.Kjelaas, P.E.Nordal, and A.Bjerkestrand, Scintillation and multi wavelength coherence effects in a long–path laser absorption spectrometer, *Appl. Opt.*, **17**, 277–284(1978).
(9)   V.I.Tatarskii, The Effects of The Turbulent Atmosphere on Wave Propagation, *Israel Program for Scientific Translation*(1971).

Section II

DIODE LASERS AND THEIR PROPERTIES

EMBOSSED GRATING DFB-BH LEAD CHALCOGENIDE DIODE LASERS

A. FACH, H. BÖTTNER, M. TACKE

Fraunhofer-Institut für Physikalische Messtechnik
Heidenhofstrasse 8, W-7800 Freiburg, Germany

ABSTRACT

The successful preparation of buried layer Double-Heterostructure Distributed Feedback Leadchalcogenide lasers using an embossing technique for the DFB-Grating is reported here for the first time. The embossing grating was formed in (111) oriented silicon by photolithography and argon ion milling. This grating was pressed into the bars of the later buried active PbSe layer. By an overgrowth process the laser structure was completed. Some lasers made from such wafers show the desired monomode behaviour.

1. INTRODUCTION

Lead chalcogenide diode lasers are the key device for high resolution spectroscopy in the mid infrared region [1]. The actual aim in the development of these lasers is mainly focussed on the improvement of the mode behaviour caused by special built-in waveguide structures like in mesa and buried heterostructure lasers (BH). The state of the art of those lasers with suppressed lateral modes is described in recent publications [2, 3, 4, 5]. In order to achieve longitudinal monomode behaviour, distributed feedback lasers (DFB) and distributed bragg reflector lasers have been developed [6 - 16], but without any advanced lateral waveguide structure as mentioned above. Here we report for the first time on the fabrication of DFB BH lead chalcogenide lasers, which should combine both advantages. The BH waveguide was made similar to that described in detail in [3]. The DFB-grating was transfered into the active layer by an embossing technique recently introduced for lead chalcogenide semiconductors [16], which was used here for the first time to corrugate lateral structured laser active semiconductor material.

Some of the resulting lasers show the desired monomode behaviour due to the DFB-BH-structure, some simultaneously emit multimode at higher and lower emission frequency around the 'DFB'-mode.

This effect is tentatively explained by end face reflections.

2. EXPERIMENTAL

The embossed grating DFB-BH lasers are based on PbSe/PbEuSe double-heterostructure (DH) lasers grown by molecular beam epitaxy (MBE) on (100) oriented PbSe substrates. Details are given in [17]. The BH-structure was formed by photolithography and ion milling processes as described in [3]. The most delicate technological step is the transfer of the submicrometer grating into the vertical laser structure. This is usually done by photolithographic and dry etching techniques. This technology treats the laser active layer by several technological steps, and the resulting grating is not well suited for the necessary MBE overgrowth process due to contours with sharp edges. Improvements are expected by a one step technology which is able to form variable grating contours without contaminating the laser wafer surface. As lead chalcogenides are easily plastically deformable, we transfered the well known embossing technique for integrated waveguides [18 - 21] to this material as a much more simple process. The embossing master sub-micrometer grating was made by conventional techniques from (111) oriented silicon wafers. The master grating has a periodicity of 0.71 μm and a grating etch depth of about 300 nm. The grating was oriented perpendicular to the BH-waveguide. The embossing can be performed with low force because only the narrow BH bars have to be structured. The overall pressure is about 2 N/mm^2 for the whole wafer. The pressure on the bars is nearly two orders of magnitude higher. The silicon sample is usually smaller than the lead chalcogenide wafer. Thus, after complete wafer processing, the properties of embossed DFB-BH lasers as well as normal BH lasers from the same wafer can be compared directly. Fig. 1 shows the SEM picture of a cross section through a complete DFB-BH-Laser.

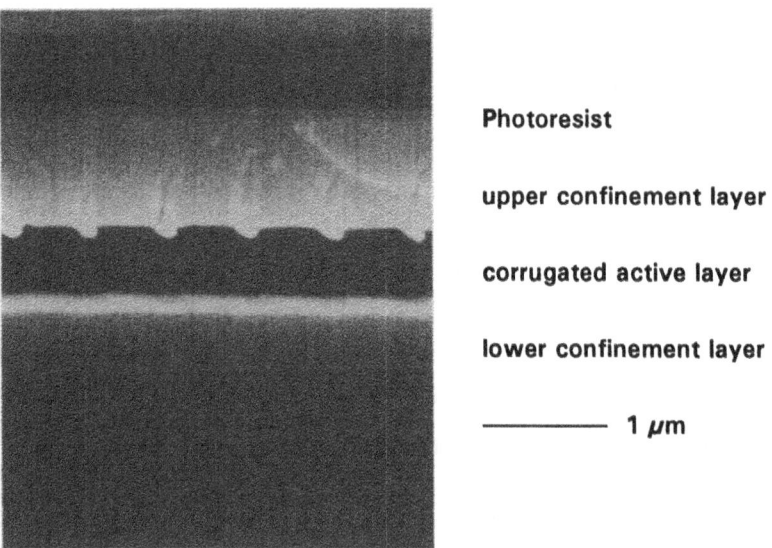

Photoresist

upper confinement layer

corrugated active layer

lower confinement layer

——————— 1 μm

Fig. 1. SEM-picture of the cross-section of the DFB-BH structure after a selective wet etching process.

The significant contrast in Fig. 1 between the different MBE-layers is caused by treatment with a special etching solution. The grating in the PbSe active layer is clearly seen with a corrugation depth of 0.1 - 0.2 μm. The overgrown sample was coverd with photoresist and suitable windows were opened over the DFB-BH-waveguides for stripe contacts. Low resistance p- and n-contacts were made simultaneously by electroplating gold. The wafer was cleaved into chips of about 500 μm length, that were mounted in a standard package by cold pressing with In.

3. OPTICAL STUDIES

In a recent publication [16] we reported on the DFB-characteristic of IV-VI DH-lasers, which were corrugated by embossing the upper confinement layer. We found no significant deterioration of electrical and optical properties compared to lasers made of the same wafer without embossed grating [16]. This quite surprising behaviour is still valid for the DFB-BH type with the embossed corrugated laser active PbSe-layer. Maximum operating temperatures were not measured in detail but up to 120 K cw operation was generally achieved.

Mode spectra with dc bias current were taken in the temperature region around 90 K, where we expected the lasers to operate as DFB-lasers. The measured mode behaviour could be classified into three types. First type the devices have no grating supported modes, but the mode spacing in the feedback wavenumber region is different from that of plain Fabry-Perot reference lasers (Fig. 2);

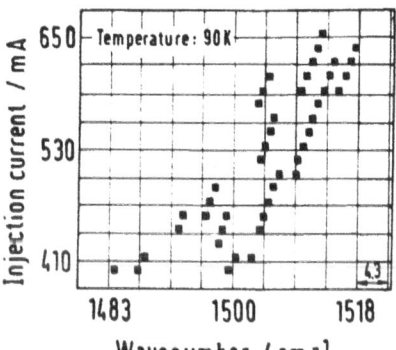

Fig. 2. Mode positions depending on injection current for embossed DFB-BH lasers classified in the text as type I.

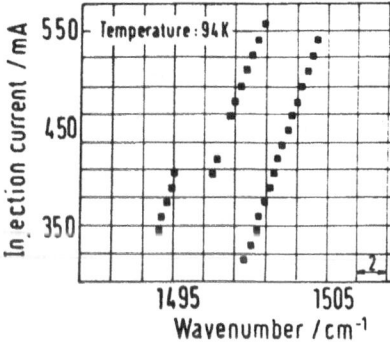

Fig. 3. Mode positions depending on injection current for embossed DFB-BH lasers classified in the text as type II.

second: devices show typically two or three modes with different mode spacing in the feedback region (Fig. 3);

third: devices show the desired monomode behaviour (Fig. 4b).

By shifting the gain curve away from the supported feedback mode, e. g. by temperature tuning, modes of the Fabry Perot Type were observed (Fig. 4a and 4c). The supported mode around 1500 cm^{-1} fits to the grating periodicity of 0.71 μm. The total tuning length of the DFB-mode is about 7 cm^{-1}. Typical values for Fabry Perot modes of this laser type are 2 - 3 cm^{-1}.

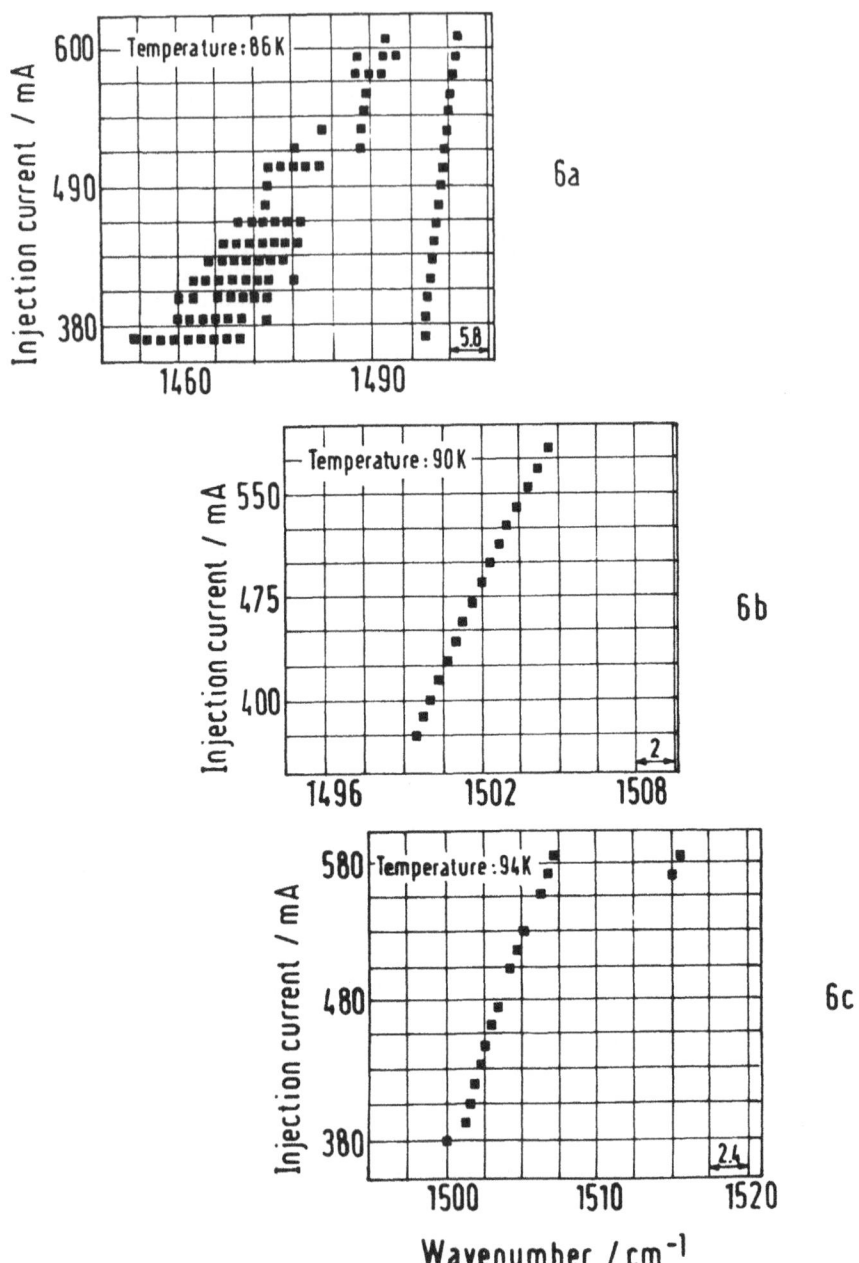

Fig. 4a-c. Mode positions depending on injection current at different operating temperatures for embossed DFB-BH lasers classified in the text as type III.

4. DISCUSSION AND CONCLUSIONS

Compared to the first embossed DFB lasers [16] the DFB-BH type with the embossed active layer shows for the first time devices with monomode behaviour, probably caused by the higher coupling efficiency due to the corrugated active layer. It is remarkable that the mode behaviour can be classified as mentioned above, which can hardly be explained by simple quality variations among the different devices. We suppose that this behaviour is caused by low coupling efficiency and individual scatter in the position of the laser resonator facets relative to the grating. A quantitative analysis is still in progress and will be published elsewhere.

5. SUMMARY

Embossed grating DFB-BH lasers with corregated grating in the active layer were made for the first time. Some devices show the expected grating supported DFB-mode without any deterioration of optical and electrical properties.

We thank A. Lambrecht for growing the epitaxial layers.

REFERENCES

[1]   GRISAR, R., SCHMIDTKE, G., TACKE, M. and RESTELLI, G.: Proceedings of the 2nd International Symposium 'Monitoring of Gaseous Pollutants by Tunable Diode Lasers', Freiburg, 17. - 18, Oct. 1988, Kluwer Academic Publishers, Dordrecht, Boston London, 1988.

[2]   SCHLERETH, K.H., BÖTTNER, H., TACKE, M.: 'Mushroom' Double Channel Doublehoterostructure Lead Chalcogenide Lasers made by Chemical Etching, Appl. Phys. Lett., Vol. 56, No. 22, S. 2169 - 2171, 28 May 1990.

[3]   SCHLERETH, K.-H., SPANGER, B., BÖTTNER, H., LAMBRECHT, A., TACKE, M.: Buried Waveguide Double-Heterostructure PbEuSe-Lasers Grown by MBE, Infrared Physics, Vol. 30, No. 5, S. 449 - 454, 1990.

[4]   FEIT, Z., KOSRYK, D., WOODS, R.J., MAK, P.: Molecular Beam Epitasxy-Grown PbSnTe-PbEuSeTe Buried Heterostructure Diode Lasers, IEEE Phot. Technol. Let., Vol. 2, No. 12, pp. 860 - 863, Dec. 1990.

[5]   FEIT, Z., KOSTYK, D., WOODS, R.J., MAK, P.: Single-mode Molecular Beam Epitaxy Grown PbEuSeTe/PbTe Buried-heterostructure Diode Lasers for CO_2 High-resolution Spectroscopy, Appl. Phys. Lett., 58 (4), pp. 343 - 345, 1991.

[6]   WALPOLE, J.N., CALAWA, A.R., CHINN, S.R., GROVES, S.H., HARMAN, T.C.: Distributed Feedback PbSnTe Double-heterostructure Lasers, Appl. Phys. Lett, 29, 307 (1976).

[7]   WALPOLE, J.N., CALAWA, A.R., CHINN, S.R., GROVES, S.H., HARMAN, T.C.: cw Operation of Distributed Feedback PbSnTe Lasers, Appl. Phys. Lett, 30, 524 (1977).

[8]   HSIEH, H., FONSTAD, C.G.: Liquid Phase Epitaxy Grown PbSnTe DFB Laser Diodes with Broad Continous Single-Mode Tuning range, 16, 1039 (1980).

[9]   KAPON, E., ZUSSMANN, A., KATZIR, A.: Distributed Bragg Reflector Lattice Matched $Pb_{1-x}Sn_xTe/PbSe_yTe_{1-x}$ Diode Lasers, Appl. Phys. Lett, 44, 275 (1984).

[10]   KAPON, E., KATZIR, A.: Distributed Bragg Refelector $Pb_{1-x}Sn_xTe/PbSe_yTe_{1-y}$ Diode Lasers, IEEE J. Quantum Electron., 21, No. 12, 1947 (1985).

[11]   SHANI, Y., KATZIR, A., BACHEM, K.-H., NORTON, P., TACKE, M., PREIER, H.M.: 77 K cw Operation of Distributed Bragg Refector $Pb_{1-x}Sn_xSe/Pb_{1-x-y}Eu_ySn_xSe$ Diode Lasers, Appl. Phys. Lett., 48, 18, 1178 (1986).

[12]   SHANI, Y., ROSMAN, R., KATZIR, A., NORTON, P., TACKE, M., PREIER, H.M.: Distribute of Bragg Refector $Pb_{1-x}Sn_xSe/Pb_{1-x-y} Eu_ySn_xSe$ Diode Lasers with a Broad Single Mode, J. Appl. Phys., 63, 5603 (1988).

[13]   SHANI, Y., KATZIR, A., TACKE, M., PREIER, H.M.: Metal Clad $Pb_{1-x}Sn_xSe/Pb_{1-x-y}Eu_ySn_xSe$ Distributed Feedback Lasers, IEEE J. Quantum Electron., 24, 2135 (1988).

[14]   SHANI, Y., KATZIR, A., TACKE, M., PREIER, H.M.: Highly Collimated Laser Beams from Grating Coupled Emission $Pb_{1-x}Sn_nSe/Pb_{1-x-y}Eu_ySn_xSe$ Diode Laser, Appl. Phys. Lett., 53, 462 (1988).

[15]   SHANi, Y., KATZIR, A., TACKE, M., PREIER, H.M.: $Pb_{1-x}Sn_xSe/Pb_{1-x-y}Eu_ySn_xSe$ Corrugated Diode Lasers, IEEE J. Quantum Electron., 25, 8, 1828 (1989).

[16]   SCHLERETH, K.H., BÖTTNER, H.: Embossed Grating Lead Chalcogenide Distributed-Feedback Lasers, to be published Jan./Febr. 1992, Journal of Vacuum Science and Technology B.

[17]   TACKE, M., SPANGER, B., LAMBRECHT, A., NORTON, P.R., BÖTTNER, H.: Infrared Double Heterostructure Diode Laser made by Molecular Beam Epitaxy of $Pb_{1-x}Eu_xSe$, Appl. Phys. Lett. 53, 23, 2260 - 2262 (1988).

[18]   ULRICH, R., WEBER, H.P., CHANDROSS, E.A., TOMLINSON, W.J., FRANKE, E.A.: Embossed Optical Waveguides, Appl. Phys. Lett., 20, 213 (1972).

[19]   SABINE, P.V.H.: Stress Induced Light Guiding in Embossed Waveguides, Electron. Lett., 11, 21, 501 (1975).

[20]   LUKOSZ, W., TIEFENTHALER, K.: Embossing Technique for Fabricating Integrated Optical Components in Hard Inorganic waveguiding Materials, 8, 10, 537 (1983).

[21]   HEUBERGER, K., LUKOSZ, W.: Embossing Technique for Fabricating Surface Relief Gratings on Hard Oxide Waveguides, Appl. Optics, 25, 9, 1499 (1986).

EPITAXIAL GROWTH OF LATERALLY STRUCTURED LEAD CHALCOGENIDE LASERS

A. LAMBRECHT, A. FACH, R. KURBEL, B. HALFORD,
H. BÖTTNER and M.TACKE

Fraunhofer-Institut für Physikalische Messtechnik
Heidenhofstrasse 8, W-7800 Freiburg, Germany

SUMMARY

Buried Heterostructure (BH-)Lasers based on PbEuSe were made using Molecular Beam Epitaxy (MBE). After growing the active layer of a DH-structure the growth was interrupted. By use of photolithography and Ar^+-ion beam etching mesa stripes were defined and overgrown with a PbEuSe clad by a second MBE step. The overgrowth was investigated by scanning electron microscopy (SEM), using a selective chemical etch to clearly resolve areas of different Eu-content. The SEM pictures indicate the crucial role of the mesa shape for the overgrowth quality. Sharp edges result in deep cracks or grooves aside a well grown ridge. To optimize the active mesa shape a chemomechanical etch process was developed after the ion milling process. The obtained smooth mesa shape results in a better overgrowth. As an alternative approach for fabrication of BH-lasers shadow masked growth of the active mesa stripe was investigated. Masks with slits in the 10 μm range and sharp edges were obtained by anisotropic etching of (100)-Si. BH-lasers with an active stripe width of 40 μm were made by this technique and showed a reduced threshold current compared to DH-lasers.

1. INTRODUCTION

Laterally structured lead chalcogenide diode lasers like buried heterostructure (BH-)lasers are expected to have superior properties for spectroscopic applications than stripe contact double heterostructure (DH-) lasers. They should not show higher order lateral modes which often cannot be suppressed by use of a monochromator. They should have a stable single emission area which is a prerequisite for an even and aperture filling farfield distribution (1). Additionally, Feit et al. (2) demonstrated that BH-lasers can achieve higher cw-operation temperatures up to 203 K plus an increased range of single mode operation. The combination of lateral structurization with the incorporation of a distributed feedback (DFB-) grating into the laser can be used to gain a predicted single mode operation in a definite spectral emission region on a routine basis (3).

Current laser development is based on the use of molecular beam epitaxy (MBE-)techniques. Therefore a process resulting in a buried lateral structure has to be compatible with MBE-growth. Most problems encountered with MBE-overgrowth are not specific to BH-lasers, but may also be found with buried ridges, when growing on structured substrates, and when growing on DFB-gratings.

In the following we report on a conventional photolithographic approach using Ar^+-ion milling for BH-stripe definition. As an alternative method we also investigated the use of shadow masks for the preparation of the active mesa stripes.

2. EXPERIMENTAL

Lateral structured lasers based on $Pb_{1-x}Eu_xSe$ active and cladding layers (with a higher Eu content x in the confinement layers) were grown using MBE. Details on the basic DH-structures and the MBE-system are given in (4) and (5). Since the increase of the lattice constant with Eu-content is quite small in the concentration range used we expect that mismatch is mainly compensated by built-in elastic strain (6). For sake of process simplicity we did not go into lattice matched structures as used by Feit et al. (7).

MBE-growth and overgrowth results of structured samples were investigated employing Scanning Electron Microscopy (SEM). A selective chemical etch was used to clearly resolve areas of different Eu-content.

a) photolithographic technique:

The MBE-layer growth is interrupted after the active layer. From the active layer a mesa stripe about 5 μm wide is formed by photolithography. The stripes are defined by Ar^+-ion beam etching (8). The Ar^+-etching rate is independent from the $Pb_{1-x}Eu_xSe$ concentration used. It is mainly shape conserving and can be used for structurization in the sub-μm range (9). The etch rates reproduce on a routine basis and can be kept uniform across the wafer. However, the ion bombardment probably creates surface defects which might influence MBE overgrowth and laser operation. The role of such defects and the influence of the subsequent annealing at a MBE-growth temperature of 400 °C is not known.

Using Ar^+-ion milling the profile of an etched stripe can be controlled by the inclination of the ion beam towards the substrate, which is mounted on a cooled rotating plate. This is shown in Fig. 1.

Unfortunately this does not form the smooth mesa shape required for optimum MBE-overgrowth of the top confining layer. Growth discontinuities arise at the steep edges of the ridge as can be seen in Fig. 2. Such discontinuities were already observed by Schlereth et al. (8). In those experiments the main PbSe flux was oriented parallel to the mesa ridge and no substrate rotation was used. Since this causes asymmetries even with slight misalignment, all samples reported here were overgrown with MBE-substrate rotation.

The optimum shape of the ridge has to be formed by an additional process step. Feit et al. (7), who use a wet chemical etch for stripe definition, reported on the use of an electrochemical treatment for this purpose. This results in an improved shape suitable for discontinuity-free MBE-overgrowth.

a) 15°

b) 30°

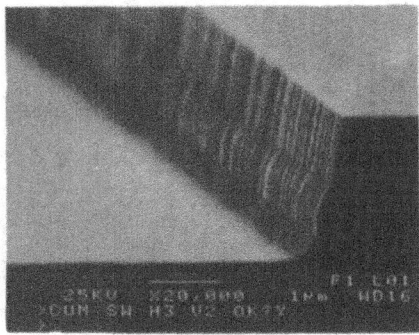

c) 45°

Fig. 1: Ion beam etching of PbSe- substrates with 15° (a), 30° (b), 45° (c) angle between the vertical Ar^+-ion beam and the (100) wafer normal. When etching, the wafer is rotated around this normal. (a) shows the trenching close to the mesa edge, (b) is an optimum angle and (c) shows the undercutting occurring at larger inclination angles.

We developed a chemomechanical step similar to the final substrate wafer preparation before MBE growth. Assisting the chemical etch by a mechanical polish gives good control of the etch rate and shape. By taking SEM-cross-sections of the mesas during and after the etching process we found that with this process the mesa height can be controlled with a resolution better than 50 nm. Even with no measurable change of the mesa the chemomechanical polish serves as a cleaning step prior to MBE growth, and removes damaged surface which may be due to the ion-milling.

The resulting smooth edges after polishing a 30° Ar^+-etched ridge are displayed in Fig. 3. The overgrowth result of such a polished structure is given in Fig. 4. The extent of growth discontinuities, especially the formation of broad and deep cracks is significantly reduced.

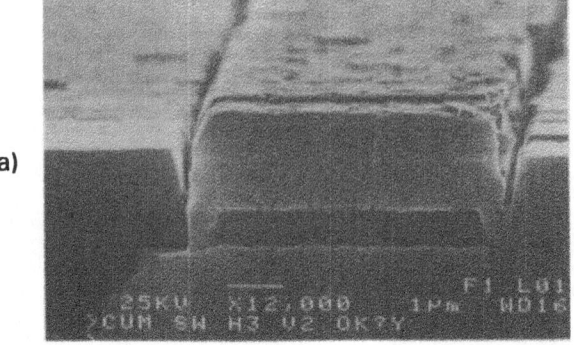

Fig. 2: Overgrowth discontinuities observed within a BH-laser (a) (30° Ar$^+$-etch) and within a 60° etched PbSe-substrate (b). Cracks arise at sharp edges of the mesa ridge.

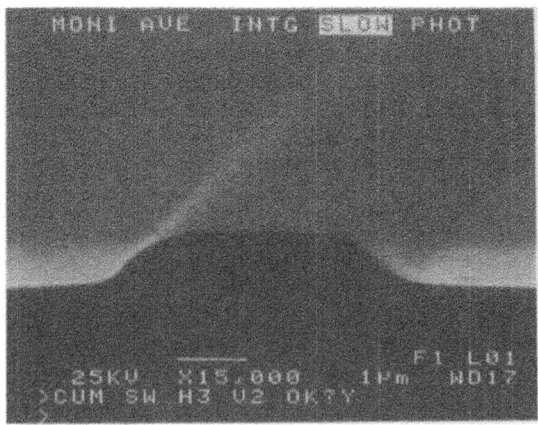

Fig. 3: 30° ion milled PbSe ridge after chemomechanically polish.

The first 5 μm BH-lasers fabricated by this method show cw-threshold currents of about 20 mA at T = 77 K and can be operated up to 170 K in cw mode. They have improved farfield properties compared to their DH counterparts (1). These lasers generally emit several longitudinal modes with modal powers in the 100 μW range.

a)

b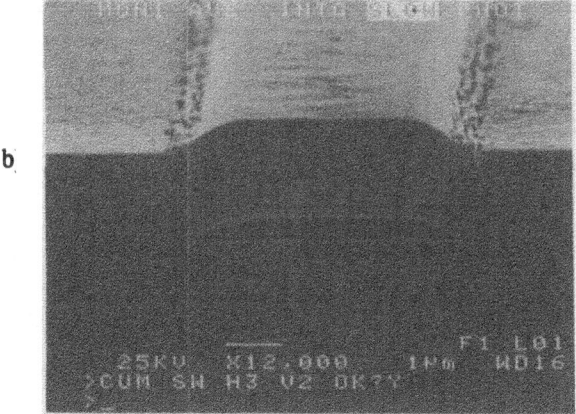

Fig. 4: (a) Overgrowth result of polished PbSe-rigde of Fig. 3. (b) Overgrowth result of a polished active layer for a BH-laser.

b) Shadow mask technique:

The above described photolithographic technique suffers from some major disadvantages:

♦ It requires many technological processes which will influence the final quality and lengthens the turn-around time. Quality control processes have to be developed adapted for the IV-VI materials.

♦ Special care has to be taken to ensure a clean surface for the overgrowth process. Especially any residual photoresist which was applied to the top layers during lithography has to be removed completely.

♦ The optimum shape for buried ridges remains a problem especially for structure widths below 5 μm. It will be difficult to reproduce a good shape.

The use of shadow masked growth as already reported for III-V materials by (10) offers a straightforward and least damaging way of fabrication of buried structures even in the μm range. By use of this technique the fabrication of a BH-laser structure may be possible within 5 to 6 hours compared to the present 2 to 4 days with photolithography. A complete in-line UHV-process with no mechanical or chemical treatment of the as-grown surfaces could be realized.

Fabrication of the Silicon shadow-mask (front view):

- thermic oxidation

-photolithographic structuring of the Si-wafer:

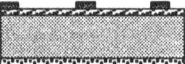

- etching of the oxide in HF:

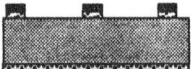

- removing of the photoresist and etching in KOH:

-removing of the oxide. Shadow mask is now ready for overgrowth

Fig. 5: Fabrication steps for Si shadow masks.

However, an optimum mask still has to be developed. There are several conditions to the mask:

The mask has to be positioned close to or in contact with the wafer. It has to withstand the growth temperatures of 400 °C with negligible movement or warping. The mask should be very clean and no outgassing should occur. It should also be rigid and easy to handle, i. e. to mount on a suitable MBE substrate holder.

All of these necessary properties are matched by use of Si as mask material. The possibility of anisotropic chemical etching of Si greatly stimulated the rapid development of Si-micromechanics (11). By anisotropic etching of Si (100) with KOH, as displayed in Fig. 5, shadow masks with slits down to 10 µm were fabricated. Fig. 6 shows a SEM picture of a partly etched sample mask.

Fig. 6: SEM picture of a sample etched for 260 min in 10 % KOH.

In Fig. 7 a 44 µm wide active stripe obtained using a Si mask is shown. A very smooth shape with strongly tailed edges is obtained.

Fig. 7: PbSe active layer grown with a MBE shadow mask.

Fig. 8 shows a SEM picture of a complete BH-laser. The width of the active layer, which was visualized by selective chemical etching, is about 40 μm. This is of course too wide for the suppression of high order lateral modes. On the other hand this laser already shows a reduced threshold current compared to DH-lasers of the same composition. This means that the effective lateral width of the shadow masked BH-laser is less than the gain width of DH-lasers. By current spreading we expect for 20 μm stripe DH-lasers a gain region of approximately 40 μm. Nearfield data (1) suggest even larger values of about 60 μm. Fig. 8 does not show any growth discontinuities due to the smooth shape of the buried region. This laser was grown in the MBE within one day.

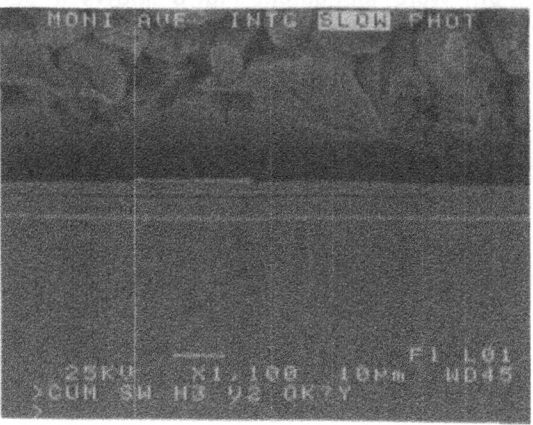

Fig. 8: SEM picture of a contacted and mounted MBE-shadow mask grown BH-laser.

3. CONCLUSIONS

We have shown that by the application of chemomechanically etching to ion milled active mesa stripes a better MBE-overgrowth for BH-lasers is achieved. Further improvement of this BH-process for stripe widths less than 5 μm must concentrate on the final preparation of the surface with special care to get an optimum shape of the buried stripe. It has to be demonstrated that this multi-step technique is suitable for different active layer materials (i. e. wavelengths) and that a reproducible fabrication with reasonable yield can be realized.

First experiments using shadow masked MBE-growth for IV-VI-BH-lasers have shown the potential of this technique for laser fabrication with structure widths above 10 μm.

Calculations by Schlereth et al. (8) show that the cut-off of the first order lateral mode for a rectangular symmetric PbSe/PbEuSe BH-laser is at a stripe width of some 3 μm. This value might even be reduced for shorter wavelength diodes. To get lateral mode suppression a softer lateral guiding seems to be advantageous without sacrificing the good thermal and carrier confining properties of a buried structure. By use of an adapted soft shape of the waveguide this could be achievable with structures 10 μm wide.

Combining both methods may be favourable too: A suitable shadow mask material can be grown epitaxially onto the first confinement layer.

After opening slits by photolithography the active layer is grown using the shadow mask technique. The shadow mask can be removed by a selective etch. A final MBE-overgrowth covers the remaining mesa structure. This method was already used for growth of III-V devices (12). Epitaxial BaF_2 (13) could be used as a suitable etch-away mask.

ACKNOWLEDGEMENTS

This work was supported by the BMFT as part of the JETDLAG program (grant No. 7EU733). The authors are greatful to their colleagues of the IPM-laser technology group. The development of shadow masks was greatly supported by the encouraging help of Christa Vogel and Gabi Uptmoor.

REFERENCES

(1)   AGNE, M., SCHIESSL, U., LAMBRECHT, A. and TACKE, M. (1991): Measurement of IV-VI Diode Laser Near and Far Field Distribution. This volume.
(2)   FEIT, Z., KOSTYK, D.,WOODS, R.J. and MAK, P. (1991): Recent Developments in MBE grown $Pb_{1-x}Eu_xSe_yTe_{1-y}/Pb_{1-v}Sn_vTe$ Diode Lasers for High Resolution Spectroscopy. This volume.
(3)   FACH, A., BÖTTNER, H. and TACKE, M. (1991): Embossed Grating DFB-BH Lead Chalcogenide Diode Lasers. This volume.
(4)   TACKE, M.,SPANGER, B.,LAMBRECHT, A., NORTON, P.R. and BÖTTNER, H.(1988).Infrared Double Heterostructure Diode Lasers made by Molecular Beam Epitaxy of $Pb_{1-x}Eu_xSe$. Appl. Phys. Lett.**53** (23),.p. 2260.
(5)   BACHEM, K.-H., NORTON, P. and PREIER, H.(1984) in: Two Dimensional Systems, Hetrostructures, and Superlattices, Eds.: Bauer, G., Kuchar, Γ. and Heinrich, H. (Springer, Berlin), pp. 147 - 156.
(6)   BÖTTNER, H.,SCHIESSL, U. and TACKE, M. (1990): Dependence of Lead Chalcogenide Diode Laser Radiation on Lattice Misfit Induced Stress. Superlattices and Microstructures 7 (2), p. 97.
(7)   FEIT, Z., KOSTYK, D., WOODS, R.J. and MAK, P. (1990): Single-Mode Molecular Beam Epitaxy Grown PbEuSeTe/PbTe Buried Heterostructure Diode Lasers for CO_2-High-Resolution Spectroscopy. Appl.Phys. Lett. **58** (4),p. 343.
(8)   SCHLERETH, K.-H., SPANGER, B., BÖTTNER,H., LAMBRECHT,A. and TACKE, M. (1989): Buried Waveguide Double-Heterostucture PbEuSe-Lasers Grown by MBE. Infrared Phys. Vol. **30** (5), pp. 449 - 454.
(9)   SHANI, Y., KATZIR, A., TACKE,M. and PREIER, H.M. (1989): $Pb_{1-x}Sn_xSe/Pb_{1-x-y}Eu_ySn_xSe$ Corrugated Diode Lasers. IEEE-QE **25**(8),p.1828
(10) TSANG, W.T., ILEGEMS, M. (1977) Appl. Phys. Lett. **31**, p. 301.
(11) HEUBERGER, A. (Ed.) (1989): Mikromechanik. Springer, Berlin, Heidelberg.

(12) DEMEESTER, P., POLLENTIER, I., VAN DAELE, P. (1990): Novel Optoelectronic Devices and Integrated Circuits Using Epitaxial Lift-off. Proceedings of the SPIE International Conference on Physical Concepts of Materials for Novel Optoelectronic Device Applications, Aachen, Germany, October 28 - November 2, 1990.

(13) ZOGG, H., VOGT, W. and MELCHIOR, H. (1985): MIS-Capacitors on BaF_2/PbSe Layers and Epitaxial Si/BaF_2/PbSe Structures for IR-Detection. Infrared Physics vol **25**, (1/2), pp. 333 - 336 (1985).

SEMICONDUCTOR LASERS AND PHOTODIODES FOR GAS ANALYSIS IN THE SPECTRAL RANGE 1.8–2.5 μm

A.N. BARANOV, A.N. IMENKOV, M.P. MIKHAILOVA, and Yu.P. YAKOVLEV

A.F. Ioffe Physico-Technical Institute,
194021 St.Petersburg, Russia

1. INTRODUCTION

In this paper semiconductor lasers and photodiodes for the 1.8–2.5 μm spectral range are described, which can be used in gas analysis systems. In spite of the fact that strongest absorption lines of many gaseous substances are concentrated at longer wavelength, the spectral range in vicinity of 2 mm seems to be more promising because powerful light sources and high efficiency photodetectors operating at room temperature can be developed in this case in contrast to the wavelengths exceeding 3 μm.

Characteristics of the lasers of several types are presented. Among them there are pulsed lasers with output optical power reaching 1 W and low threshold single mode lasers. The photodiodes posses high quantum efficiency (0.6 without antireflection coating) and high speed of response (0.5 ns at reverse bias of several volts).

The quaternary solid solutions GaInSbAs and GaAlSbAs lattice matched to GaSb substrates were used for fabrication of all the structures. The first material was used in active regions of the devices and the second one in wide gap layers — emitters in the lasers and window layers of the photodiodes. The structures were grown by liquid phase epitaxy (1).

2. LASERS

Structures of three types were used for fabrication of the lasers (Fig.1). The first one was a conventional DH structure, where the narrow-gap GaInSbAs active layer 0.25–2 μm thick was enclosed between the wide-gap GaAlSbAs layers, and the top GaSb layer was grown to provide good ohmic contact (Fig.1a). In the structures of the second (Fig.1b) and third (Fig.1c) type one or two GaSb layers were introduced into the wavegude region, which formed a staggered-lineup heterojunction with the GaInSbAs active layer (2).

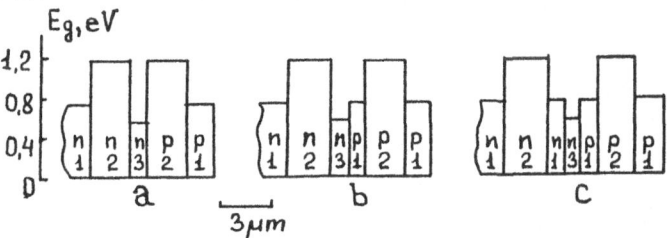

Fig. 1 : Band gap profile of the laser structures. 1 - GaSb; 2 - GaAlSbAs; 3 - GaInSbAs

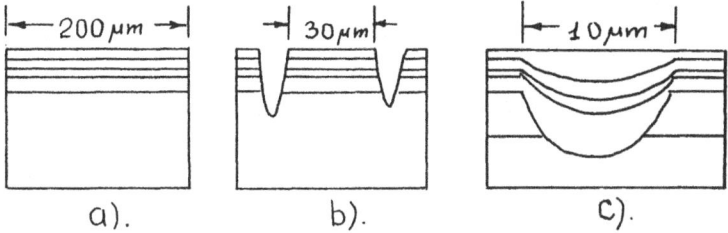

Fig. 2 : Design of the lasers.

The structures were used to fabricate broad contact lasers (Fig.2a), mesa-stripe lasers (Fig.2b), and substrate-channel buried lasers (Fig.3c). The resonator length was in the range 200–500 μm.

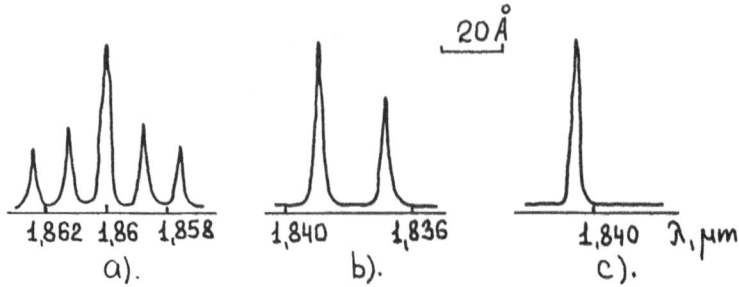

Fig. 3: Coherent emission spectra of the lasers.

The broad contact lasers are multimode (Fig.3a), threshold currents are high (2–4 A), but output optical power reaches 0.2–1 W (3). For the mesa-stripe (Fig.3b) and buried (Fig.3c) lasers there are usually 1–3 modes in the emission spectra and these lasers are more favorable for the laser diode spectroscopy. Threshold current is in the range 200–600 mA and output optical power is essentially lower — 0.2–1 mW.

Electroluminescence characteristics of the lasers depended on the internal structure of the devices. Consider now these features.

The introduction of the staggered-lineup heterojunction into the waveguide region resulted in lowering the threshold current I_{th} and made the dependence of I_{th} on the active region thickness weaker (Fig.4). The threshold current was two times lower in these devices comparing with the DH lasers with the same thickness of the active layer. These effects are due to the change in prevailing mechanism of

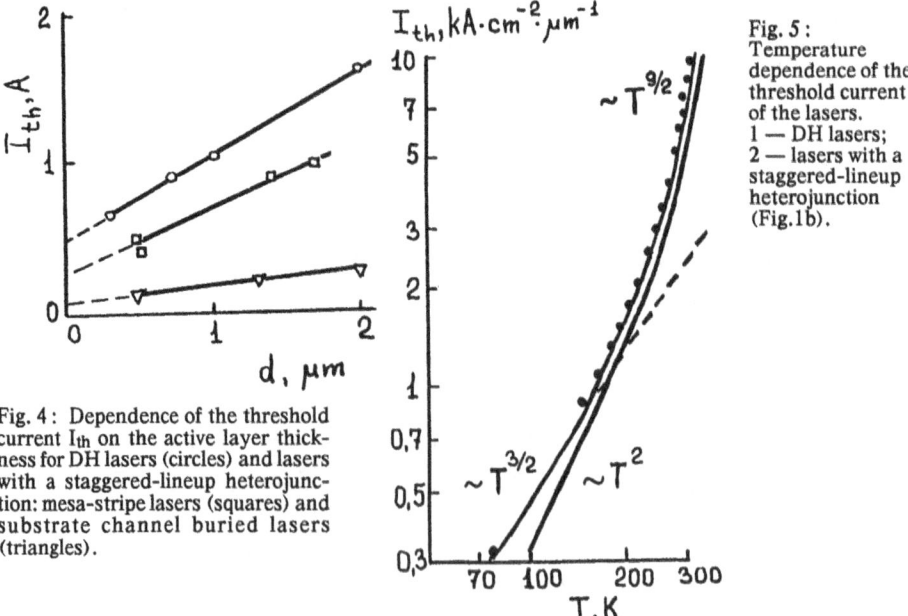

Fig. 4: Dependence of the threshold current I_{th} on the active layer thickness for DH lasers (circles) and lasers with a staggered-lineup heterojunction: mesa-stripe lasers (squares) and substrate channel buried lasers (triangles).

Fig. 5: Temperature dependence of the threshold current of the lasers. 1 — DH lasers; 2 — lasers with a staggered-lineup heterojunction (Fig.1b).

radiative recombination from the bulk one to the recombination via quantum states of the staggered-lineup heterojunction.

The temperature dependence of the threshold current was also different for the lasers with different structure.

The temperature dependence of the threshold current of the DH lasers consists of three parts. In

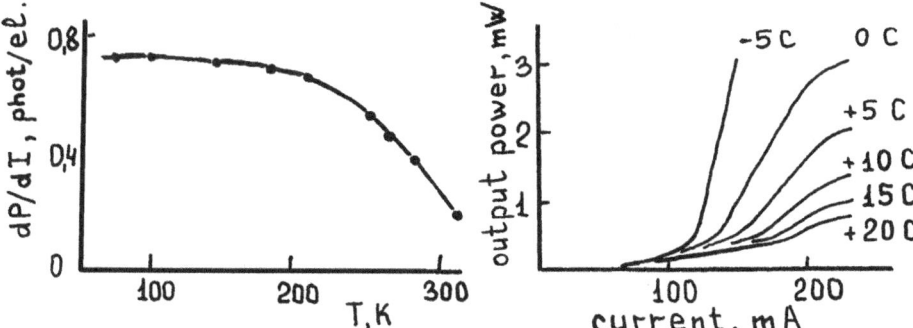

Fig. 6: Temperature dependence of differential quantum efficiency of the lasers.

Fig. 7: Output optical power versus current for a cw laser at different temperatures.

the temperature range 77–320 K I_{th} is proportional to $T^{3/2}$ and is defined by the rate of radiative recombination, for the lasers with a staggered-lineup heterojunction in this range I_{th} is proportional to T^2 (Fig.5). At higher temperatures (200–300 K) for the lasers of all the types I_{th} $T^{9/2}$ and is due to Auger-recombination (CHH-process). Above room temperature the threshold current increases exponentially and is due to the CHCC Auger process.

Differential quantum efficiency is nearly the same for all the lasers and does not depend on the temperature in the range 77–250 K (Fig.6). Above 250 K the differential efficiency decreases with increasing temperature and the point of intersect-ion of this curve with the temperature axis indicates the maximum temperature of lasing of such the wavelength lasers — 320–350 K.

The important question for many spectroscopy applications is a possibility of cw operation of a laser at room temperature. We have realized cw mode at room temperature using the Fig.1b structure and the Fig.2c design. The current dependence of output optical power of the cw laser is shown in Fig.7 at different temperatures.

These data show that the 1.8–2.4 μm lasers can be used not only in laboratory DLS installations but also in the systems for a wide use.

3. PHOTODIODES

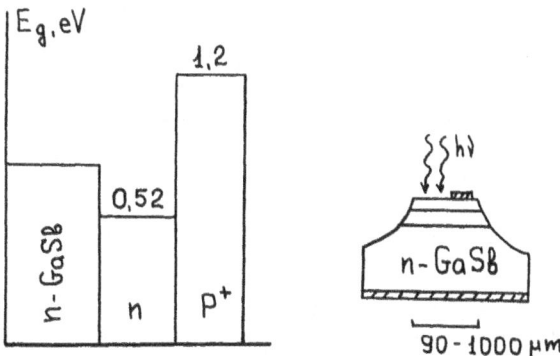

Fig. 8: Band gap profile of the photodiode structure and the photodiode geometry.

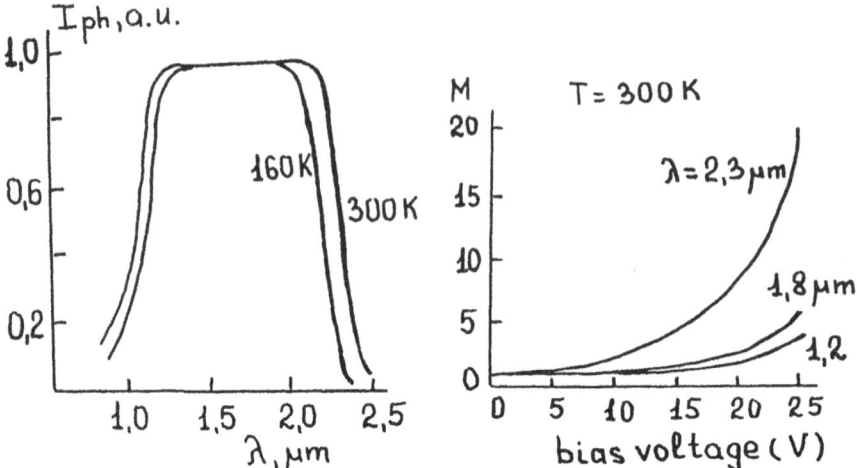

Fig. 9 : Spectral dependence of the response of the photodiodes at different reverse voltages.

Fig. 10 : Multiplication coefficient versus reverse voltage at different wavelengths.

Photodiodes were fabricated also by LPE on n-GaSb (111) Te-doped substrates (4). The structure includes the active GaInAsSb layer ($E_g = 0.52$ eV), 2–3 μm thick, and the wide band gap GaAlAsSb ($E_g = 1.2$ eV) window layer. The concentration of uncompensated donors in the n-type active layers was about $(5–7) \, 10^{15}$ cm^{-3}; the p-type wide band gap layers were doped with Ge up to a level of 10^{18} cm^{-3}. Mesa photodiodes (200–300 μm diameter) were fabricated by photolitography (Fig.8).

The photoresponse spectrum of such a photodiode is shown in Fig.9. The spectrum is typical for

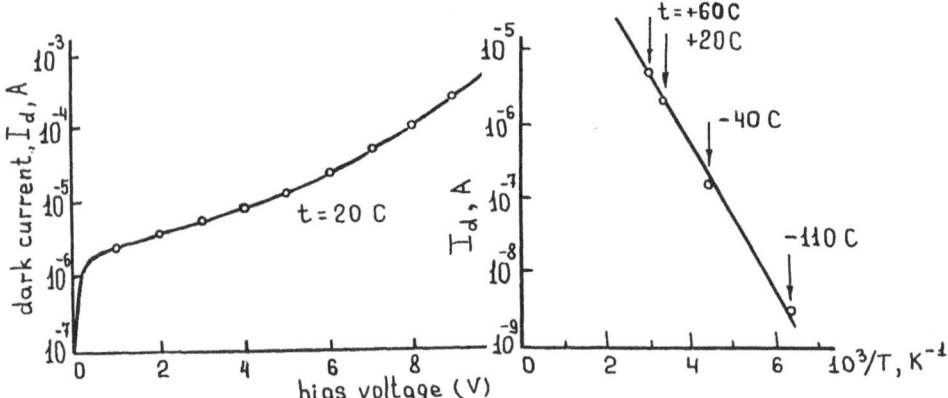

Fig. 11 : Dark current of a photodiode versus reverse voltage.

Fig. 12 : Temperature dependence of the dark current at U=–1 V.

heterephotodiodes with the wide plateau of constant sensitivity (0.6 el/photon) from 1.3 to 2.2 μm.

The photosensitivity at the 1.55 μm wavelength was compared with the sensitivity of a calibrated Ge-photodiode and occurred to be 0.6 electrons/photon at the reverse bias 2 V in the region of the plateau.

At reverse voltages higher than 5 V the increase in the photocurrent was observed due to avalanche multiplication. The spectral dependence of the multiplication coefficient indicates prevailing multiplication by holes in the active GaInSbAs layer. At the reverse bias 15 V the multiplication coefficient amounted 5–10.

Voltage-current characteristics were measured at different temperatures in the range 170–370 K (Fig.10). The activation energy for the temperature dependence of the dark current at fixed bias is 0.36 eV, that is close to a half of the active layer band gap (Fig.11). This means that the dark current at low voltages is due to generation of carriers inside the depletion layer. The effective lifetime for this process was calculated to be $6 \, 10^{-8}$ s. Reverse current density for best photodiodes was in the range $j = 6 \, 10^{-3} - 1.5 \, 10^{-2}$ A/cm−2 at the reverse bias 2 V.

At voltages exceeding 8 V the sharp exponential increase of the current and its weak temperature dependence showed that at high reverse bias the dark current is due to the band-to-band tunneling (the characteristic energy 0.5 eV is close to the band gap of the active layer).

To measure the speed of response of the photodiodes they were illuminated through an optical fiber by light pulses from the 1.3 μm InGaAsP laser. The rise and fall time of the photoresponse pulses did not exceed 0.5 ns at the reverse bias 3 V for the 200 μm diameter diodes with the 3 μm thick active layer, and the photocurrent nearly did not depend on the bias. The capacitance of the photodiodes was about 2–3 pF at the reverse voltage of several volts.

4. SUMMARY

The properties of GaInSbAs/GaAlSbAs lasers and photodiodes for the spectral range 1.8–2.5 μm are described. The single mode lasers and high-speed photodiodes can be used for laser diode spectroscopy.

REFERENCES

(1)   De Winter J.C., Pollack M.A., Srivastava A.K., Zyskind J.L., J.Electron.Mater. 14 (6), 729, 1985.
(2)   Baranov A.N., Imenkov A.N., Mikhailova M.P., Rogachev A.A., Titkov A.N., Yakovlev Yu.P., Superlat. and Microstruct. 8 (4), 375, 1990.
(3)   Baranov A.N., Danilova T.N., Dzhurtanov B.E., Imenkov A.N., Ershov O.E., Yakovlev Yu.P., Sov.Phys. Techn. Phys. 33 (8), 983, 1988.
(4)   Andreev I.A., Baranov A.N., Mirsagatov M.A., Mikhailova M.P., Yakovlev Yu.P., Sov.Techn.Phys.Lett. 12 (21), 1311, 1986.

LINEWIDTH AND NOISE OF LEAD CHALCOGENIDE DIODE LASERS

G. SPILKER, R. DADDATO, U. SCHIESSL, A. LAMBRECHT AND M. TACKE

Fraunhofer-Institut fuer Physikalische Messtechnik
Heidenhofstrasse 8, W-7800 Freiburg, Germany

ABSTRACT

A setup for the investigation of the linewidth of IV-VI lasers using a scanning confocal Fabry-Perot-Interferometer (FPS) was developed. This setup fits on to existing spectroscopic experiments and can be used on a routine basis. The scanning rate is tunable up to 75 GHz/s, so that the conditions of derivative spectroscopy can be simulated. For this application it is necessary to suppress optical feedback from the FPS. Moreover, we are able to investigate the laser line under controlled optical feedback conditions. As an example we present first results of lasers made by molecular beam epitaxy (MBE), which show linewidths below 5 MHz.

1.   INTRODUCTION

The linewidth and noise of lead chalcogenide diode lasers used in spectroscopy limit the maximum possible signal to noise ratio. Therefore linewidth and noise investigations have been performed using several different experimental techniques. P. Werle et al. (1) investigated the spectral dpendence of the amplitude noise (see also reference (2)). The noise was found to decrease strongly with the frequency. This is the main reason for the development of high frequency modulation spectroscopy. The Relative Intensity Noise (RIN) of the light emitted by the laser was investigated by H. Fischer (3). The main results were:
a)   The 1/f noise has a cutoff at a laser-specific high frequency.
b)   Sharp noise maxima occur at kinks in the current-intensity curve and at mode jumps.
c)   The RIN increases when one of several modes is selected by a monochromator or a gas absorption line.
d)   Low noise operation for spectroscopic application is preferably performed in single mode operation.
e)   Noise is increased by some orders of magnitude by unwanted optical feedback.

In spectroscopy the measured signal is a convolution (Faltung) of the gas spectrum with the laser spectrum. To resolve narrow absorption lines and to quantitatively determine the gas concentration, the linewidth of the diode laser being used has to be much smaller than the width of a doppler limited transmission line (typically 30 MHz) (4). While scanning the laser mode over the absorption line by modulation of the laser current, the linewidth of the mode should not increase.
Udo Lambrecht investigated the linewidth of two homostructure diode lasers (5) in a self-homodyne experiment. The laser beam was divided into two optical branches with a path difference greater than the coherence

length by a beamsplitter. This was realized using a White cell with a path length of 50 m. For measuring the beat signal, the two branches were superimposed on a fast photovoltaic CMT detector. Due to the limited path length of the cell this experiment yields a high frequency linewidth with a phase noise highpass of 6 MHz. Results for the linewidth at different currents are almost in accordance to Henry's formula (6):

$$\text{Linewidth} \quad 1/\text{Laser Power} + \text{const.}$$

While the theoretical work of Henry predicts a vanishing constant term, the minimum linewidth extrapolated from experimental data was 4 MHz. This is in qualitative accordance with empirical results obtained by other groups working with other diode lasers.

Complementary experiments on low frequency linewidth were made by FM/AM conversion with a Germanium etalon. The linewidth derived by this way was shown to be governed by power supply noise, cooler vibrations and optical feedback. This method, while being simple, hence gives data on the complete system rather than on the laser itself.

The goal of the present work was to employ a scanning Fabry-Perot-Interferometer as an alternative means of linewidth measurement.

2.   INSTRUMENT DESCRIPTION

The setup is designed modularly so that it can easily be changed for the analysis of spectroscopic systems. The laser module is based on a liquid nitrogen cooler to avoid vibrations. A mini monochromator (7) can be plugged into the pinhole position. A beam splitter enables the observation of the intensity with a reference detector while using the other beam for linewidth measurements. On the module for the linewidth measurement (FPS module) the light is collimated into a scanning spherical Fabry-Perot-Interferometer (FPS Burleigh CF-100 IR). The FPS has a free spectral range (FSR) of 750 MHz and a finesse of 200 between 1540 and 1740 cm^{-1}. The cavity can be scanned over 3 μm by using a piezoelectric element. The transmitted signal is observed on an oscilloscope which is triggered by the Piezo ramp (Fig. 1).

While a properly aligned plane FP, such as a Germanium etalon, reflects the parallel beam back as a parallel beam, a spherical FP does not. When the parallel beam is collimated by a coupling mirror and the FPS placed into the focus, the feedback is at first parallel but then is collimated by the coupling mirror. A small absorbing anti-iris-diaphragm placed into the focal position of the backrefelcted light suppresses optical feedback from the FPS. It causes only a minor decrease in the intensity of the light coming from the laser (Fig. 2).

A theoretical analysis of the optics yields backreflection of 10^{-6} withouth the diaphragm. However this small degree of feedback is enough to change the line shape strongly and prevent continuous tuning of the laser mode. We observed mode jumps at discrete positions within the free spectral range of the external cavity. These effects remain the same when the beam path between the FPS and the laser module is lengthened in order to achieve a back refelction of less than 10^{-7}. Only if the diaphragm is well aligned, is the mode behaviour of the laser no longer affected by feedback from the FPS.

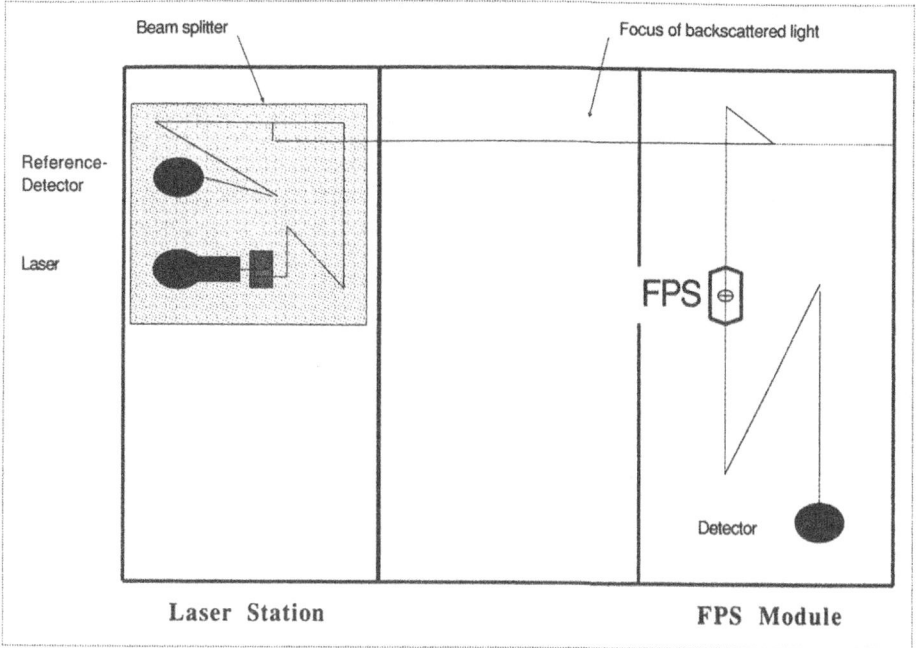

Fig. 1: Optical Setup

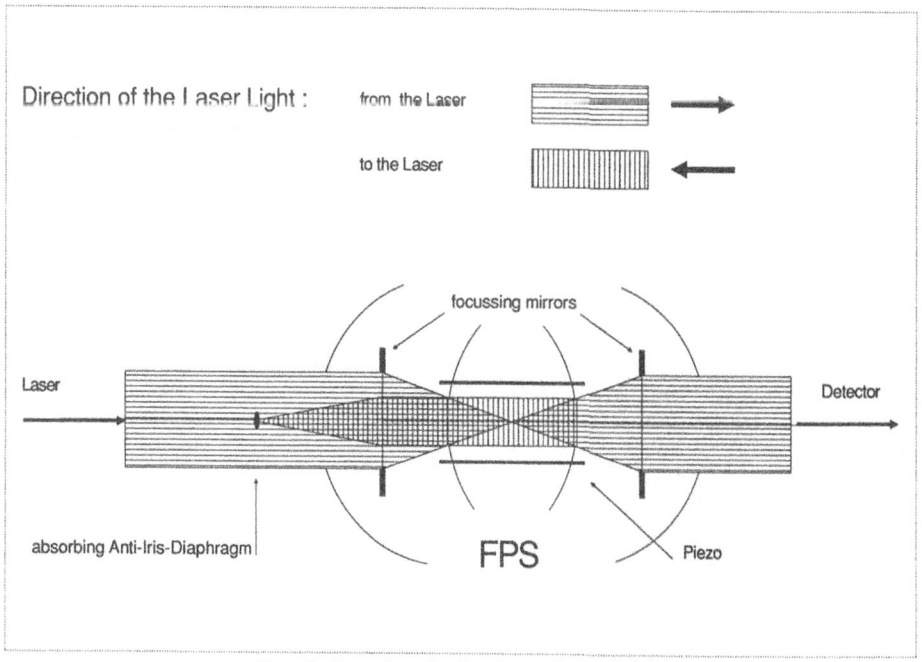

Fig. 2: Feedback Optics of the FPS

Fig. 3 shows the dependence of the linewidth (measured as in Fig. 4) on the scanning rate of the FPS. This is due to laser wavelength fluctuations that we trace to oscillations of the laser head temperature. The active temperature regulation is stabilized to some mK, but 1 mK is equivalent to a wavelength shift of about 30 MHz. An oscillation of 1 mK with 10 Hz (typical for a small cooler) leads to a laser wavelength tuning speed of 2000 MHz/s. The measurement of the linewidth has to be faster. For further experiments we choose the fastest scan to avoid this additional line broadening.

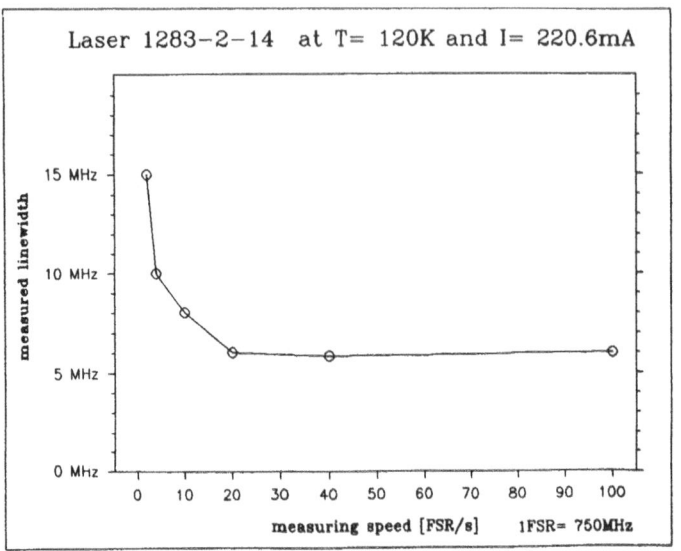

Fig. 3: Measured linewidth versus FPS scanning speed

The summarized features of our setup are as follows:

a)   The resolution is 3 MHz in the wavenumber region between 1540 and 1740 cm^{-1}. The region depends on the mirror coatings, other regions are available changing the mirror settings. This is an advantage over other experiments using a gas absorption line or a heterodyne technique that work only at a fixed, non-variable wavelength.

b)   The FPS is simple to align and has a reproducible finesse of 200.

c)   The optical feedback from the FPS can be suppressed.

d)   The laser line is directly visible on an oscilloscope. Therefore changes in laser operation due to alignment, temperature and current can be seen on-line. Optical feedback leads to an asymmetrical line shape.

e)   We operate the laser at cw current, so we can change between power supply and battery.

f)   The ramp duration of the Piezo is tunable to a minimum period of 20 ms. This gives an optical scanning speed of the FPS between 0 and 75 GHz/s.

g)   Due to its modular design, the setup is compatible with components that interface with a parallel beam, and fits to a commercial modular system (Mütek). It will be integrated into the IPM laser testing device. This allows one to take mode charts and linewidth charts at the same time.

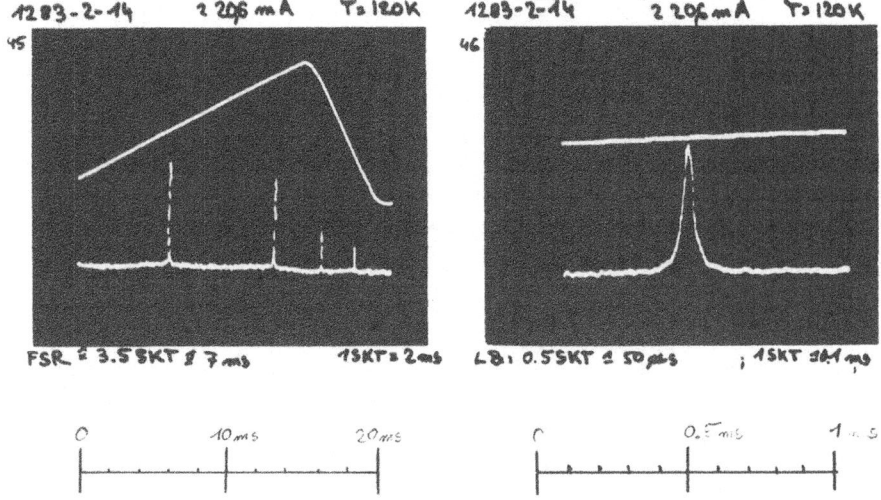

Fig. 4: A linewidth measurement of a double heterostructure IV-VI laser. The pictures were taken from the oscilloscope. On the left you see the piezo ramp and the FSR, on the right the x-axis scale is increased. The laser emits in a single mode at 1600 cm⁻¹. The linewidth is 50 μm $*$ 750 MHz/7 ms = 5.4 MHz.

3. RESULTS

An example of routine linewidth measurements as a function of operating conditions is shown in Fig. 5. The laser is a buried heterostructure MBE laser made at the IPM. Comparing the mode chart, the mode power and the measured linewidth of the strongest mode we see that the linewidth varies dramatically with the laser current. At an injection current between 260 and 313 mA the laser operates single-mode and the linewidth decreases as predicted in Henry's formula. In this range the laser can be used for a spectroscopic application.

At 314 mA the laser begins to emit a second mode. Although the intensity ratio is only 50:1 the linewidth of the stronger mode is doubled. At a ratio of 1:1 (both modes same power) the linewidth increases to more than 200 MHz (mode competition).

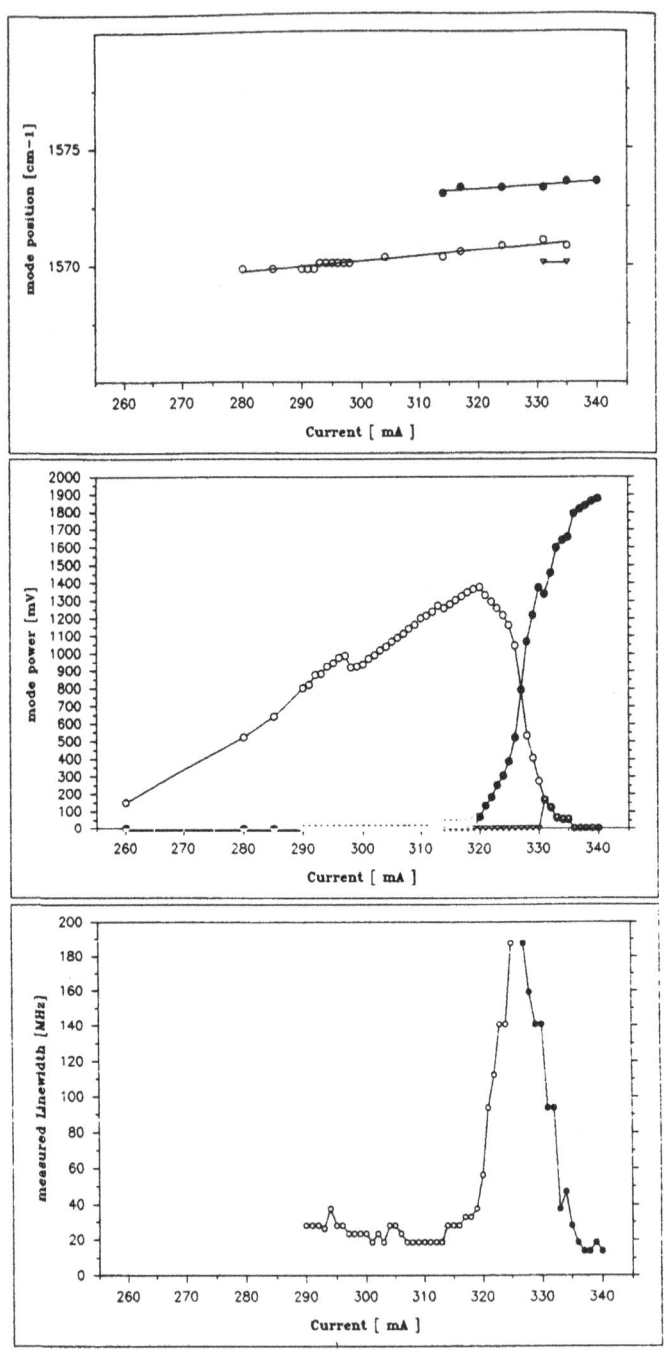

Fig. 5: A typical result: Laser 1224-1b-16 at 115 K.

CONCLUSIONS

It was shown that using an FPS, the linewidth of lead chalcogenide lasers can be measured without affecting the laser emission line itself by backreflection. Our setup allows routine investigations of the laser line, an on-line reference of the linewidth in derivative and integrative spectroscopy and sensititve investigation of optical feedback. For further experiments we will pay special attention to the influence of the internal laser structure on linewidth and feedback sensibility.

ACKNOWLEDGEMENTS

The work was supported by the BMFT as a part of the Eurotrac-Subproject JETDLAG, grant number 07Eu733. The authors want to thank A. Huber and the colleagues of the laser technology group of the IPM for cooperation.

REFERENCES

(1) WERLE, P., SLEMR, F., GEHRTZ, M. (1989). Wide-band noise characteristics of a lead-salt diode laser: possibility of quantum limited TDLAS performance. Appl. Optics 28.
(2) FISCHER, H., WOLF, H., HALFORD, B., TACKE, M. Low-frequency amplitude noise characteristics of lead salt-diode lasers fabricated by molecular beam epitaxy. IEEE J. Quant. Electron (to be published).
(3) FISCHER, H., TACKE, M. High frequency intensity noise of lead salt diode laser. J. Opt. Soc. Am. B (to be published).
(4) GRISAR, R. et al. Monitoring of Gaseous Pollutants by Tunable Diode Lasers. Kluwer Academic Publishers, ISBN 0-7923-0334-2.
(5) LAMBRECHT, U. et al. Linewidth measurement of lead-salt diode lasers using etalon and delayed self-heterodyne techniques (to be published).
(6) HENRY, C.H. (1986). Phase noise in semiconductor lasers, J. Lightwave Technol. LT-4, 298-311.
(7) RIEDEL, W.J. (1991). Optics for tunable diode laser. FTIE Proceedings

MEASUREMENTS OF IV-VI DIODE LASER NEAR- AND FARFIELD DISTRIBUTIONS.

M.Agne, U.Schiessl, A.Lambrecht and M.Tacke,
FHG-IPM, Heidenhofstr.8, 7800 Freiburg, Germany

Abstract:

At the IPM a flexible setup for routine measurements
of the near- and farfield distributions of IV-VI
laser diodes has been developed. The aperture of the
optical system is 1:2 . Measurement of absolute,
differential and noise (RIN) near- and farfield
behaviour and parallel beam intensity distibutions in
user systems is possible. With this setup the
emission patterns of MBE-DH and MBE-BH lasers with
wavelengths between 3.5µm and 7µm have been
investigated.
Series of measurements with systematic variation of
diode current and temperature allow a comparison
between the spectral behaviour and changes in the
farfield and nearfield of a laser. Simple relations
between emission pattern and modal structures have
not been found.
DH-lasers show nearfield distributions with more than
one peak, probably due to higher order lateral modes,
which result from gain guiding. This leads to multi-
lobe emission and a complicated dynamic behaviour in
the farfield.
BH-lasers have a single peak nearfield due to index
guiding in lateral direction. This results in a
relatively simple farfield and dynamic behaviour.

Introduction:

The role of the laser farfield distribution in a
spectroscopical equipment is known to be crucial for
ultimate sensitivity as required in atmospheric trace gas
analysis (1). Therefore farfield data should be regarded as
part of the laser specifications (as with III-V laser
diodes) and should be determined together with the
spectral data as part of the laser testing at the
production facility itself.

Equipment:

Since the end of 1990 a setup designed for measuring the
farfield intensity distributions of laser diodes within the
frame of our multi-purpose measurement system is available.
This setup allows also measurements of nearfield and noise
distributions and the dynamic behaviour of near- and
farfield with an aperture of 1:2. It further allows
measurements of parallel beam distributions in a user
system with a beam diameter up to 60mm.
The complete system consists of three main parts:
a) With a stirling cooler up to 16 lasers can be cooled
down to temperatures between 20K and 200K. The normal
operation temperature of MBE-grown lasers is between 80 and

120K. Our system's temperature stability in this region is of the order of some millikelvins. Most of the farfield data were taken in this temperature range.
b) The optical system combines a special off-axis ellipsoid mirror lens which was developed in our institute with another off-axis ellipsoidal mirror. It forms a real image of the light distribution on the laser facet (nearfield) in the plane "N" with a magnification of 11.0 .
The beam cross-section in the focal plane of the first ellipsoid mirror corresponds to the laser farfield (angular intensity distribution at infinity). The second ellipsoid mirror forms an image of this light distribution in the plane "F" near the focal plane. 150µm in this plane correspond to a farfield angle of 1˙.
c) A HgCdTe-detector can be scanned through the farfield plane or through the nearfield image via two computer-controlled stepmotor-driven stages. Two detectors are available. The first one is used for standard measurements and in cases where very high sensitivity is required and has a rectangular geometry with an element size of $(150µm)^2$. The data presented in figures 2,3 and 4 were taken with this device. The second detector is used for high-frequency measurements like those shown in figures 5 and 6 and has a round element with a diameter of 110µm.

The laser light can be modulated with a mechanical chopper (absolute intensity distribution) or via the laser current (differential intensity distribution). The detector signal is measured with a lock-in amplifier at the modulation frequency and read into a computer via IEC-bus. Measurements with a modulation frequency between 100Hz and 30kHz are possible; usually we work in the 1kHz range.
For nearfield measurements the image of the laser facet is scanned. Additionally the detector signal can be analysed with a frequency analyser; thereby a noise distribution at a given noise frequency (e.g.30 MHz) and bandwidth (250 kHz or 12.5 kHz) is obtained. These data can then be analysed with computer help.

Instrument testing:
The experimental setup has been tested thorougly in order to ensure the significance of the measured data:
The nearfield resolution is given by the detector element size "b",the optical magnification "m" and the radius of the Airy-disc:

$$r = b/m + 1.22*a*lambda = 13.6µm + 2.44*lambda.$$

Lambda is the wavelength of the emitted laser light and "a" the inverse of the optical aperture . Test measurements using a HeNe-laser showed that the nearfield resolution comes very close to this limit over the whole image field. This means, that at a laser wavelength of 3.0µm two emission spots 21µm apart can be clearly resolved while a wavelength of 7.0µm requires a distance of 31µm for clear separation. The image distortion caused by the mirror optics is negligible, but especially in the case of lasers with very narrow emission peaks the geometry of the detector element leads to artefacts in the observed emission patterns (fig.2).

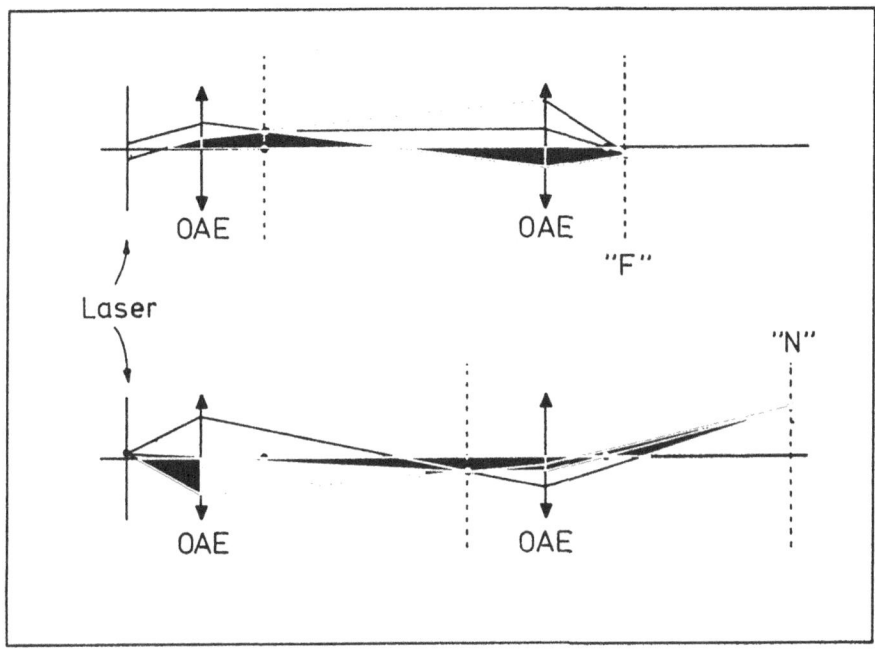

Fig 1:
The optical setup
(a) for farfield measurements,
(b) for nearfield measurements

These artefacts are due to backscattered light from the direct surroundings of the detector element and have an intensity of typically 1-5% of the maximum signal (Fig.2), depending on the detector and the wavelength. However, since they have a very distinct geometry, they are clearly distinguishable in the field measurements and cause no serious problems.
The lock-in technique enables us to measure intensities of a few nW/mm² so that observation of the nearfield below threshold is possible. The calibration of the nearfield measurements (i.e the optical magnification) is calculated from the focal length of the mirrors and their distance. It has been checked experimentally to an accuracy of about 2% using a glass scale and HeNe-laser light. Test measurements of a commercially available laser diode (Toshiba, 670nm) agreed well with the farfield data given by the manufacturer.

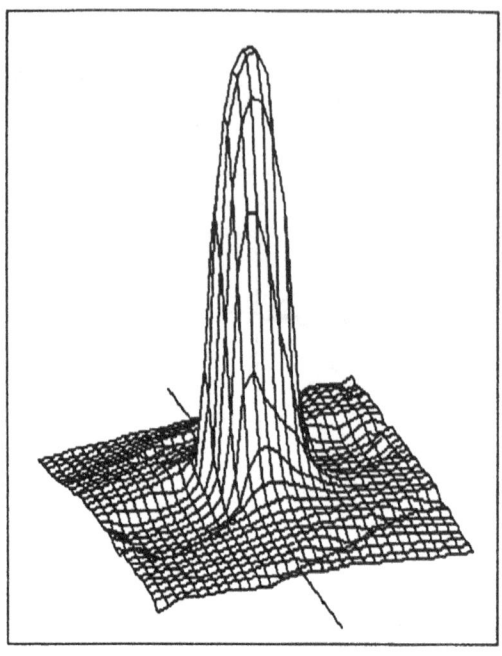

Fig. 2
A measurement of
the near-pointlike
farfield of a HeNe-
Laser demonstrates
the small artefacts
caused by the detector
element.
Scansize is 31*31 steps
à 25µm.

Special care was taken to achieve an accurate angular
calibration of the setup; the mirror optic allows an
alignment in the visible and thereby an accuracy better
than 1˙. The calibration was calculated and checked
experimentally using a HeNe-laser which could be tilted
against the optical axis.
The aperture of 1:2 limits the accepted farfield to a cone
with a semi-angle of 15˙. All lead-chalcogenide lasers that
have been characterised show wider farfield patterns so
that no full information about the laser emission is
obtained. However, our farfield measurements deliver
reasonable physical results especially for the shorter
wavelength lasers. Since many user systems work with
apertures about 1:2, our system is suitable for routine
characterisation of lasers which shall be delivered to
external groups.
For research purposes a setup without this aperture limit
has been developed and first measurements have been
performed. These measurements already proved that the
farfield distributions can be reproduced in another system
and after temperature cycling of the laser.
We could also show that the measured farfield patterns are
similar to the parallel beam distributions obtained under
the same laser conditions. Figures 3a and 3b give a
comparison of a parallel beam distribution (beam diameter
= 18mm, beam length = 1.30m) which has been measured in
our standard system and the corresponding farfield.

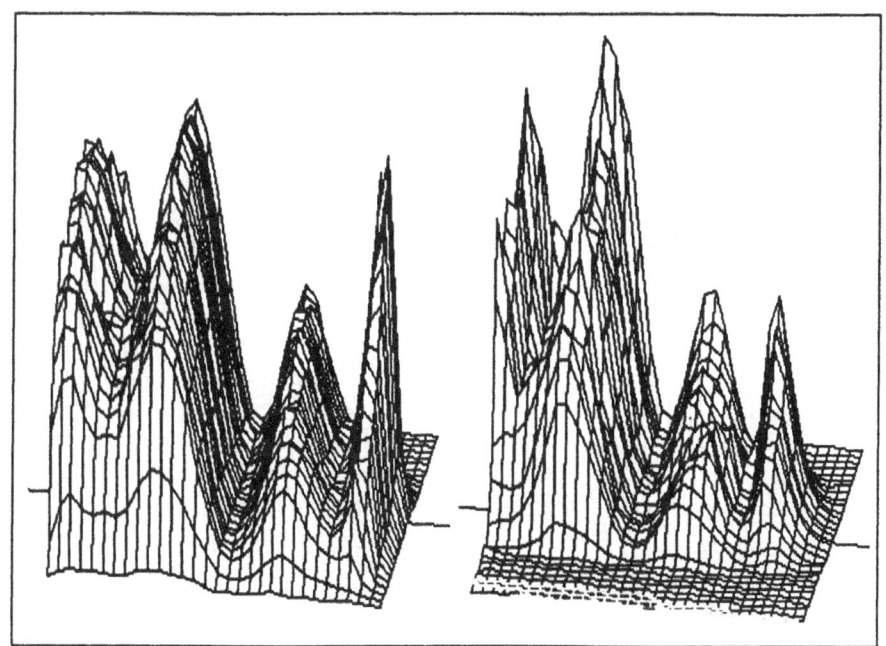

Fig 3:
(a) The farfield of a 3.5µm MBE-DH laser diode and
(b) the corresponding parallel beam intensity
    distribution.

Results:

Parallel to the test measurements, investigations of near-
and farfield patterns of many lasers have been performed.
The following general results were obtained:
1) Double heterostructure (DH) lasers with wavelengths
shorter than 5µm usually show more than one peak in the
nearfield pattern (Fig. 4.b). Lasers with longer wavelengths tend to have "shoulders" in the nearfield due to
the decreasing resolution with increasing wavelength. These
observations are interpreted as higher order lateral modes
caused by gain guiding, which results from the relatively
wide pumped region -the width of the stripe contact is
20µm. Typically a MBE-DH laser with a 20µm stripe contact
has a nearfield width of about 60µm. The dynamic behaviour
and the spatial noise distribution are relatively complex
and show lateral mode competition effects. The farfield
patterns of DH laser diodes with wavelengths between 2.9µm
and 4.5µm always exhibit stripe patterns (Fig. 4) where the
relative intensity of the stripes depends on diode current
and temperature.

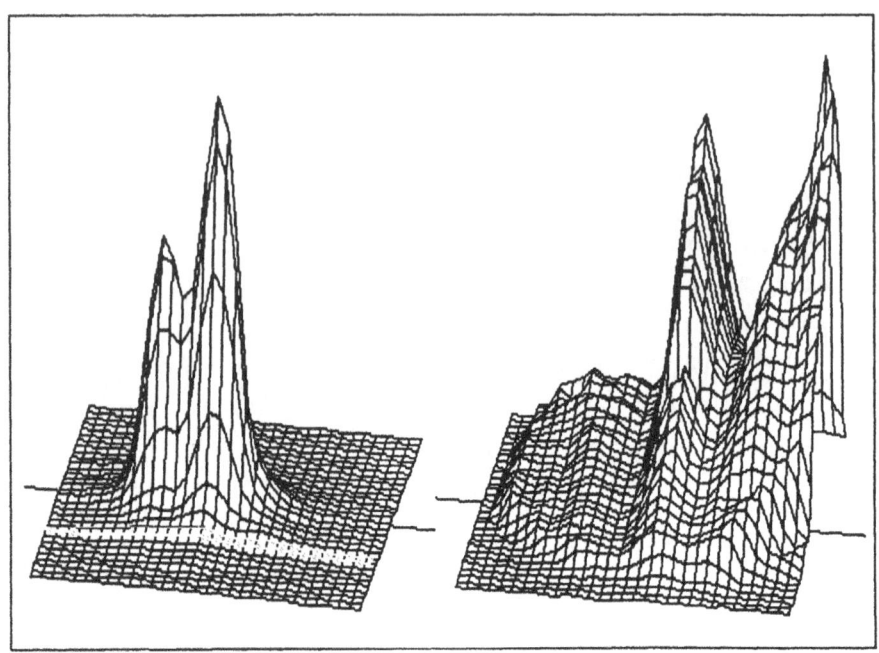

Fig 4:
(a) The nearfield of a 3.5µm MBE-DH laser diode and
(b) the corresponding farfield.
The scansizes correspond to 141µm*141µm in the nearfield
and to 31'*31' in the farfield scan.

Whilst the details of the farfield distribution of a given
laser may change very rapidly with diode current, the
general farfield structure (stripe pattern) is the same
even for laser diodes from different wafers. Fast changes
in the farfield are usually related to nearfield changes
and high total intensity noise and kinks in current-power
curve. However, as a comparison of figure 5a and 5b shows,
there may be modes which do not contribute to the increased
noise.
2) Buried heterostructure (BH) lasers with wavelengths
about 6.9µm and 4.5µm and a BH-stripe width of 5µm always
show very narrow nearfield distributions with a width of
typically 20µm. This comes close to the resolution of our
experimental setup in this wavelength range, so that no
inner structure could be detected. However, due to the
index guiding no higher order lateral modes and therefore
no complicated nearfield structure is expected. No BH-
lasers in the 3.0-4.0µm wavelength range are currently
available for measurements.
All measured BH diode lasers show only minor changes of the
near- and farfield structure with current and temperature,
a relatively smooth farfield distribution and a simple

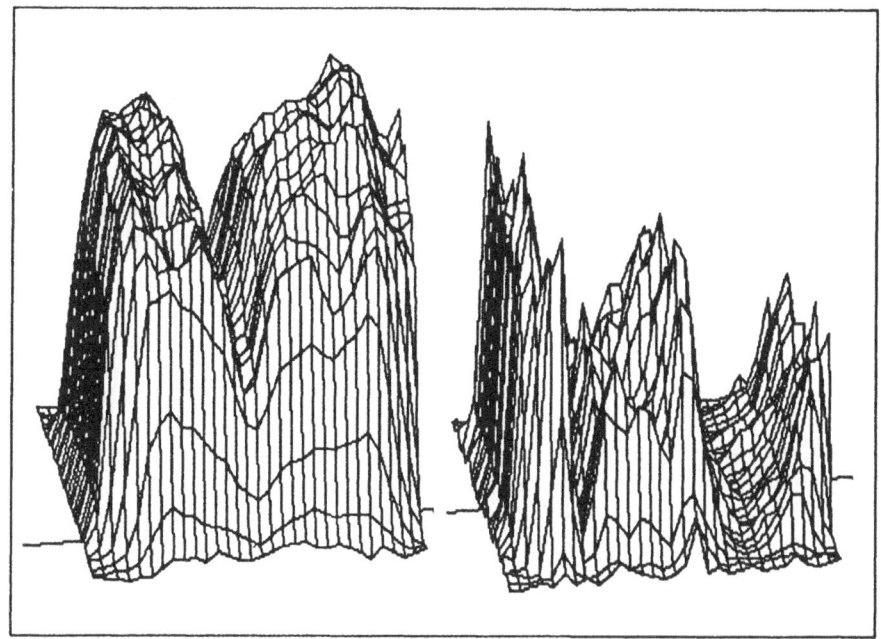

Fig 5:
(a) The farfield of a 3.5μm MBE-DH laser diode and
(b) the corresponding noise (RIN) distribution.
The arrow indicates the farfield structure corresponding to
the mode with the lower intensity noise.
The scansizes correspond to an angle of to 31'*31'.

dynamic behaviour. Fig.6 shows the near- and farfield of a
4.5μm BH-type laser diode. The total intensity of the
nearfield changes with current, but its structure is
preserved over a big current range. Since these lasers
should emit in the lowest order lateral and transverse mode
these results are quite reasonable. The farfield of this
laser changes slowly from the nearly ideal one-lobe
emission to the two-lobe emission .This might be explained
with an increase in phasefront curvature with increasing
current (2).

Conclusions:
Our experimental equipment allows routine measurement of
the near- and farfield disributions of laser diodes and of
their noise characteristics and dynamic behaviour.
Extensive testing proved the reliability of the
measurements. The farfield structure is characteristic for
a given laser and can be reproduced after temperature
cycling of·the laser.
With this setup MBE-DH and -BH laser diodes were
characterised. The DH structured lasers showed stripe

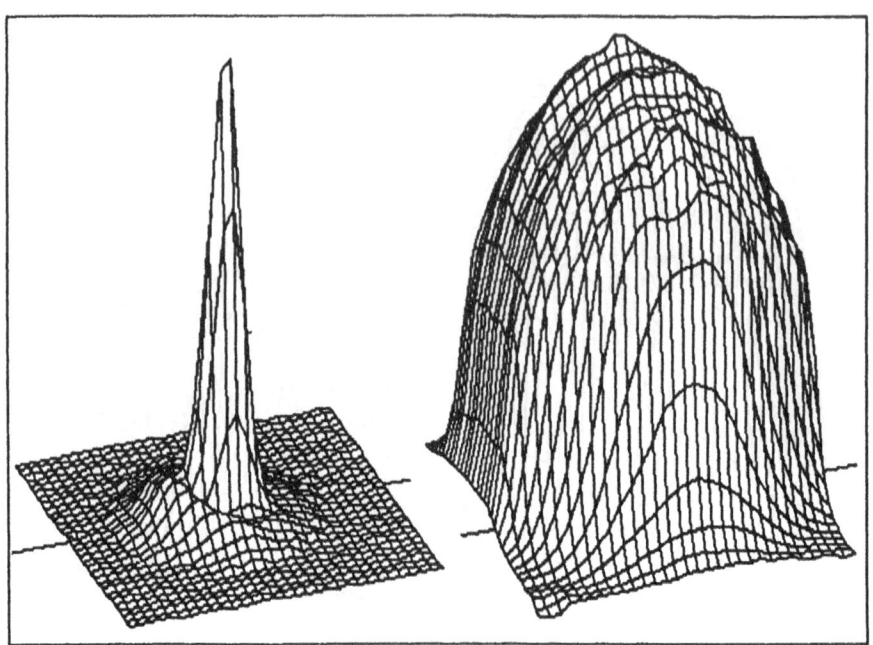

Fig 6:
(a) The nearfield of a 4.5μm MBE-BH laser diode and
(b) the curresponding farfield.
The farfield structure of this BH laser is smooth and
narrow compared with that of a DH-laser.
The scansizes correspond to 141μm*141μm in the nearfield
and to 31'*31' in the farfield scan.

patterns in the farfield which depended strongly on diode
current and a spatially structured noise distribution. The
nearfield of these lasers was usually double- or triple-
peaked. Both effects can be explained qualitatively with
higher order lateral laser modes. The BH-structured laser
diodes with BH widths about 5μm did not show any sign of
higher order lateral modes in the near- or farfield. The
farfield emission of BH laser diodes was distributed more
evenly over the observed angular range.

It has yet to be shown that the delivery of farfield data
together with a laser leads to an increased sensitivity in
the user apparatus. However, the farfield data are
certainly an important information for improvement of the
laser structures and a useful tool for laser selection.

Large aperture measurements of the farfield of selected
lasers are currently under way. This is not intended to
become a standard characterisation procedure, but for laser
development a complete farfield knowledge is important.

As with the III-V laser diodes the farfield structure of a laser should follow directly from the waveguide properties of the buried heterostructure. First attempts to model the light distribution inside the laser turned out to be not sufficient to explain the observed farfield distributions.

Acknowledgements:
The authors wish to thank M.Knothe, W.Riedel and A.Anders for their beautiful design and realization of the mirror optics used in our setup, which make accurate alignment and calibration procedures easy and comfortable.
This work was supported by the BMFT-JETDLAG project (grant no. 07EU733).

References:
(1) G.Schmidtke, W.Kohn, U.Klocke, M.Knothe, W.Riedel, J.Wolf: Diode laser spectrometer for monitoring up to five atmospheric gases in unattended operation, Applied Optics Vol 28, No.17 (1989).

(2) U.Schießl, A.Lambrecht, M.Tacke: Far field and Phase Front Curvature of Index Guided Diode Lasers, Submitted to IEEE Journal of Applied Physics.

PROGRESS IN IV-VI PHYSICS AND OPTOELECTRONIC DEVICES

(Proceedings of the Freiburg IV-VI Colloquium:
H. Böttner, A. Lambrecht, M. Tacke Editors)

Recent Developments in MBE Grown Pb$_{1-x}$Eu$_x$Se$_y$Te$_{1-y}$/Pb$_{1-v}$Sn$_v$Te Diode Lasers For High Resolution Spectroscopy.

Z. Feit, D. Kostyk, R.J. Woods, P. Mak
Laser Photonics Analytics Division, Bedford,Ma. 01730.

SUMMARY

Buried heterostructure (BH) tuneable diode lasers with PbEuSeTe confinement layers and PbSnTe,PbEuSeTe active layers were fabricated using a two stage molecular beam epitaxy growth procedure. The buried heterostructure technology was also used to prepare quantum well Pb$_{1-v}$Sn$_v$Te diode lasers (BQW) with PbEuScTe cladding/confinement layers. These lasers exhibit improved performance characteristics and significantly higher operating temperatures in continuous mode. Continuous wave (cw) operating temperature of 204 K was realized for a BH diode laser with a PbTe active layer, and 189 K for a Pb$_{0.932}$Sn$_{0.068}$Te BQW with a 1000 Å thick active layer. Considerably good spectroscopic properties have been routinely observed at operating temperatures exceeding 100 K. Mode tuning in excess of 1 cm^{-1} and tuning rates smaller than 1.8 Ghz/mA were measured in the majority of the diode lasers.

1. INTRODUCTION

Semiconductor diode lasers constructed of PbEuScTe/PbTe compounds are mostly used in spectroscopy-related applications throughout the infrared region of 3-30 μm (1). In the past , widespread utilization of Pb-salt lasers has been limited by their very low operating temperatures typically below 77 K. The introduction of molecular beam epitaxy (MBE) growth technique and rare-earth alloys such as PbEuSeTe and PbEuSe contributed to rapid advances in diode laser properties. The higher operating temperatures, routinely achievable at present, extend the temperature tuning range for laser spectrometry applications while simplifying the cooling apparatus requirements. Most noticeable has been the introduction of a new generation of a smaller and cheaper TDL spectroscopy systems based on LN$_2$ cooled sources, which are gradually replacing the bigger, noisier and more expensive He cooled systems. Extending cw operating temperatures towards thermoelectric cooling capability while preserving high quality modal characteristics and low threshold currents will be highly useful for example in fixed wavelength monitoring instruments to be used for instance in air pollution monitoring, trace gas analysis and medical diagnostics. This paper presents recent improvements in BH technology which enabled us to achieve 204 K cw operation for PbEuSeTe/PbTe diode lasers. We will also present recent developments in PbEuSeTe/ PbSnTe buried heterostructure (BH) and buried quantum well (BQW) diode lasers.

2. EXPERIMENTAL

PbEuSeTe, PbTe and PbSnTe epitaxial growth is carried out in a standard Varian Gen II MBE system which contains PbTe,Eu,PbSe, $Pb_{0.7}Eu_{0.3}Te$ and Te effusion sources whose fluxes were combined to grow $Pb_{1-x}Eu_xSe_yTe_{1-y}$ and $Pb_{1-v}Sn_vTe$ compounds. Tl_2Te and Bi_2Te_3 effusion sources are used for p and n doping, respectively (2). Buried heterostructure $Pb_{1-x}Eu_xSe_yTe_{1-y}$/PbTe lasers were prepared in a two stage MBE growth. In the first stage a PbTe buffer layer is grown on a (100) PbTe substrate followed by the cladding and active layers. The wafers are then removed from the MBE system for microfabrication of the buried stripes using standard photolithography. The growth is completed in the second stage with the growth of the second cladding and contact layer.

3. RESULTS AND DISCUSSION

Electrical and optical measurements as a function of temperature were carried out with the devices mounted in a closed-cycle He refrigerator capable of providing stable temperatures within the range of 10 to 300 K. The dependence of the cw threshold current on heat sink temperature for diode lasers 225.85 and 113.85, both with a binary PbTe active layer, is presented in figure 1. The main difference between these otherwise identical devices is the lack of growth discontinuities at the active stripe in laser 225.85 compared to laser 113.85 (3). Laser 225.85 exhibited cw operating temperature of 203 K , which in fact enables extended wavelength coverage from 4.2 μm to 6.4 μm by temperature tuning of the laser. Figure 2. exhibits modal characteristics of the above diode laser observed at heat sink temperature of 100 K. The threshold current at 100 K is 1.9 mA and the emission spectrum in figure 2 is plotted versus various injection currents. The injection currents are multiples of 5xI_{th} . As evident from figure 2 operation in single transverse and longitudinal modes is preserved up to injection currents as high as 30xI_{th} where the emission power at single mode reaches 0.18 mW (0.1 mW is sufficient for most spectroscopy applications). We have been able to demonstrated spectroscopic operation on a low pressure CO_2 gas samples as using diode laser 225.85 in cw operation when stabilized at 200K (3). Generally, the optical output characteristics of lattice matched PbEuSeTe BH lasers operating above liquid nitrogen temperature are suitable for molecular spectroscopy applications, as demonstrated in Table I for two devices with different active layer compositions. These results were consistently found in most high temperature devices tested so far.

Table I - Optical output characteristics for lasers 113.85 and 9121.07

Laser No.	I(th) [80 K]	I(th) [140 K]	cw Single Mode [100 K]	Mode Tuning	Tuning Rates	Max. Power
113.85 [PbTe]	3.2 mA	36.2 mA	8.5xI(th) [0.22 mW]	4.4 cm^{-1}	0.79 GHz/mA [190 K]	1.22 mW
9121.07 [0.22 at.% Eu]	2.9 mA	23.4 mA	11xI(th) [0.28 mW]	2.4 cm^{-1}	1.0 GHz/mA [180 K]	1.22 mW

Since the usefulness of the BH configuration for the lattice matched PbEuSeTe/PbTe system was demonstrated we have chosen to examine this approach in a non-matches PbEuSeTe/PbSnTe system. In this case the PbEuSeTe active layer is replaced with a PbSnTe active layer which does not match the cladding layers which have the same lattice constant as the PbTe substrate. Optical test data measured in different PbSnTe devices is presented in Table II.

TABLE II - Optical Test Results for BH and BQW PbTe and PbSnTe TDLs

Device type	ActiveLayer Composition	Max cw T [K]	Tuning Rate [GHz/mA]	Mode Tuning [cm^{-1}]	Latt.Mismatch $\Delta a/a$
BH	PbTe	203	0.36 [at 200 K]	2.05[at 100 K]	$< 10^{-4}$
BH	$Pb_{0.96}Sn_{0.04}Te$	176	2.1 [at 100 K]	2.2[at 100 K]	8.2×10^{-4}
BH	$Pb_{0.95}Sn_{0.05}Te$	174	1.44 [at 80 K]	2.65[at 60 K]	1.0×10^{-3}
BH	$Pb_{0.932}Sn_{0.068}Te$	179	1.15 [at 100 K]	2.0[at 80 K]	1.4×10^{-3}
BH	$Pb_{0.904}Sn_{0.096}Te$	178	1.58 [at 80 K]	1.65[at 80 K]	1.96×10^{-3}
BQW	$Pb_{0.932}Sn_{0.068}Te$	189	0.57 [at 180 K]	1.15[at 100 K]	1.4×10^{-3}

Molecular spectroscopy applications require laser output with mode tuning bigger than 1cm^{-1} and tuning rates smaller than 1.8 GHz/mA. It is therefore, very encouraging to find that even at elevated operating temperatures the devices presented in Table II perform very well. In comparing PbSnTe BH diode lasers having various degrees of lattice mismatch we have found almost no difference in performance. However, we have kept the cladding layer composition the same for all PbSnTe which, in a way , can contribute to better electrical and optical confinement as the band gap of the active layer becomes more narrow with the increase in Sn content of the active layer. The improvement in confinement can compensate for the increase in lattice mismatch as the Sn content is increased. The lattice matched PbTe device and the BQW PbSnTe device performed considerably better.

The usefulness of laser 225.85 with a PbTe active layer, when operating cw at 200 K , was well demonstrated by the ability to perform high resolution spectroscopy of a low pressure CO_2 gas sample . Single mode operation and tuning rates as low as 360 Mhz/mA obviously contributed to the observed high quality molecular spectrum obtained when the laser was operating cw at 200 K.

In conclusion, BH PbEuSeTe/PbSnTe diode lasers have been manufactured in a two stage process. These lasers operate cw at high operating temperatures between 176 - 204 K. We have found it encouraging that even at high operating temperatures mode tuning exceed 1cm^{-1} and tuning rates are considerably bellow 1.8 GHz. We strongly believe that Pb-salt diode lasers can be operated when thermoelectrically cooled for molecular spectroscopy applications.

REFERENCES

(1) K.J. Linden, "Tunable diode lasers for 3-30 microns infrared operation," SPIE Conf. Tunable Diode Laser Develop. Spectroscopy Appl., vol.438,p.p 2-9, 1983.
(2) Z. Feit, D. Kostyk, R. Woods and W. Jalenak, "Low threshold PbEuSeTe double heterostructure lasers grown by molecular beam epitaxy," Appl. Phys. Lett., vol. 55,pp. 16-18,1989.
(3) Z. Feit, D. Kostyk, R. Woods and P. Mak, "Single-mode molecular beam epitaxy grown PbEuSeTePbTe buried-heterostructure lasers for CO_2 high resolution spectroscopy," Appl. Phys. Lett., vol. 58,pp.

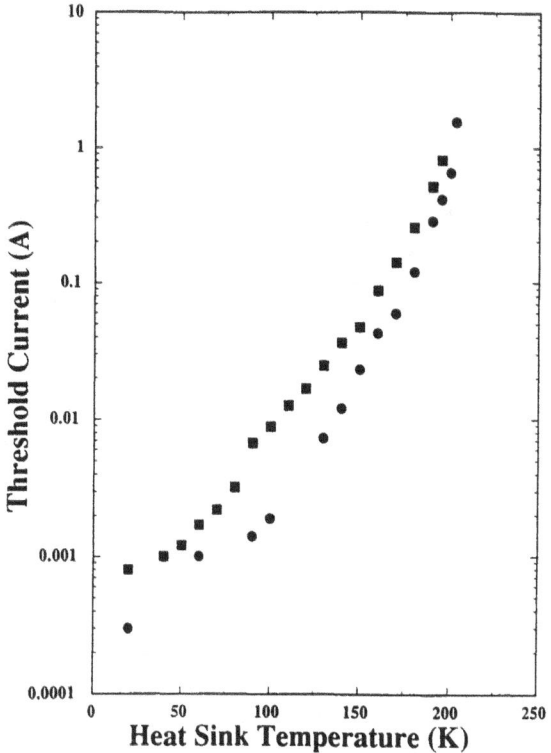

FIG. 1 cw threshold currents vs heat sink temperature for diode lasers
113.85 (■) and 225.85 (●) active layers.

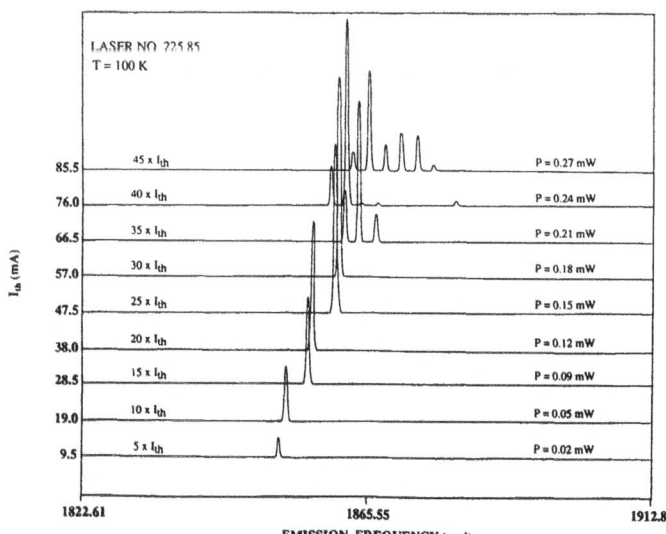

FIG. 2 Emmission spectrun for buried-heterostructure diode laser 225.85
as measured for various injection currents at 100 K in cw mode.

CONTRIBUTIONS TO THE THEORY OF AUGER RECOMBINATION
IN LEAD CHALCOGENIDES

M. MOCKER and F. LEMKE
Department of Physics,
Humboldt-University Berlin,
Invalidenstraße 110, O-1040 Berlin,
Germany.

SUMMARY

Theoretical models, e.g. with consideration of nonparabolicity, anisotropy and many-valley structure, respectively, enable us to get at least a qualitative understanding of the Auger recombination effect in dependence on temperature and carrier concentration for several classes of mixed crystal systems. For PbSnTe the theory is extented to quasi-two-dimensional (Q2D) and quasi-one-dimensional (Q1D) systems.

1. INTRODUCTION

A number of experimental and theoretical results indicate the significance of the Auger recombination in narrow gap semiconductors for the understanding of the radiative properties of optoelectronic devices[1,2]. However, getting experimental data of Auger recombination is difficult because lifetime measurements only yield integral information about recombination. It is known that extrinsic recombination is more important than the Auger effect in some cases. However, the Auger process as a technologically unavoidable intrinsic recombination process gives the lower limit for the total nonradiative effect.

The basic quantities describing theoretically the Auger process are the net recombination rate R or the recombination lifetime $\tau = \delta n/R$ with $\delta n = n - n_0$, where n and n_0 are the electron concentration and its equilibrium value, respectively. According to a many particle theory of the spectral recombination rate $R(q,\omega)$ [3] for one particle excitations we can approximate the expression for R by a Beattie–Landsberg formula [3] provided we replace the static dielectric constant by a high frequency one. For small carrier concentrations MAXWELL-BOLTZMANN-statistics (MBS) applies and the Auger rate R_A^{eeh} takes the form

$$R_A^{eeh} = C_A^{eeh} \, n \, (np - n_0 p_0) \tag{1}$$

in the case of an electron-electron-hole process. The Auger coefficient C_A^{eeh} depends only on the transition matrix elements M_{if} and the band structure. It is given by

$$C_A^{eeh} = \frac{2\pi}{\hbar} \frac{1}{N_c^2 N_v} \frac{1}{\Omega} \exp\left(-\frac{E_A}{k_B T}\right) \sum_{if} |M_{if}|^2 \delta\,(E_1 - E_f) \, \exp\left(-\frac{E_{2'} - \mathrm{Min} E_{2'}}{k_B T}\right). \tag{2}$$

Here we use common notations. N_c and N_v are the effective densities of states. At low temperatures the activation energy E_A determines the order of magnitude of the Auger effect. The expression for the activation energy follows directly from the band structure. Usually it is investigated an one-valley-process, i.e. all states involved in the recombination act are localized in the same $\vec{k}$-space-valley. For parabolic bands and direct gap we have

$$E_A = E_{2',0} - E_g + \Delta E \left(\frac{m_1 + m_2 + m_{1'}}{m_1 + m_2 + m_{1'} - m_{2'}} \right),$$

$$\Delta E = E_g + E_{1,0} + E_{2,0} + E_{1',0} - E_{2',0} .$$

(3)

The $E_{i,0}$ are the band edge energies and the m_i are the effective masses in the band of the state i. From formula (3) we find out that with increasing mass ratio m_c/m_v the activation energy of the eeh process increases and for nearly mirror symmetrical bands like in lead chalcogenides E_A gets the maximum value $E_g/2$.

A more complicated theory yields the results for the Kane bands. The activation energy in the case of ideally hyperbolic and mirror symmetric bands diverges [2], and generally we find a reduced Auger effect considering more realistic models of nonparabolicity. We conclude a small efficiency of the one-valley-process in lead chalcogenides. However, a two-valley-process effectively works in those compounds with a small mass ratio $r=m_\perp/m_\parallel$, i.e. with strong anisotropy of the band structure[2]. In this case E_A approaches the values $r/2 \cdot E_g$ or rE_g for parabolic or Kane bands, respectively.

2. RECOMBINATION LIFETIME IN BULK STRUCTURES

A detailed discussion of the lifetime dependence on doping concentration N_D and temperature T is given in [5] for different mixed crystals on lead salt base. It turns out that the Auger recombination is dominated by the two-valley-process and that the activation energy of this most probable Auger process increases with increasing gap energy and decreasing anisotropy of the band structure,i.e. it increases in the series PbTe, PbSe, PbS. Thus, the small signal lifetime $\tau_{SA}(N_D)$ exceeds the corresponding radiative lifetime in $Pb_{0.78}Sn_{0.22}Te$ only for a doping level of $N_D < 10^{16} cm^{-3}$ and in $PbS_{0.1}Se_{0.9}$ for a doping level of $N_D < 10^{18} cm^{-3}$.

The small density of states and the large carrier concentration in lead salts require the consideration of degenerate statistics. Clearly, that the n^2p-dependence of the Auger rate results from MBS and holds no longer if degeneracy comes into play. A simple power law like $\tau_{SA} \sim N_D^a$ or $\tau_A^h \sim \delta n^a$ does not exist when FDS has to be used (τ_A^h denotes the high injection lifetime). Only numerical calculations can help to get detailed information about lifetimes and recombination rates in the case of FDS. However, some qualitative statements are possible [2]. For the transition region from MBS to FDS we find for the lifetimes $\tau_{SA}(N_D)$ and $\tau_A^h(\delta n)$: (i) if $E_A \ll kT$ the slope of these functions is somewhat weaker than that for MBS, and (ii) the slope of these function can be much stronger than that for MBS, if $E_A \geq 10 \, kT$. In this transition case we propose approximate formulae for the Rate R_A^{eeh} and for the lifetime τ_{SA}^{eeh}

$$R_A^{eeh} = C_A N_c^2 N_v \exp\left(\frac{2F_c + \tilde{F}_v}{k_B T} \right), \qquad \tau_{SA}^{eeh} = \frac{1}{C_A} N_c^2 \exp\left(-\frac{2F_c}{k_B T} \right),$$

(4)

which replace the conventional formulae $R_A^{eeh} = n^2 p C_A$ and $\tau_{SA}^{eeh} = 1/(C_A N_D^2)$. F_c, $\tilde{F}_v$ are the quasi Fermi levels of electrons and holes, respectively.

The slope of these functions decreases for strong degeneracy ($F_c, \tilde{F}_v \geq E_A$) in any case.

3. AUGER RECOMBINATION IN LOW-DIMENSIONAL STRUCTURES

Considering Auger recombination in low dimensional structures, some new aspects have to be taken into account [6]. The activation energy is changed by three facts: (i) new recombination channels appear due to the subband splitting, (ii) the anisotropy parameter r can change because of the restricted free electron motion, and (iii) the gap increases due to the additional subband energies. For small structures where only the lowest subbands are occupied the alteration of the Auger rate is mainly due to the increasing gap energy. The modifications of the activation energy for some special planes and directions caused by the changed anisotropy are listed in table 1 for PbSnTe (bulk value of r=0.1). Except for [100]-direction, these changes modify the increasing gap effect only slightly. For the [100]-axis, the Q1D band structure is isotropic and the Auger rate for a Kane band model vanishes.

plane	2D		Q2D		direc-tion	1D		Q1D	
	r	E_A/E_g	r	E_A/E_g		r	E_A/E_g	r	E_A/E_g
{100}	0.4	0.54	0.14	0.15	[100]	1	∞	1	∞
					[110]	0.4	0.54	0.14	0.15
{110}	0.11	0.11	0.11	0.11	[100]	1	∞	1	∞
					[110]	0.4	0.54	0.14	0.15
					[111]	0.11	0.11	0.2	0.23
{111}	0.2	0.23	0.11	0.11	[110]	0.4	0.54	0.14	0.15
					[211]	0.2	0.23	0.11	0.11

Table 1: Anisotropy parameters r_{v0} and corresponding activation energies E_A for different low-dimensional systems. Contrary to [6,7] we have to distinguish between 2D and Q2D or 1D and Q1D systems, respectively. The latter one are the physically interesting systems.

The temperature dependence of the activation energy is given in Fig. 1(a). The reason for the strong variation at high temperatures is the considerable temperatur dependence of the gap energy.

Further, the density of states for reduced dimensionality, the wave function form factors, and related effects can modify the Auger coefficient. Making some suitable approximations[7], we derive an analytical expression for C_A^{eeh} for v-dimensional structures (bulk: $v=3$; Q2D: $v=2$; Q1D: $v=1$):

$$C_A^{vD} = \mu \left(\frac{3}{2}\right)^{2(3-v)} \frac{\sqrt{\pi}}{2} \left(\frac{e^2}{\epsilon_0 \epsilon_\infty}\right)^2 \frac{p_\perp^4 (k_B T)^{\frac{1}{2}}}{E_g^{\frac{11}{2}} \hbar} \times$$

$$\times \frac{\left(\frac{v-1}{2} + I_{1x} I_{2x} \frac{E_g}{k_B T}\right)}{\left(1 + \gamma_v \frac{k_B T}{E_g}\right)^3} \frac{\sqrt{I_{vD} I_{HvD}}}{(F_{val}^v)^3} \exp\left(-\frac{E_A}{k_B T}\right).$$

(5)

The symbols are the same as in [7]. Here, a quasi-Kane model has been used. In the bulk this leads to $C_A = 0.2\ C_A^{Kntage}$[8].

The temperature dependence of the Auger coefficient given by (5) is shown in Fig. 1(b) for several particular planes. A comparison with Fig. 1(a) shows that the rate is mainly influenced by the increasing gap energy. Formula (5) enables us to compare the Q2D and the Q1D Auger rate with the bulk one. At 77K we find a reduction in all Q2D structures by about one order of magnitude. In Q1D we expect a further reduction, the lowest rate should appear in an [100]-directed quantum wire.

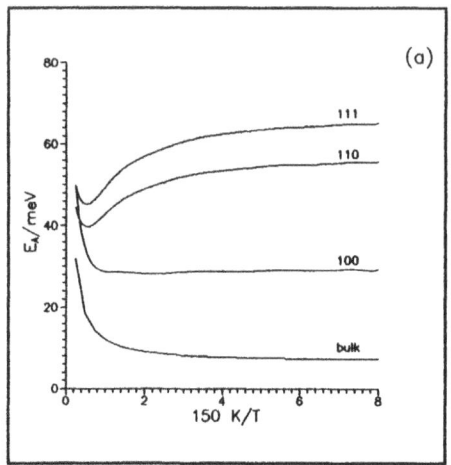 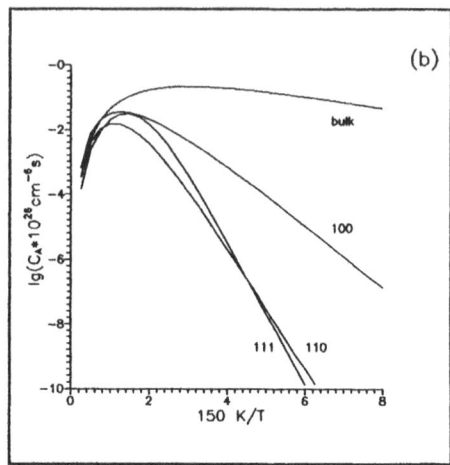

Fig. 1: a) Activation energy E_A and b) Auger coefficient C_A^{eeh} as a function of the inverse temperature for bulk and for some special Q2D PbSnTe-structures.

REFERENCES

[1] S. D. BENESLAVSKII, A. V. DIMITRIEV, Solid State Communications **39**, 811(1981).
   J. W. TOMM, K. H. HERRMANN, and A. E. YUNOVICH, phys. stat. sol. (a)**122**, 11(1990), and references therein.
[2] O. ZIEP, M. MOCKER, D. GENZOW, K. H. HERRMANN, phys. stat. sol. (b)**90**, 197(1978).
   M. MOCKER, O. ZIEP, phys. stat. sol. (b)**115**, 415(1983).
[3] O. ZIEP, and M. MOCKER, phys. stat. sol. (b)**98**, 133(1980).
   O. ZIEP, and M. MOCKER, phys. stat. sol. (b)**119**, 299(1983).
[4] A. R. BEATTIE, and P. T. LANDSBERG, Proc. R. Soc. (London) **A 249**, 16(1959).
[5] O. ZIEP, M. MOCKER, D. GENZOW, Wiss. Z. Humboldt Univ., Math.-Nat.R. 30, 81(1981).
[6] M. MOCKER, F. LEMKE, and P. SELBMANN, phys. stat. sol. (b)**151**, K151(1989).
   M. MOCKER, F. LEMKE, Superlattices and Microstructures 10, 231(1991).
[7] F. LEMKE, and M. MOCKER, phys. stat. sol. (b)**167**, 219(1991).
[8] P. R. EMTAGE, J. Appl. Phys. **47**, 2565(1976).
   M. MOCKER, O. ZIEP, and M. BEILER, Wiss. Z. Humboldt Univ., Math.-Nat. R. 35, 165(1986).

THERMODYNAMICS OF QUASITERNARY SYSTEMS CONTAINING IV-VI-SEMICONDUCTORS

VOLKMAR LEUTE and DIETER MENGE

INSTITUTE OF PHYSICAL CHEMISTRY, UNIVERSITY OF MÜNSTER

SCHLOSSPLATZ 4, 4400 MÜNSTER, GERMANY

SUMMARY

Phase diagrams of quasiternary systems are calculated from data of the quasibinary subsystems on the basis of a cluster model. The thermodynamic factor of interdiffusion is explained by special cluster reactions.

As to the crystal structures of the binary constituents of these quasiternary systems the IV-VI-components crystallize in the rocksalt structure, except for SnSe, which crystallizes with an orthorhombic lattice. The II-VI-components occur with zincblende or wurtzite structure.

If we want to use a cluster model for the calculation of phase diagrams, we have to choose suitable cluster sizes. In the rocksalt lattice of the IV-VI-compounds the octahedrons would be the natural clusters. But these 7-particle clusters are too big to be handled effectively in calculations. Therefore we subdivide these octahedrons into 8 pyramidal 4-particle clusters with a particle of one sublattice at the apex and three particles of the other sublattice at the base plane of the pyramid. Analogously, the tetrahedron clusters of the wurtzite or zincblende lattices can also be subdivided into 4 pyramidal clusters. This procedure has the advantage that we can compare the cluster energies of thetrahedrally and octahedrally coordinated lattices.

In order to calculate the enthalpy of a solid solution, one has to know the probabilities and the energies of all clusters. As long as we can work with the so-called random distribution assumption, i.e. as long as we can assume, that in both sublattices the particles are randomly distributed, the cluster probabilities for a given composition of a quasiternary system $M_{(k)}N_{(1-k)}X_{(1)}Y_{(1-1)}$ are simple products of the corresponding mole fractions, for example:

$$p(MXXY) = b(MXXY) \cdot k \cdot 1 \cdot 1 \cdot (1-1). \qquad (1)$$

The degeneracy factor b(MXXY) considers the multiple possibilities of cluster orientation with respect to the crystal lattice.

If the cluster energies are known, the mean molar excess enthalpy of the solid solution can be calculated by summing up over the energy contributions of all clusters, considering their probabilities. This procedure yields an expression of the following structure:

$$h^E(k,1) = h(k,1) - h^0(k,1)$$
$$h^E(k,1) = k(1-k)[\alpha_{MN} + k\beta_{MN} + 1\tau_{MN} + k1\delta_{MN}]$$
$$+ 1(1-1)[\alpha_{XY} + 1\beta_{XY} + k\tau_{XY} + 1k\delta_{XY}] + k(1-1)\Delta_R H^0 \qquad (2)$$

The coefficients α, β, τ, and δ in this expression are simple functions of the cluster energies. On the other hand, we have shown [1], that, within the scope of the random distribution assumption, the cluster energies can be calculated from the interaction parameters of the quasibinary subsystems, provided the dependence of these parameters on the mole fractions is not higher than linearly.

$$a_{[MN]X} = \alpha_{[MN]X} + k\beta_{[MN]X} \tag{3}$$

Thus the quasiternary interaction parameters can be given as linear combinations of the quasibinary parameters, e.g.:

$$\alpha_{MN} = \alpha_{[MN]Y}; \quad \beta_{MN} = \beta_{[MN]Y}; \tag{4a,b}$$
$$\tau_{MN} = \alpha_{[MN]X} - \alpha_{[MN]Y}; \quad \delta_{MN} = \beta_{[MN]X} + \beta_{[MN]Y} \ldots \tag{4c,d}$$

The only intrinsic quasiternary parameter, $\Delta_R H^0$, in expression (2) can be interpreted as the standard reaction enthalpy of the double conversion, $MX + NY \dashrightarrow MY + NX$, between the pure binary components of the quasiternary system.

According to the random distribution assumption, the entropy contribution to the mean molar free enthalpy of such a quasiternary solid solution corresponds to the configuration entropies of the metal and chalcogen sublattices. Thus, the function $g(k,l)$ for a given phase can be calculated, if the standard reaction enthalpy for this phase, as well as the interaction parameters for the quasibinary subsystems, are known.

In the context of the cluster model we can explain the thermodynamic factor for interdiffusion as being caused by reactions between clusters. In this regard, we have investigated the interdiffusion of sulfur and tellurium in the quasibinary system PbS/PbTe. Between the critical point of the spinodal miscibility gap (T = 1073 K) and the azeotropic point (T = 1043 K) there is a region of complete solid solubility [2]. Nevertheless, this solid solutions will show a distinct deviation from ideal behaviour that will influence the diffusion properties.

The cluster reactions (r1) and (r2) can be interpreted as pair formation reactions, describing the formation of YY-pairs within an X-matrix and of XX-pairs within a Y-matrix [1]:

$$2\ M(XXY) \dashrightarrow M(XXX) + M(X\underline{YY}) \tag{r1}$$
$$2\ M(XYY) \dashrightarrow M(YYY) + M(\underline{XX}Y) \tag{r2}$$

The reaction enthalpies for these processes can be derived in the usual way from the individual cluster energies:

$$\Delta_{(r1)} H^0 = E_{M(XXX)} + E_{M(XYY)} - 2 \cdot E_{M(XXY)} \tag{5a}$$
$$\Delta_{(r2)} H^0 = E_{M(YYY)} + E_{M(XXY)} - 2 \cdot E_{M(XYY)} \tag{5b}$$

On this basis, the thermodynamic factor can be expressed

as a linear function of these reaction enthalpies.

$$F = 1 + [6(1-1))/(RT)] \cdot 8[1 \cdot \Delta_{(r1)}H^0 + (1-1) \cdot \Delta_{(r2)}H^0] \quad (6)$$

The more exothermic these pair formations are, the higher the deviation of the thermodynamic factor from the ideal value $F = 1$. As F decreases for an exothermic pair formation the inter-diffusion coefficient must also decrease for this case. Provided the component diffusion coefficients are independent on composition, then, according to Darkens' equation [3], $\tilde{D} = 1*D(S) + (1-1)*D(Te)$, the interdiffusion coefficient should depend linearly on the mole fraction. The experiments, however, show a distinct minimum for $\tilde{D}$ at intermediate composi-tions. If Darkens' coefficient is modified by multiplying with the thermodynamic factor, the solid lines in Fig.1 are ob-tained in good accordance with the experimental results.

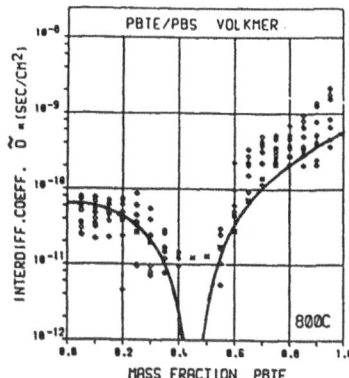 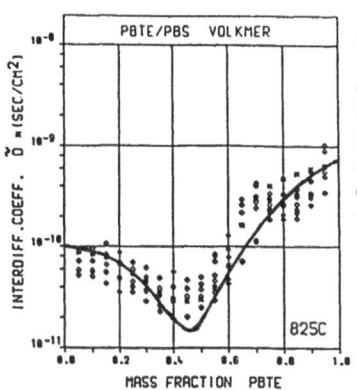

Fig.1
coefficient for inter-diffusion of S and Te in Pb(S,Te) as function of composi-tion

Thus, it is shown, that the deviation of the diffusion from ideal behaviour for the system PbS/PbTe can be attributed mainly to the effect of exothermic pair formation. At even lower temperatures the same effect causes the spinodal misci-bility gap.

If phase diagrams of systems with different phases are to be calculated, the free enthalpy of transformation between these phases has to be considered [4].

$$g^{*I} = RT \cdot \Sigma x^I(i) \cdot \ln(a^I(i)) \quad (7a)$$
$$g^{*II} = RT \cdot \Sigma x^{II}(i) \cdot \ln(a^{II}(i)) + \Sigma x^{II}(i) \cdot \Delta_{Tr(I \to II)}G^0(i) \quad (7b)$$

As long as we can assume, that the enthalpy and entropy of transition are independent of temperature, the free enthalpy of transition can be expressed as a function of the transition entropy and the transition temperature.

$$\Delta_{Tr(I \to II)}G^0(i) = \Delta_{Tr(I \to II)}S^0(i) \cdot [T_{Tr(I \to II)}(i) - T] \quad (8)$$

The comparison of a series of chalcogenide systems has shown, that the transition entropy is independent of the chemical nature of the transforming components, but depends on the structures of the phases, between which the transformation occurs. Moreover, we found a general sequence of structures with increasing temperature for all binary chalcogenides that we have investigated till now:

zincblende --> wurtzite --> (orthorhombic) --> rocksalt.   (9)

If the system PbS/PbTe is doped with the corresponding Cd-chalcogenides, we have to treat this new system thermodynamically as the quasiternary system (Cd,Pb)(S,Te). Besides the phase with rocksalt structure, in this system also phases with wurtzite or zincblende structure can occur. Among the three other quasibinary phase diagrams of this quasiternary system the diagrams of the systems PbTe/CdTe and PbS/CdS (Fig.2) show an eutectic behaviour with extended solid solubility regions on the Pb-chalcogenide rich side, but nearly no solubility on the Cd-chalcogenide rich side [5]. The system CdS/CdTe shows an interesting behaviour in so far as in the interesting temperature region there exist simultaneously a spinodal miscibility gap and a small structural miscibility gap [5]. From the phase diagrams of the four quasibinary subsystems the transition data for the pure binary compounds, as well as most of the interaction parameters, that are needed for the calculation of the quasiternary g-functions, can be derived. The remaining parameters, which can not be determined from the experimentally accessible quasibinary subsystems, have to be determined by adjusting calculated quasiternary phase diagrams to experimental data [5]. If the $g(k,l)$-functions for a given temperature for all participating phases have been determined, an isothermal section of the quasiternary phase diagram can be constructed. For this pur-

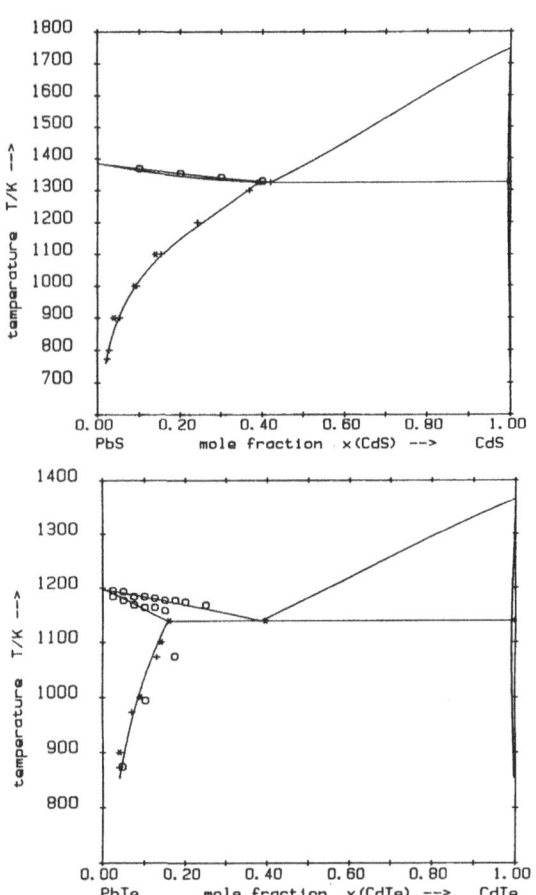

Fig.2 full lines = calculated,
o [6, 7], + [8], * [5].

pose the tie lines, connecting equilibrium points in different phases or, in the case of spinodal miscibility gaps, connecting equilibrium points whithin the same phase, have to be calculated. The envelope of the tie line fields yield the solubility limits of the solid solutions. A point, where two tie line fields intersect becomes one of the vertices of a three phase triangle; the other two vertices, in most cases, are known from the quasibinary subsystems.

In the special case of the system $Cd_k Pb_{(1-k)} S_l Te_{(1-l)}$, Fig.3, there are two types of three phase triangles: one type in which all three phases have different structures, namely zincblende, wurtzite and rocksalt, and an other type where two phases belong to one and the same structure, but differ in composition. This last type results, whenever a spinodal miscibility gap is involved into the three phase region. This is the case for the three phase regions 'rocksalt, rocksalt, wurtzite' and 'rocksalt, wurtzite, wurtzite'.

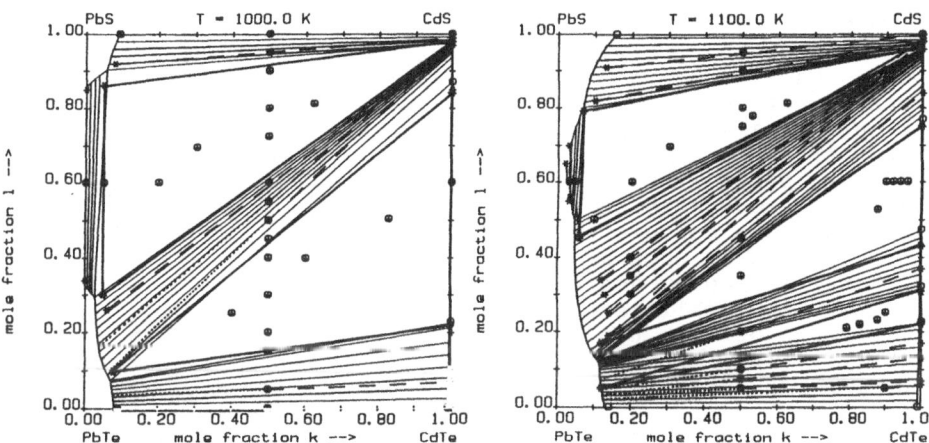

Fig.3 Phase diagram for (Cd,Pb)(S,Te); full lines = calculated tie lines; --- = experimental tie lines.

The formation of a three phase region including a spinodal miscibility gap can be demonstrated by the two intersecting g-surfaces for the rocksalt and the wurtzite structure at T = 1000 K (Fig.4). The projection of the triangle, connecting the three points of contact between the tangential plane and the two g-surfaces onto the l,k-plane yields the corresponding three phase region for the quasiternary phase diagram. As to the spinodal miscibility gap in Pb-chalcogenide-rich solutions, one can observe a shifting of the critical point of demixing along a critical line from the quasibinary edge at T = 1073 K to higher temperatures with increasing Cd-chalcogenide content of the solid solution. This feature can best be visualized by a sequence of g-surfaces in the interesting temperature region (Fig.5).

Fig.4

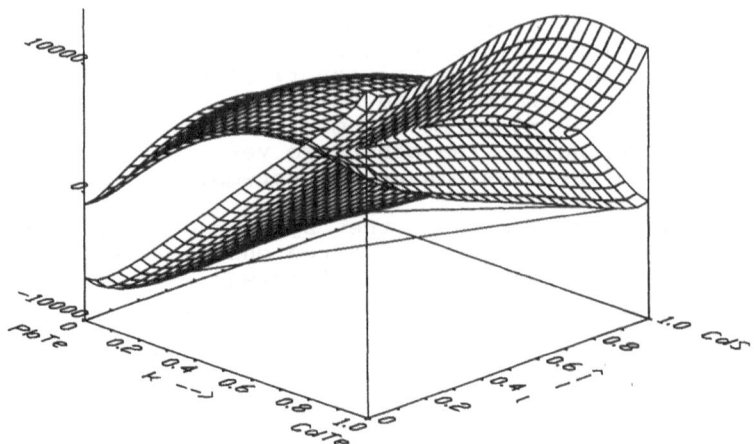

One could argue, that
the calculated g-surfaces
would have to be incorrect,
because of the random distribution assumption used
for the calculation of the
cluster probabilities. In
order to check this argu-
ment, we have calculated for
two quasibinary edge systems
the real, energy dependent
cluster probabilities. The
comparison of both types of
probabilities shows, that
the differences between
them, in these cases, are
smaller than 0.02 (Fig.6).

As to the quasiternary
system $Cd_k Sn_{(1-k)} Se_l Te_{(1-l)}$
the procedures, used to determine the phase diagram,
in principal are the same as
already described. Most of
the quasiternary interaction
parameters can be derived
from the phase diagrams of
the quasibinary edge sys-
tems. In this special case 4
different structures, namely
zincblende, wurtzite, rocksalt, and the orthorhombic
structure have to be considered.

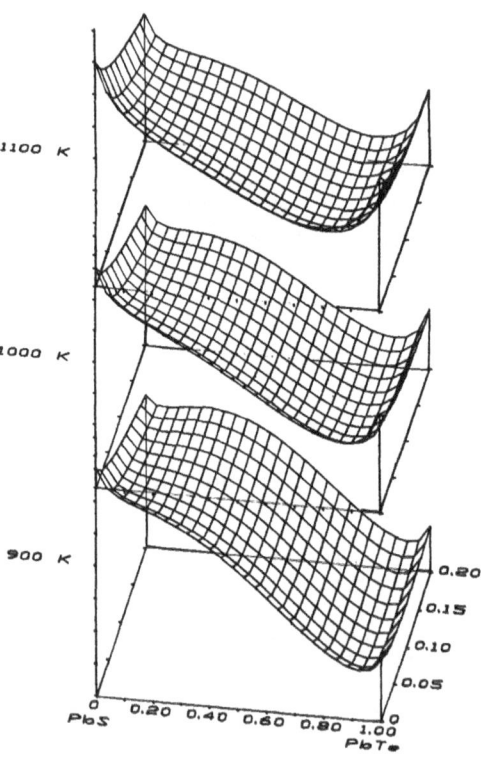

Fig.5 Pb-chalcogenide rich
part of the g-surfaces for
the system (Cd,Pb)(S,Te)
for several temperatures.

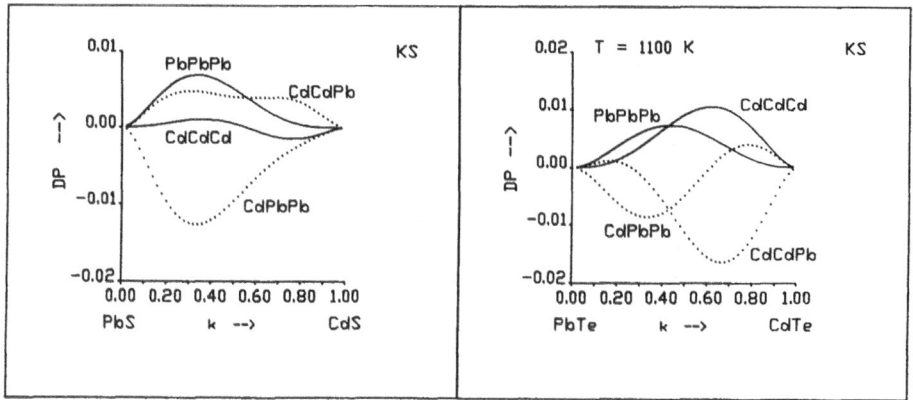

Fig.6 Difference between energy dependent probabilities and "random distribution probabilities" for pyramidal clusters.

The subsystem SnTe/ SnSe (Fig.7a) shows an eutectic behaviour with broad solubility regions for both, the phase with rocksalt structure and the phase with the orthorhombic structure. The selenide rich phase is further characterized by a high temperature phase transition between two orthorhombic structures. But for the calculation of the quasiternary phase diagram, we can here neglect this transition, as we are interested in that temperature region, in which only the high temperature phase is stable.

The subsystem CdTe /CdSe (Fig.7b) [9] shows

Fig.7 Phase diagram for Sn(Se,Te) and Cd(Se,Te); full lines = calculated;
a.) +, *, ⊕ from DTA-measurements, o, ⊛ from x-ray measurements;
b.) +, * [10]

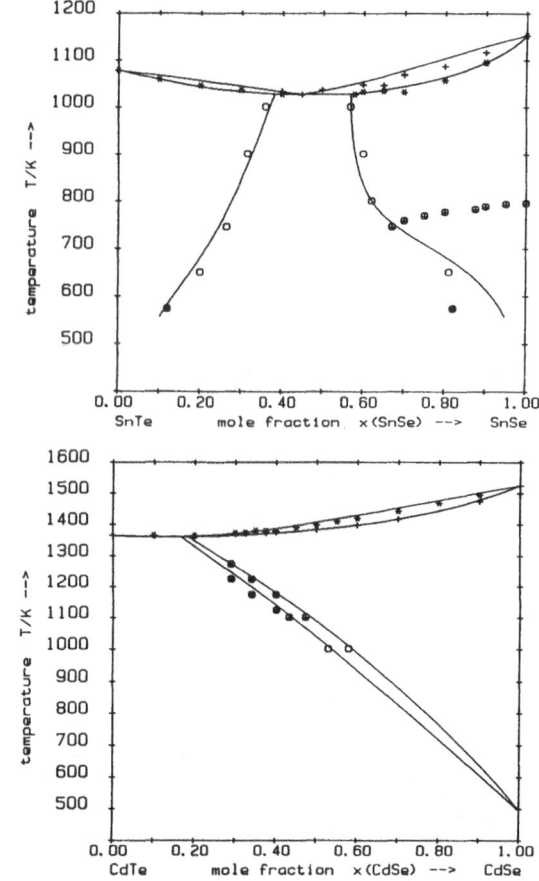

broad regions of solid solubility and is characterized by a small structural miscibility gap between the phase with wurtzite and the phase with zincblende structure.

The two quasibinary subsystems SnSe/CdSe (Fig.8a) and SnTe/CdTe (Fig.8b), in which IV-VI-components are combined with II-VI-components, show the same topology as the equivalent systems with Pb-chalcogenides.

The quasiternary phase diagram shows two 3-phase regions at both temperatures T = 800 K and T = 1000 K (Fig.9). But it is astonishing, that for the 3-phase region that includes the miscibility gap between wurtzite and zincblende structure, at the lower temperature, the third point of the triangle belongs to the phase with the orthorhombic structure, whereas at the higher temperature the third point belongs to the phase with rock-salt structure. Thus we have to conclude, that at an exactly defined intermediate temperature the two 3-phase triangles are united, forming a region with 4 solid phases. Including the gas phase, this is the maximum number of phases that can coexist in equilibrium in a ternary system. The temperature of this quintuple point must be at about T = 850 K.

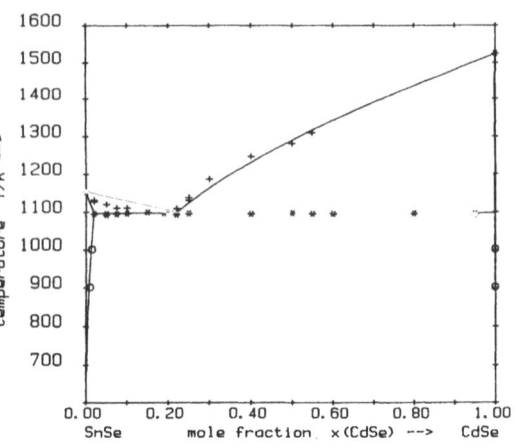

Fig.8a Phase diagram (Cd,Sn)Se;
full lines = calculated;
+, * from DTA-measurements;
o from X-ray measurements.

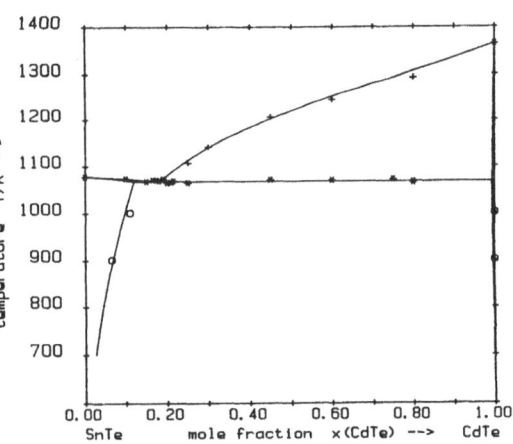

Fig.8b Phase diagram (Cd,Sn)Te;
full lines = calculated;
+, * from DTA-measurements;
o from X-ray measurements.

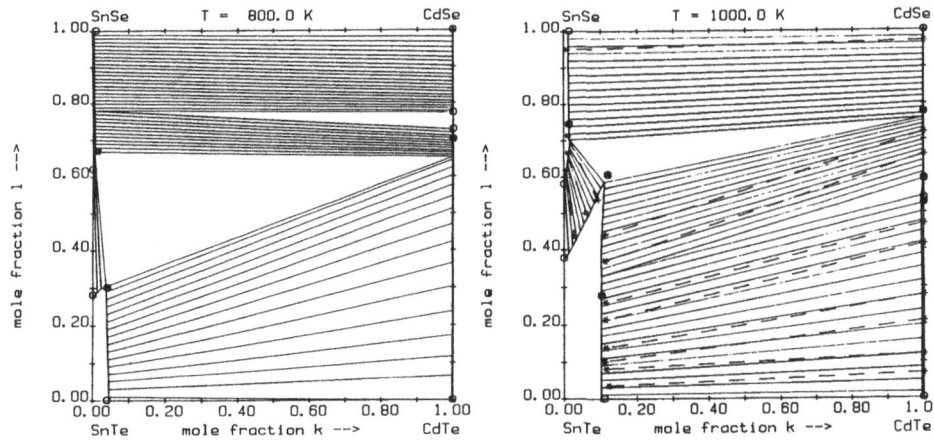

Fig.9 Phase diagram for (Cd,Sn)(Se,Te); full lines = calculated tie lines; --- = experimental tie lines.

ACKNOWLEDGEMENTS
    This research was supported by the Deutsche Forschungsge-
meinschaft and by the Fonds der Chemischen Industrie.

LITERATURE:
[1] V.Leute,"Thermodynamic Description of Quasibinary Systems
     Based on a Four Particle Cluster Model",
     Ber.Bunsenges.Phys.Chem. 93 (1989) 548-555
[2] V.Leute and N.Volkmer, "A Contribution to the Phase Dia-
     gram of the Quasibinary System $Pb(S_{(1-x)}Te_x)$",
     Z.Physikal.Chem.N.F. 144 (1985) 145-155
[3] L.Darken, Trans.Met.Soc.AIME, 174 (1948) 184
[4] V.Leute and H.-J.Köller, "The Four Quasibinary Phase Dia-
     Diagrams of the Quasiternary System (Hg,Pb)(Se,Te)",
     Z.Physikal.Chem.N.F. 149 (1986) 213-227
[5] V.Leute and R.Schmidt, "The Quasiternary System
     $(Cd_k Pb_{1-k})(S_l Te_{1-l})$", Z.Physikal.Chem.N.F.(1991) in press
[6] Z.F.Tomashik and V.N.Tomashik, "The CdS + PbSe --> CdSe +
     PbS Ternary Reciprocal System",
     Inorganic Materials 20 (1984) 486
[7] A.J.Rosenberg, R.Grierson, J.C.Woolley and P.Nikolic,
     "Solid Solutions of CdTe and InTe in PbTe and SnTe",
     Trans.Met.Soc.AIME 230 (1964) 342
[8] P.M.Bethke and P.B.Barton, "Sub-Solidus Relations in the
     System PbS-CdS", The American Mineralogist 36 (1971) 2034
[9] V.Leute and B.Wulff, "The Phase Diagram of the Quasiter-
     nary System $(Zn_k Cd_{1-k})(Se_l Te_{1-l})$",
     Ber.Bunsenges.Phys.Chem. in press
[10] A.J.Strauss and J.Steininger, "Phase Diagram of the CdTe
     -CdS Pseudobinary System", J.Electrochem.Soc. 117 (1970)
     1420

DEVELOPMENT OF LEAD-CHALCOGENIDE TUNABLE DOIDE LASERS FOR 3 TO 4 μm SPECTRAL REGION AT LEBEDEV PHYSICAL INSTITUTE

A.P. SHOTOV

Lebedev Physical Institute, 117924 Moscow, USSR

ABSTRACT

Characteristics and structures of tunable diode lasers based on Pb-salt crystals PbSnSe, PbSSe, PbCdSe are presented. Main parts of the research have been directed to the improvement of long-term reliability and achievement of higher temperature operation. In emission characteristics we found that broad-area lasers emit in periodically localized channels (filaments), which are connected with the carrier concentration modulation by autowaves generated by an injection current. Operating parameters of the traditional diffusion lasers and some improved structures with controlled profile of carrier concentration and SCQW heterostructures prepared by HWE are discussed.

1. INTRODUCTION

Diode lasers fabricated from IV-VI narrow gap Pb-salt semiconductor compounds and alloys are used as current or temperature tunable sources in 3 - 40 μm spectral region for high resulution spectroscopy, monitoring of atmospheric pollution, gas analysis and other applications (1 -4). IV-VI Pb-salt family semiconductors include PbS, PbSe(Te) and their alloys with SnSe, CdS and SnTe compounds. The band-gap Eg of Pb-salts occur at L-point of Brillouin zone (instead of center (K-point) as for III-V compound) and varies from 0.3 eV to zero, depending on the composition of the alloy. Lasers made of these alloys cover 3 to 40 μm infrared spectral region. These types of lasers usually require cryogenic cooling with dewar or modern closed cycle He gas refrigerators, which allow the laser temperature to be varied from about 10 K to room temperature. For diode lasers operated at a higher temperature T > 200 K (usually pulse operated) a thermoelectric cooling can be used.

The main problems for Pb-salt diode lasers are the improvements of the operating temperature, the threshold current, a quality of emission and long time stability. During the last years many improvements have been made in designing and performing of the lasers. Different types of heterostructures, efficient optical and carrier confinements, extremely thin quantum well active layers were used as well as the new composition of the alloys (PbSrSe, PbEuSe). Now the highest operating temperature of up to 290 K has been reached for a small part of the IR spectral region (3 - 5 μm) (5). Typical value of the emitting power of order 0.1-10mW. This power together with a narrow linewidth ($\leq 10^{-4}$ cm^{-1}) give very high spectral brightness, which is a very important characteristic for many spectroscopic applications. Emission wavelengths of the lasers can be tuned during operation by variation of a number of parameters on which the bandgap and refractive index depend on (3 - 5). The most widely used are both the temperature and current control tuning. A very important problem is thermal stability of the laser devices. For most lead salt materials used for the diode laser $dE_g/dT > 0$ (instead of $dE_g/dT < 0$ for III-V materials). This means that some nonuniformities of the current density connected with imperfection of p-n junction are being damped. This improves thermal stability of the diode lasers. The reliability of the lasers is mainly connected with thermal cycling degradation of laser operating characteristics. In our work some stable laser structures have been developed as well as good and stable multilayer (Au/Pd/In) electrical contacts were made.

The next sections summarize the most representative types of diode lasers.

2. DIFFUSED LASERS

Ternary compounds PbCdS, PbSSe and PbSnSe are used for diffusion-based technology to cover 3 - 40 μm spectral range. The crystals were grown using a closed-tube directional vapor-phase technique, described previously (6). To prepare the p-n junction n-type crystals with carrier concentrations from 5 x 10^{17} to 5 x 10^{18} cm^{-3} were used. Slices were cut from crystals with a wire saw. P-n junction was formed by interdiffusion process from vapor phase using Se-rich $Pb_{0.49}Se_{0.51}$ source for all compositions. Wafers (n-type) and $Pb_{0.49}Se_{0.51}$ source are sealed in an evacuated quartz ampule. Heating of the ampule for 0.5 - 1 hour at 400 - 450 °C (depending on the material) creates a p-n junction at the depth 10 - 20 μm (7 - 8). In order to improve the long term stability of the lasers, much work was done on electrical contacts. We found that three layers In/Pd/Au plated produce low resistance (of $< 10^{-4}$ ohmcm^2) and stable electrical contacts to the p- and n-side of the p-n junction. After depositing the Au and the Pd layers (of thickness a ~ 0.2 μm), a thick (> 1 μm) layer of In was added. Evaporated Au layers (instead of plated) were also used. The stripes, 150 - 200 μm wide, were separated and the laser end faces were formed by cleaving. The distances between end faces usually varied from 400 to 600 μm. The devices are packaged by In-welding to In-plated copper heat sink. Typical characteristics of the laser are presented in Table 1.

Material		Wavelength (μm)		I_{th}(KA/cm^2)	
		4.2 K	77 K	4.2 K	77 K
$Pb_{1-x}Cd_xS$	$0 < x < 0.03$	3.0 - 4.3	2.9 - 4.0	0.1 - 0.2	2 - 5
$PbS_{1-x}Se_x$	$0 < x < 1.0$	4.3 - 8.5	4.0 - 7.3	0.05 - 0.2	0.8 - 2
$Pb_{1-x}Sn_xSe$	$0 < x < 0.12$	8.5 - 4.0	7.3 - 24	0.1 - 0.5	0.7 - 5

Table 1

This type of diffused lasers demonstrates good thermal stability and long time reliability, and is now used in most spectroscopic applications.

Mode and intensity characteristics of these broad-area lasers were studied. The near field distribution of intensity and spectral mode structure of p-n junction was measured by adjusting the optics in front of the laser, each spot along the p-n junction plane could be imaged separately on the entrance slit of the nonochromator. Pulse current (τ ~1 μs, 170 Hz) was used for injection of carriers. It was found (Fig. 1) that intensity of emission and also wavelength and mode separation have some oscillation which correspond to a period of 15 - 20 μm.

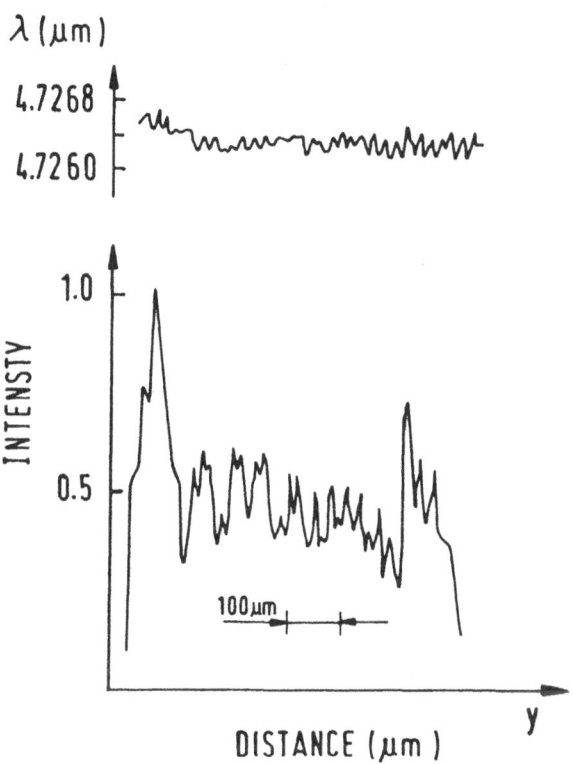

Fig. 1: Distribution of intensity and wavelength of the emisssion in the plane parallel to the junction for a $PbS_{0.67}Se_{0.33}$ broad-area ~ 500 μm wide laser at 77 K.

These emission channels (or filaments) are probably connected with corresponding oscillation of carrier concentration and refractive index along the junction plane. We connected these channels with the carrier concentration modulation by some autowaves, generated by the injection current (10). Usually the peak of emission originated from the corner of the junction. Intensity of laser emission and mode structure can be improved by controlling surface condition of lateral plate of the laser cavity. We put a copper film and electrically insolated layers on the surface of the cavity. In this case, one intensive peak of emission was obtained at the center of the p-n junction plate if the size of the stripe was less than 200 μm (Fig. 2).

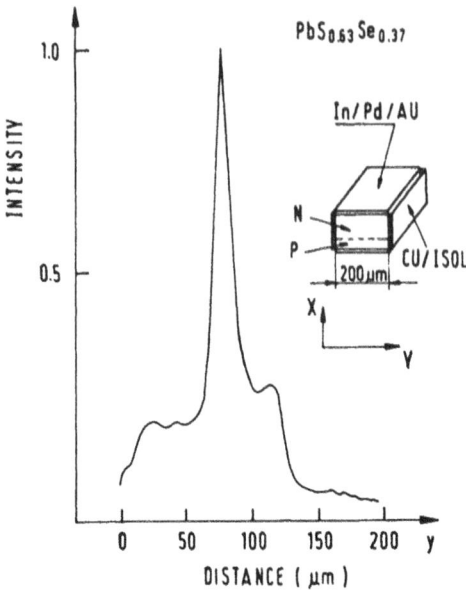

Fig. 2: Distribution of intensity of the emission in the plane parallel to the junction for a PbSn$_{0.67}$Se$_{0.33}$ 200 μm wide laser with metallized lateral cavity 77 K.

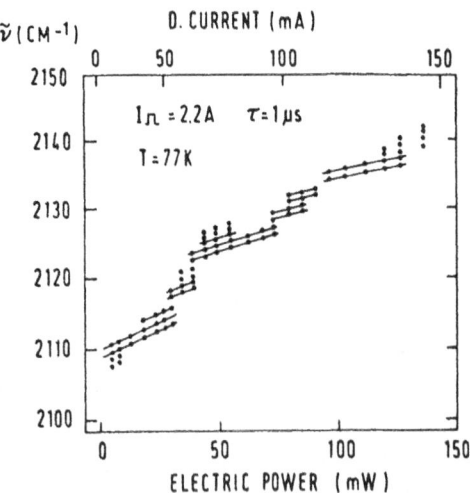

Fig. 3: Spectral modes for a PbSn$_{0.67}$Se$_{0.33}$ 200 μm wide laser versus addition D.C. power (D. current) at pulse current of 2.2A at 77 K.

Mode structure of the laser is shown in Fig. 3. for pulse injection current of I = 1.5 x J_{th} = 2.0 A. It was found that the additional DC power (D. current) is quite effective for mode tuning. Current of temperature continuously mode tuning usually has a limited range of 1 - 2 cm². In this case, larger tuning of up to 5 cm⁻¹ was observed (11).

3. PbSSe DIODE LASERS WITH CONTROLLED CARRIER CONCENTRATION

Higher operating temperature has been achieved by using a double heterostructure (12 - 13). But the lattice mismatch of the DH is often the reason for the degradation after the following thermal cycling between low and room temperature.

The problem of lattice mismatch can be avoided, and efficient confinement can be simultaneously achieved in a homostructure with controlled carrier concentration profiles (14). The carrier and optical confinement is enhanced due to the potential barrier of n^+pp^+ structure and due to a rather strong carriers' concentration dependence of the refractive index of the narrow gap IV-VI semiconductors (15 - 16).

From Fig. 4., one can see PbSSe diode lasers with controlled carrier concentration profile. The figure shows the refractive index versus the carrier density for $PbS_{0.65}Se_{0.35}$ at photon energy close to the energy gap. For the concentration step from ~ 10^{17} cm⁻³ to 2×10^{18} cm⁻³, the relative refractive index step $\Delta n/n$ 10 % can be obtained which is enough for optical confinement. Calculation results of the confinement factor show that 80 % of the radiation should be located in 1.5 - 2 μm active layer.

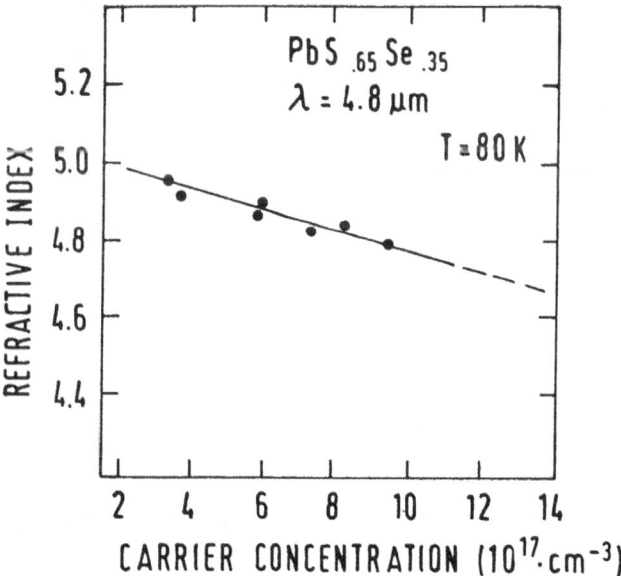

Fig. 4: Dependence of the refractive index in $PbS_{0.65}Se_{0.35}$ on carrier concentration.

The calculations of refractive positions of Fermi and quasi-Fermi levels (Fig. 5) show that potential barriers of n^+pp^+ structure (with the carrier concentration for 1×10^{17} cm^{-3} in the active layer and 2×10^{18} cm^{-3} in the n^+ and p^+ layers) can be ~ 10 KT at 4.2 K and several KT at 80 K. This is also enough for the carrier confinement.

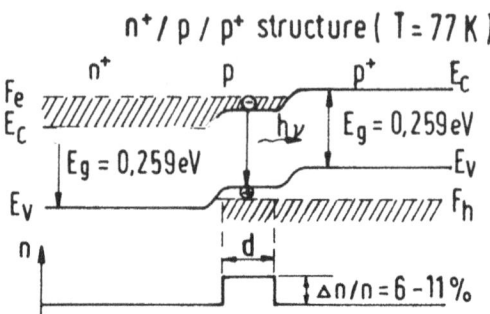

Fig. 5: Schematic n^+pp^+ laser structure at forward bias that shows carrier and optical confinement.

The laser n^+pp^+ homostructure (Fig. 6) was grown by hot wall molecular epitaxy (17). PbSSe n^+ substrate with the thickness of about 100 μm and carrier concentration of 2×10^{18} cm^{-3} was grown on KCl wafer oriented in (100) direction. KCl was desolved in water after the growing process. All the following layers were grown on the KCl side of the PbSSe substrate.

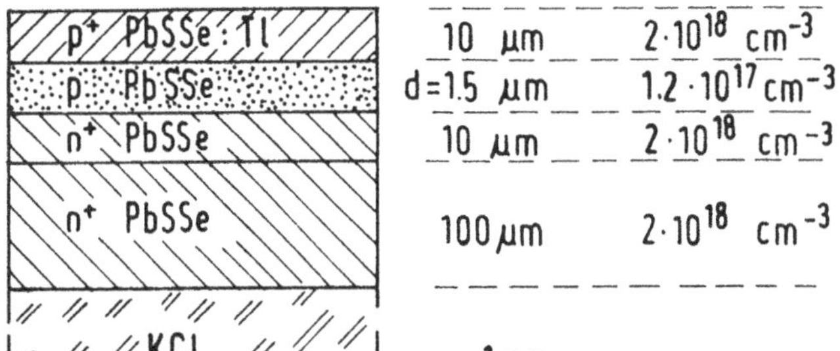

Layer	Thickness	Concentration
p^+ PbSSe : Tl	10 μm	$2 \cdot 10^{18}$ cm^{-3}
p PbSSe	d = 1.5 μm	$1.2 \cdot 10^{17}$ cm^{-3}
n^+ PbSSe	10 μm	$2 \cdot 10^{18}$ cm^{-3}
n^+ PbSSe	100 μm	$2 \cdot 10^{18}$ cm^{-3}
KCl	$\sim$1mm	

Fig. 6: n^+pp^+ laser structure grown by hot wall molecular epitaxy.

Good quality n^+ layer ($n = 2 \times 10^{18}$ cm^{-3}), thickness of about 10 μm , n-p junction and active p-layer were grown at the rate of 3 μm/hour by proper adjustment of the compensating selenium vapour pressure and growing temperature. Contact Tl doped p^+ layer with the carrier concentration of about

2×10^{18} cm^{-3} and thickness 5 - 10 μm was grown during the second deposition step. All the structure was grown without breaking the vacuum between the two deposition steps.

The best results were obtained for the active p-layer with the thickness of about 1.5 μm, and the carrier concentration of 1.2×10^{17} cm^{-3}, estimated from C-V characteristics of the p-n junction. Low carrier concentration of the active layer is very important to decrease the free carrier absorption.

The threshold current density for the best Pb$_{0.65}$Se$_{0.35}$ lasers was about 25 A/cm^2 at 4.2 K and 700 A/cm^2 at 80 K.

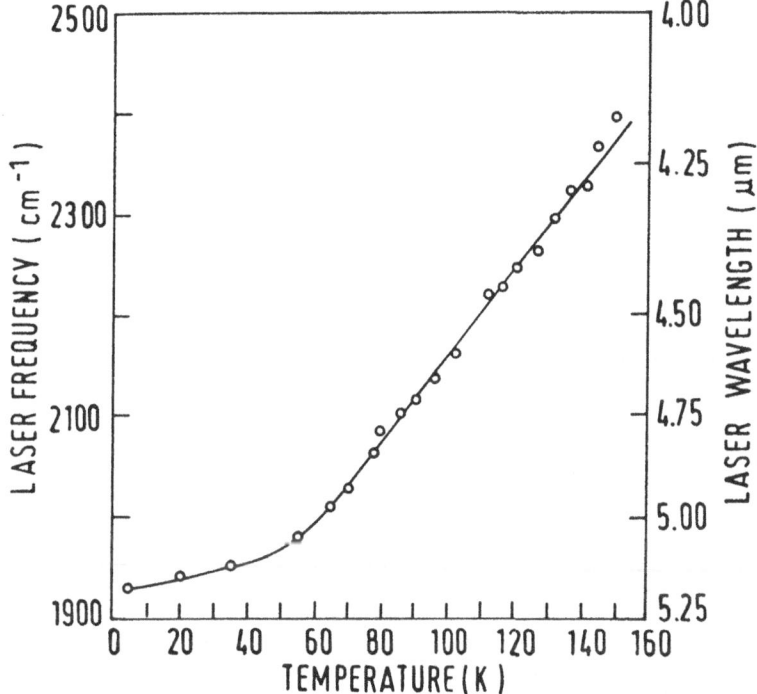

Fig. 7: Temperature dependence of the center frequency of emission for PbS$_{0.65}$Se$_{0.35}$ laser.

Fig. 7 shows the center emission frequency of PbS$_{0.65}$Se$_{0.35}$ diode laser as a function of temperature. The temperature tuning range is rather large: about 500 cm^{-1} (from 5.3 to 4.1 μm) for the 4.2 - 150 K range. The maximum power of 22 mW at 4.2 K was obtained (Fig. 8). This value corresponds to internal differential quantum efficiency of about 20 % which is rather high for this type of lasers.

Temperature stability of this laser is much higher than that of DH lasers. After more than 100 temperature cycles between 300 and 4.2 K, most of the lasers showed no change in the threshold current.

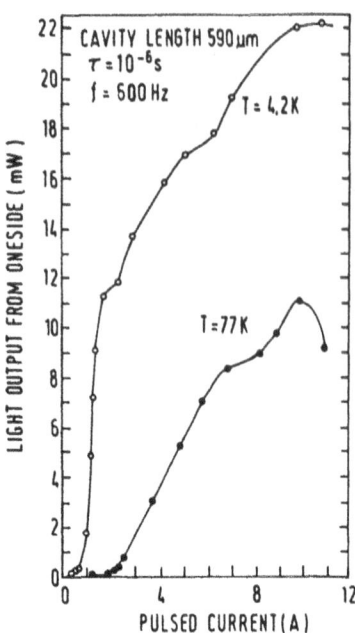

Fig. 8: Output power versus injection current of a $PbS_{0.65}Se_{0.35}$ laser in pulsed operation.

4. PbS/PbSnSe HETEROSTRUCTURE LASERS WITH QUANTUM-WELL ACTIVE REGION

To improve the theshold current and operating temperature of the lasers PbS/PbSSe/PbSnSe, double heterostructures were fabricated by hot wall epitaxy (18 - 20) with separate electron and photon confinements and a quantum-well in the active region (SQ SQW DHS).

The composition of seven layers of the structure is shown in Fig. 9. The single-quantum $Pb_{0.95}Sn_{0.05}Se$ active region and the two $PbS_{0.4}Se_{0.6}$ wave-guiding layers near the active region result in an increase in the energy band gap of $\Delta E = 109$ meV at 77 K. The n-type layers were doped with bismuth. The selenium concentration was adjusted to obtain p-type conductivity. The quantum-well active region was undoped. The thickness (L_z) of the active region was 400, 500, 1000, 2000 Å and up to 4 μm.

We observed that the photon energy for the lasers with $L_z = 1000$ Å corresponds to the band gap E_g of the active region. For the lasers with $L_z = 400$ and 500 Å, two different laser emission lines with energies greater than E_g were observed. For $L_z = 400$ Å, this increase amounts to 9.7 meV for the line with $h\nu = 127$ meV ($\lambda = 9.8$ μm) and 38 meV for the line with $h\nu = 155$ meV ($\lambda = 8$ μm) at 77 K (Fig. 10). These shifts are found from the difference between the photon energy of the QW laser and that of the DH laser with a thick active region ($L_z > 1$ μm) in pulsed operation (to eliminate heating effects).

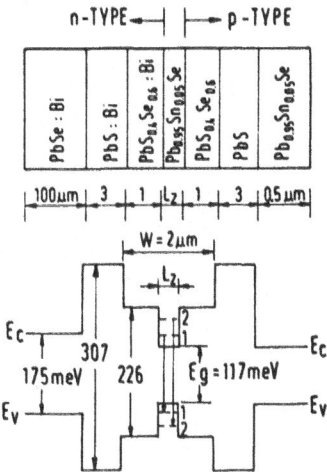

Fig. 9: Schematic energy band diagram and laser structure with single-quantum-well PnSnSe active region.

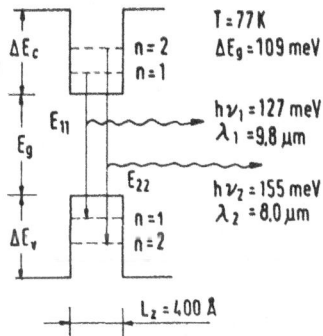

Fig. 10: Measured emission energy.

The energy levels in the quantum well (L_z = 400 Å) were calculated from the model square potential well of finite depth. Since the exact value of the band discontinuities in PbSnSe/PbSSe is unknown, we use $\Delta E_c \sim E_v$. In our case, ΔE_g = 109 meV and m_e = 0.035 m_0, m_h = 0.031 m_0. The fact that the effective masses of the electrons and holes are very nearly equal leads to a weak dependence of E_{11} and E_{22} on $\Delta E_c/\Delta E_v$. It was estimated that the observed photon energies correspond to optical transitions between localized states in the potential well with n = 1 (E_{11}) and n = 2 (E_{22}). The selection rule

$\Delta n = 0$ holds for these transitions. Similar results were observed for the laser with $L_z = 500$ Å.

Fig. 11 shows the temperature dependence of the threshold current density J_{th} for the lasers with $L_z = 400$ and 1000 Å at pulsed current ($1\ \mu s$, 600 Hz). At low temperature, $T < 60$ K, the laser with $L_z = 400$ Å operates at low J_{th} for optical transitions between states with $n = 1$ ($h\nu = E_{11}$), because only states which are close to the band edge are filled by carriers. Transitions between states with $n = 2$ require a higher pumping current. As was noted above, at $T = 60 - 120$ K two laser emission lines were observed simultaneously. At $T > 125$ K, only the $h\nu_2 = E_{22}$ transition between higher density of states at $n = 2$ was observed. In part of the temperature range, the threshold current for transition E_{11} and E_{22} is practically independent of temperature. This fact reflects the step form of the density of states for the quasi-2D system in the potential well. Similar behaviour is found for the laser with $L_z = 500$ Å. For the laser with $L_z = 1000$ Å, the J_{th} (T) curve has a structural feature which indicates a switching of the laser operation from $n = 1$ to $n = 2$ optical transitions. However, the photon energy in this case is quite close to E_g at the active layer. The temperature dependence of the threshold current for the laser with $L_z = 2000$ Å is of the standard type for lead-salt diode lasers (12 - 14).

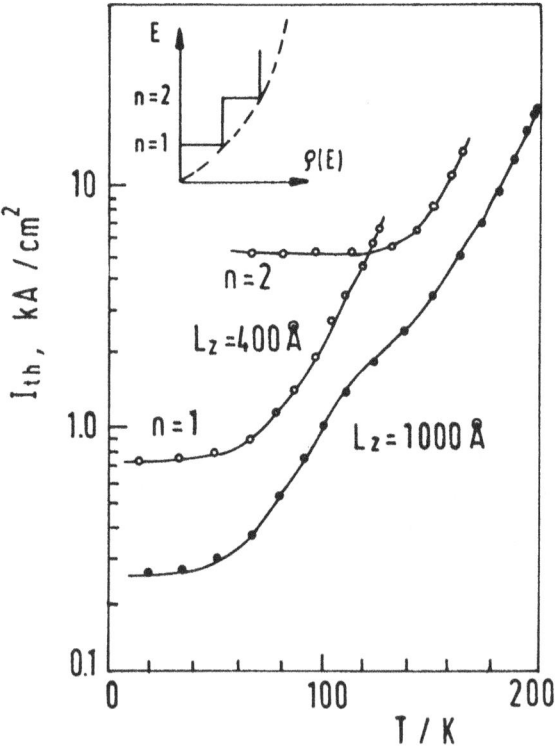

Fig. 11: Temperature dependence of the threshold current density.

Fig. 12 shows the relation between the theshold density and the active layer thickness for these SDCH lasers and also for the DH $PbS_{0.4}Se_{0.6}/Pb_{0.95}Sn_{0.05}Se$ at T = 77 K. The effect of separate confinement for carriers and photons is clearly demonstrated. The lowest threshold current of 230 A cm^{-2} at 77 K and the highest operating temperature of 218 K (λ = 6.5 μm) were obtained at pulsed current for the SCDH laser with L_z = 2000 Å. In the single-quantum-well lasers with L_z = 1000 Å, the threshold current increases with decreasing thickness of the QW active region. We suppose that the main reason for this is a leakage of carriers from the potential well due to small values of the ΔE_0 in these structures. Larger gap material (like PbSrSe) for barrier-layers can be useful in this case.

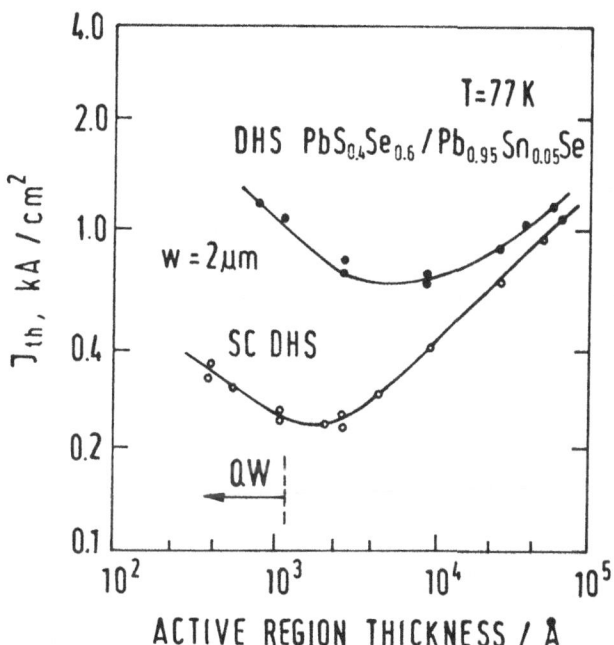

Fig. 12: Threshold current density against active layer thickness at 77 K.

ACKNOWLEDGEMENTS

The author is grateful to I.I. Zasavitsky, Yu.G. Selivanov, E. G. Chizhevsky, M.S. Murashov and my other colleagues at the narrow-gap laboratory for their cooperation.

REFERENCES

(1) SHOTOV, A.P., (1973). Tunable semiconductor infrared lasers. Proc. of the 4th Conf. on Solid State Devices, Tokyo, 1972. Suppl. to the J. Jpn. Soc. Appl. Phys. 42, 282-288.

(2) SHOTOV, A.P., (1986). Physical Research of narrow-gap semiconductors and development infrared tunable diode lasers, Vestnik Ak. Nauk SSSR, No 6, 3-9.

(3) PREIER, H., (1979). Recent advances in lead-chalcogenide diode lasers. Appl. Phys., 20, 189-206.

(4) ENG, R., BUTLER, J., LINDEN, K., (1980). Tunable diode laser spectroscopy: an invited review. Optical Engineering 19, 945-960.

(5) SPANGE, B., SCHIESSL, U., LAMBRECHT, A., BÖTTNER, H., TACKE, (1988). Near room temperature operation of $Pb_{1-x}Sr_xSe$ infrared diode lasers using molecular beam epitaxy growth techniques, Appl. Phys. Lett., 53(26), 2582-2583.

(6) SHOTOV, A., KUCHERENKO, I., KOROLEV, Yu, CHIZHEVSKY, E., (1972). Electrical properties of $Pb_{1-x}Sn_xSe$, grown by vapor transport method, Physics and Technics of Semiconductors (USSR) 6, 1508-1513.

(7) ZASAVITSKY, I, MATSONASHVILI, B., POGODIN, V., SHOTOV, A., (1974). Diode laser spectra at the hydrostatic pressure. Physics and Technics of Semiconductors (USSR) 8, 732-736.

(8) ZASAVITSKY, I., MATSONASHVILI, B., SHOTOV, A., (1975). Continuous wave operation PbSe diode laser, Pis'ma Zh. Tech. Fiz. 1, 341-343.

(9) ZASAVITSKY, I., CHIZHEVSKY, E., SHOTOV, A., (1978). Pressure tunalbe CW-PbSe diode laser Quantum Electronics (USSR) 5, 692-694.

(10) MURASHOV, M., SHOTOV, A., (1990). Carrier concentration autowaves in PbSSe diode lasers. Quantum Electronics (USSR) 17, 2426-2432.

(11) MURASHOV, M., SHOTOV, A., to be submitted for publication.

(12) SHOTOV, A., VYATKIN, K., SINYATINSKII, A., (1980). Double Heterostructure PbSnSe diode lasers growth by hot wall epitaxy. Pis'ma Zh. Tech. Phys. 6, 983-986.

(13) SHOTOV, A., VYATKIN, K. (1980). DHS PbSnSe diode lasers CW operated at 80K. Pis'ma Zh. Tech. Phys. 6, 1199-1202.

(14) SHOTOV, A., SINYATINSKy, A., (1983). PbSSe lasers with controlled carrier concentration, grown by hot wall epitaxy. Pis'ma Zh. Tech. Phys. 9, 881-884.

(15) VYATKIN, K., SHOTOV, A., (1980). Optical properties PbSe epitaxy layers. Physics and Technics of Semiconductors 14, 1331-1334.

(16) SINYATINSKY, A., SHOTOV, A., (1982). Optical properties PbSSe epitaxy layers. Physics and Technics of Semiconductrs 16, 2187-2190.

(17) VYATKIN, K., SHOTOV, A., URSAKI, V., (1981). PbSnSe thin films, grown by hot wall epitaxy, Izv. Akad. Nauk USSR, Neorg. material 17, 24-27.

(18) SHOTOV, A., SELIVANOV, Yu., (1986). DHSC PbS/PbSSe/PbSnSe lasers, grown by hot wall epitaxy, Pis'ma Zh. Tech. Phys. 12, 1386-1389.

(19) SHOTOV, A., SELIVANOV, Yu., (1987). PbS/PbSSe/PbSnSe heterostructure laser with a quantum-well active region. Pis'ma Zh. Eksp. Teor. Fiz., 45, 5-7.

(20)	SHOTOV, A., SELIVANOV, Yu. (1990). PbS/PbSSe heterostructure lasers with a quantum-well active region, Semicond. Sci. Technol., $\underline{5}$, 927-929.

OPTICAL PROPERTIES OF HIGH ENERGY GAP LEAD SALTS

Herrmann, K.H., Möllmann, K.-P., Tomm,, J.W.

Humboldt-Universität zu Berlin
Institut für Festkörperphysik, FB Physik
Invalidenstrasse 110, O-1040 Berlin, Germany

Böttner, H., Lambrecht, A., Tacke, M.

Fraunhofer-Institut für Physikalische Messtechnik
Heidenhofstrasse 8, W-7800 Freiburg, Germany

By alloying the narrow-gap semiconductor PbSe with the wider-gap indirect semiconductors EuSe and SrSe one obtains so-called high energy gap lead salt mixed crystal systems which meet demands as
♦ barriers in Q2D-structures,
♦ confinement layers in heterostructures and, possibly as
♦ optically active layers in diode lasers for the 3 μm wavelength region.

Because SrSe and EuSe exhibit very similar band structures (with the exception of the valence band formed by the 4f-states of Eu in EuSe) a comparison of both mixed crystal systems gives insight into questions of rather physical interest, too.

In the present study we compare results obtained from both systems applying the following methods:
♦ Transmittance Spectroscopy (FT-IR),
♦ Photocurrent Spectroscopy (stationary as well as transient PC) and
♦ Infrared Photoluminescence.

For the (Pb,Eu)Se-system we found for all samples band-like states within the forbidden gap, which might be attributed to the band formed by the Eu 4f-states. Nevertheless the main contribution to the luminescence arises from direct interband transitions.

For the (Pb,Sr)Se-samples long tails, possibly due to alloy disorder dominate the optical spectra.

Both, the occurence of well-pronounced levels as well as the tailing in (Pb,Eu)Se and (Pb,Sr)Se, resp., cause the present limits of the layers as radiation emitters.

PICOSECOND INFRARED SPECTROSCOPY OF LEAD CHALCOGENIDE SEMICONDUCTORS

Klann, R., Buhleier, R., Elsaesser, T.,[*] Lambrecht, A.[**]

[*]Technische Universität München, Physik Department E11
James-Franck-Straße, W-8046 Garching, Germany

[**]Fraunhofer-Institut für Physikalische Messtechnik
Heidenhofstrasse 8, W-7800 Freiburg, Germany

Optical nonlinearities in lead-selenide (PbSe) are studied in pump-probe experiments with picosecond pulses tunable from 4 to 8 μm. The pump pulse creates an electron-hole plasma giving rise to temporal and spectral changes of transmission close to the band gap. The latter are directly monitored by tunable probe pulses. We observe a blueshift of the absorption edge and a strong decrease of the refractive index. For carrier densities of several 10^{17} cm^{-3}, the index of refraction changes by $\Delta n = -0.1$ due to band filling by both electrons and holes. Theoretical calculations of the optical nonlinearities are in agreement with the experimental data.

The temporal evolution of the absorption change for excitation densities higher than 10^{18} cm^{-3} shows a partial recovery within the first 100 ps. This behavior which is studied for a wide range of carrier densities and temperatures, gives evidence of enhanced recombination rates. Different mechanisms of recombination will be discussed.

OPTICAL NONLINEARITIES OF FREE CARRIERS IN PbSe

Leidig, K.

Universität Würzburg, Physikalisches Institut, Abt. EP IV/1
Am Hubland, W-8700 Würzburg, Germany

Four-wave mixing has been observed in PbSe at 4.2 K as a function of the difference frequency of two Q-switched CO_2 laser beams. Intraband energy relaxation times τ_E of free carriers were derived from a fit to the experimental data. These results in $\tau_E = 1$ ps for all epitaxial material whereas bulk material yields $\tau_E = 5$ ps. Thereby the tendency holds on that the lower the temperature the greater the difference between bulk and epitaxial PbSe.

LIMITATIONS OF COMPOUND SEGREGATION IN THE VAPOUR CRYSTAL GROWTH OF SOLID SOLUTION CRYSTALS

Szczerbakow, Andrzej

VIGO Ltd.
ul. Radiowa 3, 00-908 Warszawa 49, Poland

Assuming total irreversibility of mass transport, the traditional approach to the evaporation-condensation procedure leads to the conclusion that the condensing solution becames substantially enriched with the more volatile component. However, crystallisation via the vapor phase in a sealed ampoule held almost isothermal ought to be considered nearly reversible i. e. with restricted segregation. This is justified by the high homogeneity of vapour grown solid solution crystals, observed commonly in the pseudobinary $A^{IV}B^{VI}$ materials. Consequently, different evaporation-condensation systems are compared in relation to the component segregation, where the case of a solid solution in a nearly isothermal ampoule is interpreted considering segregation as an effect of deviation from equilibrium of the system as a whole. Maximum segregation is estimated as a function of the mass-driving temperature difference. Consistency with experiment is discussed - also with regard to chemical reactions linked up with the phase transitions.

PbSnMnTe SEMIMAGNETIC SEMICONDUCTORS: CARRIER CONCENTRATION CONTROLLED MAGNETIC PROPERTIES

Story, T.

Polish Academy of Sciences, Institute of Physics
Al. Lotnikow 32/46, 02-668 Warsaw, Poland

Magnetic properties of the semimagnetic $Pb_{1-x-y}Sn_yMn_xTe$ semi-conductors depend on the concentration of carriers. Crystals with carrier concentration $p < p_c = 3 \cdot 10^{20}$ cm^{-3} are paramagnets, whereas crystals with carrier concentration $p > p_c$ are ferromagnets. In the case of the samples with very high carrier concentration ($p > 10^{21}$ cm^{-3}) and low concentration of magnetic ions ($x \leq 0.03$) a spin-glass phase is also observed. The type of magnetism and the critical temperature of magnetic phase transistion of the crystals of PbSnMnTe can be reversibly controlled by simple annealing processes. No changes of chemical composition of the sample are necessary. It offers a unique possibility to generate e. g. a required Curie temperature of a ferromagnetic sample or a magnetic structure grown in the same piece of IV-VI semimagnetic semiconductor.

STRUCTURAL AND ELECTRICAL PROPERTIES OF PbSe GROWN BY THE VERTICAL BRIDGMAN METHOD

Post, Ralph, Mülberg, Manfred

Humboldt Universität zu Berlin
Institut für Kristallographie und Materialforschung
Invalidenstrasse 110, O-1040 Berlin, Germany

Large PbSe crystals (> 10 cm^3) were grown by the vertical Bridgman method.

Recently, a renewed attention is directed to the Bridgman method. Especially from the III-V and II-VI compounds is known that optimized growth conditions, mainly temperature profiles with low thermal gradient (≤ 10 K/cm), yield crystals with a high structural quality.

On the basis of thermal modelling studies some parameters were fixed: the furnace axial temperature gradient in the range of 5 K/cm, the growth velocity 1 mm/h and the aspect ratio (crystal diameter/crystal length) < 0.16.

The crystals are not completely monocrystalline. Predominantly, in the first-to-freeze region 2 - 3 grain boundaries are present. A low-angle grain boundaries substructure is also observed with a cell diameter between 0.5 and 1 mm. A higher dislocation density is caused by the influence of the so-called end effects. In the middle part of the crystals the etch pits density is about 10^5 - 10^6 cm^{-2}.

Caused by the large extended stability region of PbSe a segregation behaviour with respect to the Pb/Se ratio is observed. The variation of the carrier concentration is a strong function of the melt composition. A stoichiometric composition gives initially p-type PbSe ($p_{77\,K}$ (max) = $4 \cdot 10^{18}$ cm^{-3}) and finally n-type PbSe ($n_{77\,K}$(max) = $5 \cdot 10^{18}$ cm^{-3}).

MAGNETICALLY TUNED $Pb_{1-x}Mn_xSe$ AND $Pb_{1-x-y}Mn_xSn_ySe$ DIODE LASERS

Kowalczyk, L.

Polish Academy of Sciences, Institute of Physics
Al. Lotnikow 32/46, 02-668 Warsaw, Poland

The magnetic field dependence of laser emission has been studied in $Pb_{1-x}Mn_xSe$ and $Pb_{1-x-y}Mn_xSn_ySe$ diodes at liquid-He temperature. Spectra were obtained in the energy range from 130 meV to 220 meV for $x \leq 0.02$ and $y \leq 0.08$. The spectral positions of the peaks were recorded as a function of magnetic field intensity. From measurements of splitting, rate of shift to higher energies and polarization, emission peaks were identified with transitions between sublevels of spin-split zero-order Landau levels. Magnitudes of effective g-factor have been deduced from the dependence of the emission spectra on magnetic field. The results provide evidence that the values of g-factor for $Pb_{1-x}Mn_xSe$ as well as for $Pb_{1-x-y}Mn_xSn_ySe$ are much bigger than for PbSe. The enchancement of g-factor is connected with exchange interaction between Mn ions and the band electrons. It leads to greater magnetic-tuning rates for $Pb_{1-x}Mn_xSe$ and $Pb_{1-x-y}Mn_xSn_ySe$ lasers than for other lead salt diode lasers in this range of emission energy. Furthermore in $Pb_{1-x-y}Mn_xSn_ySe$ there exists a possibility of independent regulation of the value of energy gap by means of the tin content and the magnitude of the exchange interaction by the manganese amount. It seems at present that magnetic-field-tuned $Pb_{1-x}Mn_xSe$ and $Pb_{1-x-y}Mn_xSn_ySe$ lasers can be utilized for high-resolution spectroscopy.

Se-DIFFUSED $PbS_{1-x}Se_x$-HOMOLASERS IN THE TEMPERATURE RANGE FROM 5 TO 100 K

Iwaschkina, Diana A., Siche, Dietmar

Humboldt-Universität zu Berlin, FB Physik
Abteilung Molekül- und Photobiophysik
Invalidenstrasse 110, O-Berlin 1040, Germany

For the theoretical characterization of the temperature-dependent laser parameters (threshold current density, tuning rate, mode power) in Se-diffused Pb(S,Se)-homojunction lasers the exact knowledge of the position of the optical active medium is necessary. In layer structured homolasers and in heterolasers the confinement of the optical wave field is controlled technologically by the layer growth processes. In diffused homolasers the diffusion profile is responsible for the existence, position and structure of the optical active layer.

For laser fabrication n-type crystals were grown from the gaseous phase and diffused with Se isothermally. To correlate measured laser para-meters with the depths and structures of pn-junctions the Se-diffusion times have been varied. For the characterization of the properties of the pulsed diode lasers the threshold current density in the temperature range from 5 to 100 K was used. A dependence on the diffusion parameters was established. For the known diffusion model the obtained positions of the optical wave field lead to non-lasing diodes. The model for a realistic concentration profile near the pn-junction needs some modifications.

PHOTOVOLTAIC INFRARED SENSOR ARRAYS IN MONOLITHIC LEAD CHALCOGENIDES ON SILICON

Zogg, H., Maissen, C., Masek, J., Hoshino, T., Blunier S.

Arbeitsgemeinschaft für Industrielle Forschung (AFIF)
Swiss Federal Institute of Technology
ETH Hönggerberg, CH-8093 Zürich, Switzerland

We review MBE growth of epitaxial IV-VI layers on Si(111) substrates and fabrication of photovoltaic infrared devices in the layers. We have fabricated sensor arrays in PbS, PbTe, $PbS_{1-x}Se_x$, $Pb_{1-x}Eu_xSe$ and $Pb_{1-x}Sn_xSe$, i. e. with cut-off wavelengths ranging from 3 μm to > 12 μm. An intermediate epitaxial stacked CaF_2-BaF_2 bilayer of 200 nm thickness serves to overcome the large lattice and thermal expansion mismatch between the Si-substrate and the infrared sensitive layers.

Advantages of lead salts compared to $Hg_{1-x}Cd_xTe$ are the much easier growth by MBE, less severe homogeneity problems, and the higher optical absorption (3 μm thick layers are already sufficient to absorb most of the incoming IR-radiation). The high permittivities of IV-VI materials restrict the bandwidths of the devices to the 100 MHz range, which is much more than needed for focal plane arrays. On the other hand, these high permittivities efficiently shield the electric field caused by charged defects. IV-VI materials are fault tolerant.

IV-VI-on-Si IR-sensors therefore offer the potential for a low cost technique of large IR focal plane arrays both for the 3 - 5 μm and 8 - 12 μm range, but with comparable sensitivities as in $Hg_{1-x}Cd_xTe$. The arrays may be hybridized with a Si read-out chip without thermal mismatch problems, or the read-out electronics may even be integrated in the Si-substrate.

THERMAL MISMATCH STRAIN RELAXATION OF EPITAXIAL IV-VI LAYERS ON Si(111) SUBSTRATES AT CRYOGENIC TEMPERATURES

Maissen, C., Sultan, A., Teodoropol, S., Zogg, H.

Arbeitsgemeinschaft für Industrielle Forschung (AFIF)
Swiss Federal Institute of Technology
ETH Hönggerberg, CH-8093 Zürich, Switzerland

Epitaxial IV-VI layers have been grown by MBE on Si(111) substrates with the aid of an intermediate CaF_2/BaF_2 bilayer in order to get high quality epitaxy. We found that the thermal mismatch strain in epitaxial PbSe layers on Si(111) (which is expected because of the higher thermal expansion coefficient of IV-VIs than of Si) is relieved by (111)-glide. This strain relief is near complete down to 100 K and even after multiple thermal cycles.

It is well known that binary lead salts are rather soft, while ternary compositions like $PbSe_{1-x}Te_x$ are much more brittle. We therefore have grown $PbSe_{1-x}Te_x/PbSe$ and $PbSe/PbSe_{1-x}Te_x/PbSe$ stacks on fluoride covered Si(111) to study their strain relaxation properties. The strains of the individual parts of the stack were determined with X-ray diffraction or RBS angular scans. As expected, the ternary $PbSe_{1-x}Te_x$ intermediate layers are elastically strained at cryogenic temperatures and influence the strain state of the top layer.

MBE GROWTH AND CHARACTERIZATION OF PbSe LAYERS ON Si (111) USING (Ba, Ca)F$_2$ AS BUFFER LAYERS

Nguyen-Van-Dau, F., Crété, D.G., Mathet, V.

Thomson CSF/LCR, Laboratoire Central de Recherches
Domaine de Corbeville, F-91404 Orsay Cedex, France

The heteroepitaxy between a narrow gap semiconductor and silicon have received a growing attention in the past few years. One of the main reasons is that it could lead to the possibility of monolithically integrating infrared sensor arrays with their read-out electronics on the same silicon wafer. For that purpose, lead salts compounds are of particular interest. Most of them can be epitaxially grown on Si (111) using (Ba, Ca)F$_2$ buffer layers in order to overcome the important lattice and thermal expansion mismatches.

We have used molecular beam epitaxy to study the growth of PbSe on Si (111) using these fluoride buffer layers. In this contribution, we will present an overview of our present knowledge.

An optimization of the fluoride growth process have been performed, which is based on metallographic analysis and Rutherford backscattering in channeling geometry. We have performed the growth of n and p type PbSe layers. Structural characterizations of the samples are performed by RHEED, X-ray diffraction and scanning electron microscopy. Electrical and electro-optical characterizations will also be presented.

RHEED INVESTIGATIONS OF PbTe AND $Pb_{1-x}Eu_xTe$ SURFACES DURING MBE GROWTH

Springholz, G., Bauer, G.

Johannes-Kepler-Universität Linz, Institut für Halbleiterphysik
Altenberger Strasse 69, A-4040 Linz, Austria

Extensive RHEED (Reflection High Energy Electron Diffraction) studies of the PbTe surface during MBE growth were performed. We report on the depedence of the observed RHEED oscillations of different diffraction features (i. e. specular spot, integral order streaks and diffuse scattering) on various diffraction conditions such as angle of incidence and azimuth and determine their phase relationship. Studies of the temperature dependence of the RHEED oscillations show that a 2-dimensional growth of PbTe can be achieved over almost the whole substrate temperature range from room temperature to the reevaporation temperature. In fact, the temperature for optimized RHEED oscillations ranges from about 100 °C to 200 °C where up to N = 250 oscillation periods ($\approx$ 100 nm) can be observed. The characteristics of samples grown under such conditions are presented.

Adsorption and desorption studies of the PbTe or $Pb_{1-x}Eu_xTe$ (111)-surfaces indicate that two well definded surface states of this (111) surface can be prepared, namely a surface which consists only of Pb or of Te atoms. RHEED oscillations for the growth of PbTe and PbEuTe starting from each of the different surfaces show a quite different behaviour. Also they depend on the composition of the impinging $PbTe/Te_2/Eu$ flux from the effusion cells which can be either metall rich or Telluride rich. This behaviour can be used to optimize the beam flux ratios of Tellurium to Europium for the MBE growth of $Pb_{1-x}Eu_xTe$ films.

SCANNING TUNNELING MICROSCOPY ANALYSIS OF SINGLE CRYSTAL PbSe-SURFACES AND EPITAXIAL LAYERS ON BaF_2

Stocker, W., Magonov, S.N., Cantow, H.-I.

Freiburger Material-Forschungszentrum (FMF)
Stefan-Maier-Strasse 31a, W-7800 Freiburg, Germany

Böttner, H., Schelb, S., Tacke, M.

Fraunhofer-Institut für Physikalische Messtechnik
Heidenhofstrasse 8, W-7800 Freiburg, Germany

The topography and atomic arrangement on the (001) surface of n- and p-PbSe single crystals have been studied in air by scanning tunneling microscopy (STM). Various types of prepared surfaces, chemically polished, cleaved and naturally and epitaxially grown surfaces were investigated with STM. The regular patterns of atomic scale STM images correspond well with the crystallographic positions of atoms on the (001) plane. Several types of surface defects were observed.

The atomic structure of epitaxial layers of PbSe on BaF_2 substrate have been visualied. It appeared, that with certain substrate temperatures (001) and (111) oriented grains are formed. Various growth orientations of the PbSe layer were found even in different locations in the same sample.

STM was also applied to investigate the phase boundaries of double hetero structures PbS/PbSe. The STM-images of a cleaved wafer-surface show the interfaces of the epitaxial layers with atomic resolution. The substrate and the interfaces of the epitaxial layers is inspected for dislocations using STM. The peculiarities of the epitaxial overgrowth of PbS on PbSe are discussed.

Section III

TECHNIQUES, SYSTEMS, COMPONENTS

STATE OF THE ART OF TDLS IN THE USSR

A.I.Nadezhdinskii
General Physics Institute Ac. Sc. USSR, Moscow, USSR

ABSTRACT

State of the art in Tunable Diode Laser Spectroscopy in the USSR is considered. More than twenty groups in SU use in their work Tunable Diode Lasers. Diode Lasers produced in SU cover spectral region from 0.6 to 46 microns. In all spectral region under consideration similar setup and software can be used to realize high resolution molecular spectroscopy. New generation of automated TDLS was developed using IR fibers. Different applications of TDLS in fundamental investigations are considered as well as TDL uses in environment protection, high technologies, medicine, etc.

1. INTRODUCTION

Now Diode Laser Spectroscopy (DLS) becomes one of the most intensively developing area of high resolution molecular spectroscopy (Fig.1). In USSR more than 20 scientific groups in different institutes, in different towns use Diode Lasers (DL) in their work. Table 1 presents classification of their activity. Between them are those which work in fundamental spectroscopy, intermolecular interaction, multi-photon dissociation, high pure material technology, plasma-physics, nonstationary process diagnostic, medical and metrology application and so on. Very important part of these works concerns DL application in environment protection.

This paper presents some topics which include fundamental spectroscopy of atmosphere pollutants, development and manufacturing of DL based analytic systems, atmosphere pollutants monitoring. These results were obtained in General Physics Institute of the USSR Academy of Sciences in collaboration with colleges from other institutes and organizations of the USSR.

2. NEW DIODE LASER SPECTROMETERS GENERATION

It seems that recent DLS progress allows us to speak about new generation of DL based systems. That means:
- very wide spectral region from visible to far IR;
- IR fiber optics use;
- fully automated DL systems operation.

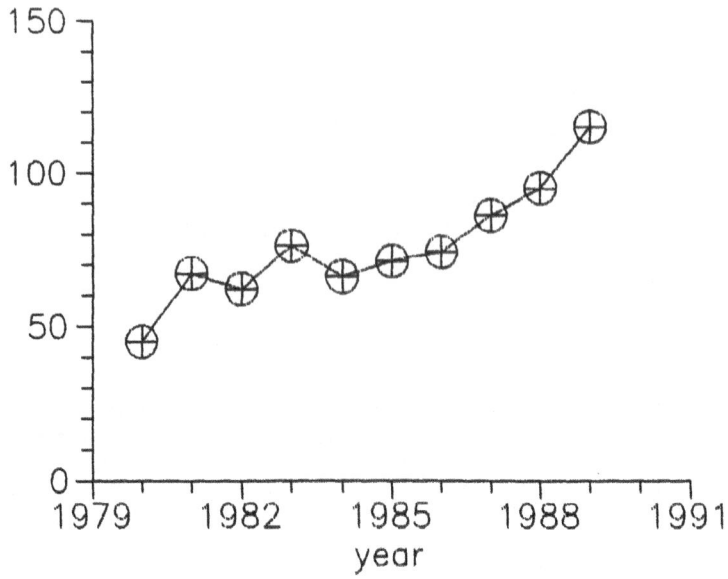

Fig.1 Number of papers concerning TDLS per year

Table 1. Diode Laser Spectroscopy in USSR

Organization	Activity			
	Laser	Spectr.	Appl.	Design
Appl.Geophys.Inst. Moscow			Y	
Atom.Energy Inst. Moscow		Y	Y	
Central Aer.Observ. Moscow			Y	Y
Cent.Bur.for Uniq.Instr.Ac.Sc. Moscow				Y
Gen.Phys.Inst.Ac.Sc. Moscow		Y	Y	Y
Inst.Atm.Opt.Ac.Sc. Tomsk		Y		
Inst.Atm.Phys.Ac.Sc. Moscow		Y	Y	
Inst.Chem.Pure Mat.Ac.Sc. N.Novgorod			Y	
Inst.of Physics Minsk			Y	
Inst.Spectr.Ac.Sc. Troitsk		Y	Y	
Leningrad Univ.		Y	Y	
Karadag National Park, Crimia			Y	
NPK Optronics Moscow				Y
Phys.Inst.Ac.Sc. Moscow	Y	Y	Y	
Phys.Tech.Inst.Ac.Sc. Leningrad	Y		Y	
Phys.Instr.Center Ac.Sc. Troitsk				Y
Surface and Vacuum Inst. Moscow			Y	Y
Tbilisy Univ.	Y			
Technol.Laser Inst. Shatura	Y		Y	

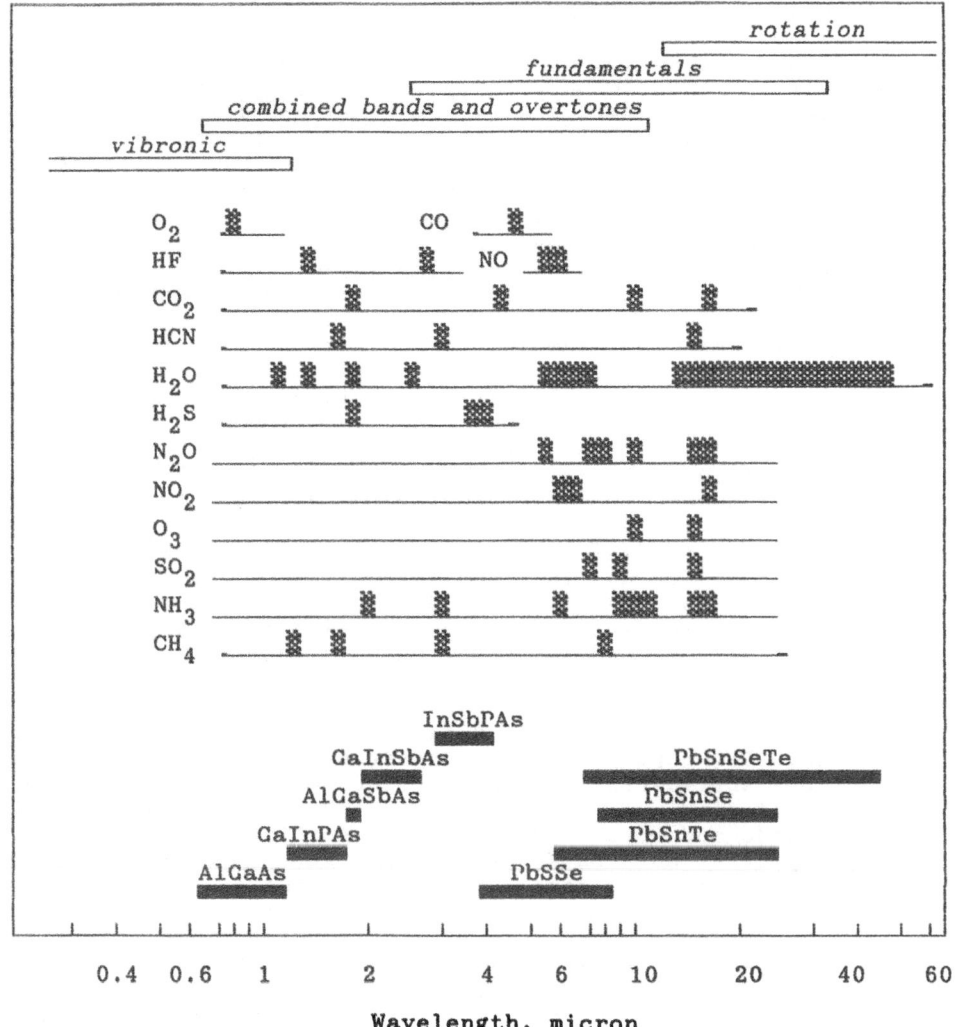

Fig.2   Spectral regions of verious types diode lasers

    ⊏⊐   Molecular spectra classification

    🎇   Location of several molecular bands

    ▬   Spectral regions of different types diode lasers

Conventionally in high resolution spectroscopy, mid-IR diode lasers were used operating in spectral region 3-30 microns. Stretching and bending fundamentals of most molecular objects lie in this spectral region. Recently diode lasers appear operating in near and far IR regions. Thus electronic, combined, overtone, rotation bands of different molecules may be investigated by DLS. Fig.2 shows spectral regions of DL which are available now in the USSR. It is of great importance that the same method and the same setup can be used to overlap very wide spectral region from 0.67 to 46 microns where different absorption bands of various objects are located.

DL with wavelength less than 2.5 micron can operate near room temperature. The fact is of great importance for DLS applications. As the examples of this type DL use, electron concentration measurement (Stark broadening) (1) and spectroscopy of metastable N_2 in plasma (2) as well as atmosphere humidity measurements can be mentioned.

Important step of DL systems simplification and improvement is connected with IR fiber optics use (3-4). Now IR fibers are produced from different materials and have optical losses as small as 0.1-1 dB/m in spectral region up to 25 microns (this is sufficient for most DLS applications).

Simultaneous DL and IR fiber optics use gives possibility for developing of really new diagnostic systems such as multicomponent, multiplexed, distributed and so on. Recently first model of DL and IR fiber based diagnostic system was developed and manufactured in Physics Instrumentation Center Ac. Sc. of the USSR.

Now it is necessary to speak about new level of DLS automation. Development of new expert systems is in progress now for DL application in environment protection, medicine, high technologies. In such systems check of suitable DL operation mode and data processing is realized automatically. Nowadays spectral data bank information should be included in mentioned expert systems for pollutant under study identification and measured concentration calibration.

Fig.3 presents scheme of TDL spectrometer. The scheme is used both in mid and near IR regions. The spectrometer consists of four channels to stabilized frequency tuning curve, to realize simultaneous frequency scale calibration and to achieve high precision in line shape measurements. Table 2 summarizes its parameters.

3. HIGH RESOLUTION MOLECULAR SPECTROSCOPY

Several groups in SU use TDL to investigated high resolution spectra of different molecules. As example investigation of fine structure of vibrationally exited nv_3 energy levels of SF_6 can be considered. This work was done by groups in General Physics Institute (S.S.Alimpiev and coworkers) and Institute of Nuclear Energy (G.S.Baronov and coworkers) and leads to explanation of spectral selectivity of multi-photon dissociation.

Phase transition in spherical top molecules (CF_4, SnH_4) spectra being theoretically predicted was then experimentally observed in Institute of Spectroscopy (Yu.A.Kuritsin,

Fig.3 DIODE LASER SPECTROMETER

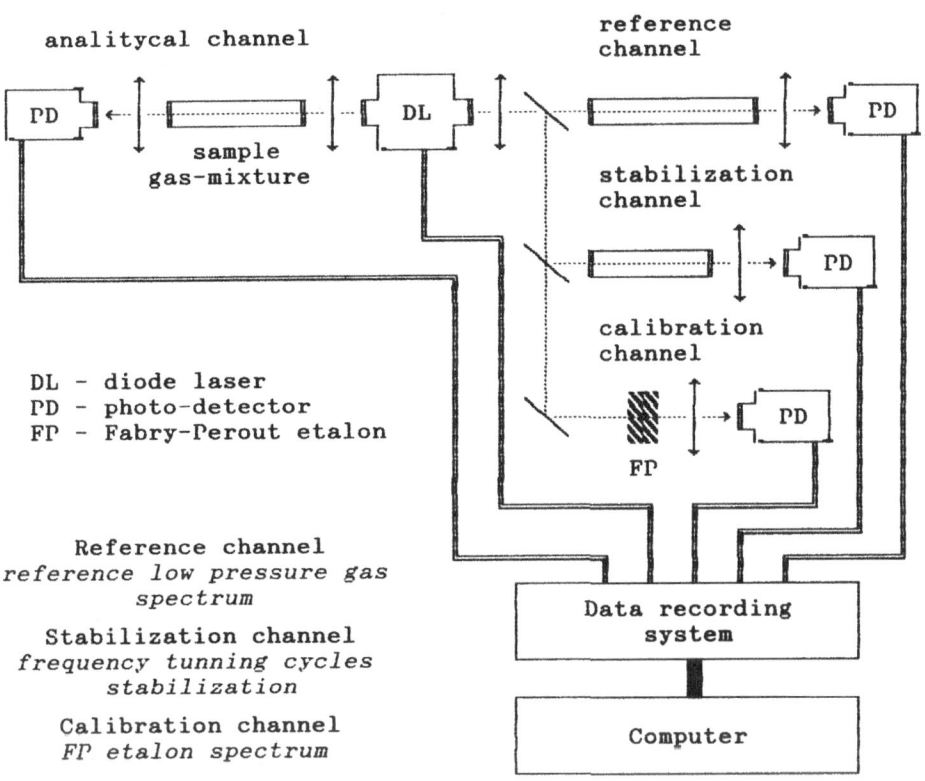

DL - diode laser
PD - photo-detector
FP - Fabry-Perout etalon

Reference channel
*reference low pressure gas
spectrum*

Stabilization channel
*frequency tunning cycles
stabilization*

Calibration channel
FP etalon spectrum

Table 2 Diode Laser Spectrometer parameters

Spectral range, microns	$0.7 - 16$
Frequency tuning rate, cm^{-1}/s	$10^2 - 10^5$
Tuning curve stability, cm^{-1}/h	$1 - 2 * 10^{-4}$
Frequency calibration accuracy, cm^{-1}	$0.5 - 5 \ 10^{-4}$
Laser linewidth, MHz	$1 - 50$
S/N (bandwidth 1 MHz)	$10^3 - 10^5$
Precision of absorption measurement	$0.3 - 3 \%$

V.M.Krivtsun and coworkers). Authors discovered changes of fine cluster structure type related to stable axes change of molecular rotation.

Fig.4 presents typical spectra obtained by DLS. Spectrum fragment near 1368 cm^{-1} belongs to v_3 fundamental of the SO_2 molecule. In near IR region combined band $v_2 + v_3$ of CO_2 molecule was investigate near 5110 cm^{-1}.

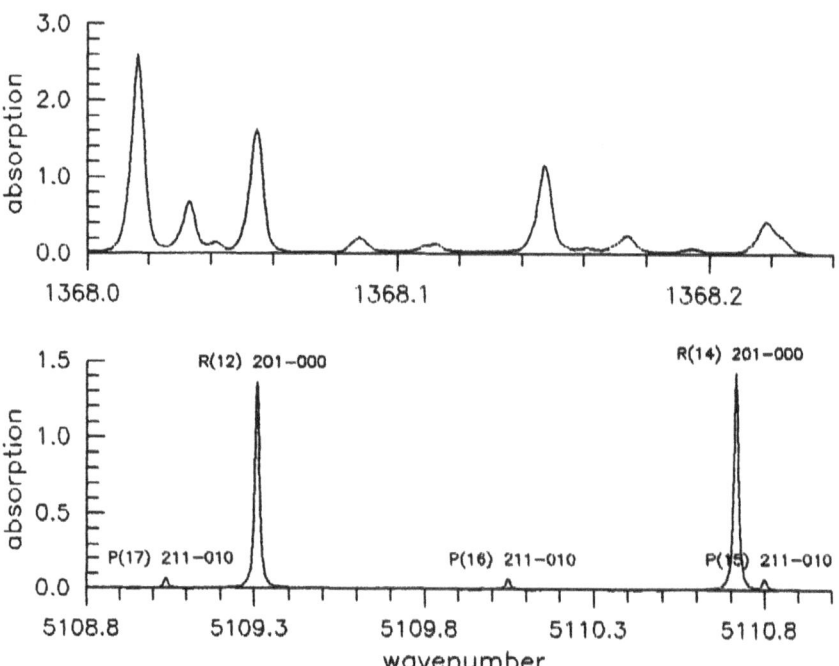

Fig.4 Typical spectra obtained in near and mid IR regions

We compared transition frequencies of v_1 and v_3 bands measured by DLS (E.V.Stepanov) with those obtained in FTS experiments (5) (Table 3). In all spectral regions under study (accept one where our precision was worse) observed rms differences between these two experimental methods were of the order of 10^{-4} cm^{-1}. Mean rms value (0.00031 cm^{-1}) agrees with our experimental accuracy (see Table 2) and is several times smaller than discrepancy between calculated (6) and measured frequencies (0.00115 cm^{-1}). Agreement between our results and FTS line intensity measurements were worse. We obtained very large discrepancy for strong lines (because of absorption saturation in FTS experiments) and about 20 % for weak lines. When we compared experimental and calculated values of spectral lines intensities it appeared to be obvious that it is necessary to take into account molecule nonrigidity and different vibrational band interaction.

SPECTRAL REGION (cm^{-1})	DLS - FTS (cm^{-1})	OBS - CAL (cm^{-1})
1166		0.00166
1175		0.00253
1321	0.0001	0.00128
1328	0.0001	0.00127
1332	0.00008	0.00089
1335	0.00072	0.00145
1338	0.00015	0.00086

Table 3

Comparison between experimental and calculated frequencies for v_1 and v_3 fundamentals of SO_2

DLS - FTS — RMS discripancy between DLS and FTS results

OBS - CAL — Results of fitting for proper spectral regions

Fragment of measured transitions' list

№	Frequency (cm^{-1}) Experiment	Calcul	O-C 10^{-4} cm^{-1}	Band	Upper state J	K_A	K_C	Lower state J	K_A	K_C	Intensity Exp. 1/cm Atm	Calc orb. un.
1	1166.7510	.74800	-30	c	14		4	14		6	1.85	5030
2	.7851	.78282	-23	c	27		26	26		26	2.18	9731
3	.8114	.80938	-20	c	13		4	13		6	1.97	4792
4	.8551	.84947	-56	c	25		22	24		22	4.52	6793
5	.9019	.90207	2	c	11		2	11		4	1.81	4174
6	.9421	.94077	-13	c	10		0	10		2	1.15	3790
		.94300	9	c	38		12	39		16		136
7	.9601	.96038	3	c	31		28	31		30	0.62	1556
		.96066	6	c	29		8	30		10		373
8	.9775	.97608	-14	c	9		0	9		2	1.97	3352
9	1167.0055	.00653	10	c	8		2	8		0	1.40	2856
10	.0335	.03315	-3	c	7		2	7		0	1.23	2294
11	.0563	.05586	-4	c	6		4	6		2	1.15	1651
12	.0729	.07498	21	c	5		4	5		2	0.86	904
13	.1401	.14180	17	c	50		42	50		44	0.90	449
		.14437	43	c	58		48	58		50		132

Important application of TDL is related to line shape measurements. Several groups in the world now declare accuracy at % level. To certify it international inter-comparison of TDL spectrometers is now discussed. Precision of spectral line intensity measurements in our experiments were checked through CO_2 combined band recording (A.N.Khusnutdinov). Ratio of measured lines intensities (Fig.4) coincided within few percents when pressure was changed more than one order of magnitude. Good agreement between measured values and HITRAN data base was observed.

Polyatomic molecules spectroscopy is of the great importance for atmosphere optics. These type molecules have spectra with high spectral line density. Even for low pressure mean distance between lines is comparable with linewidth. Naturally line broadening parameters evaluation for such spectra becomes complicated problem.

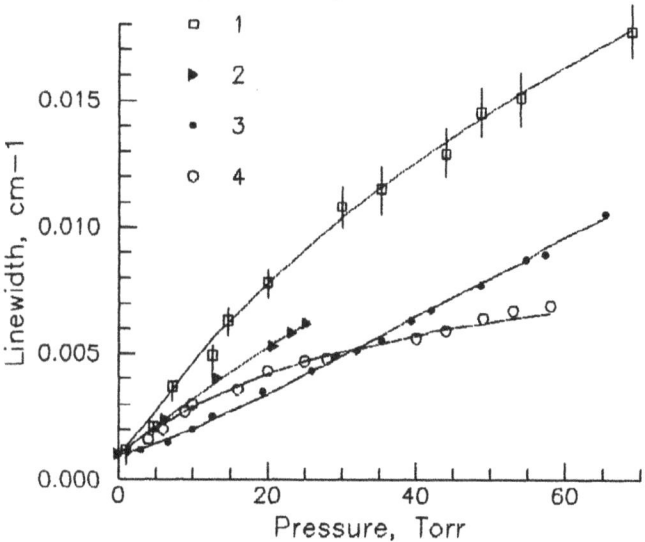

Fig.5 Pressure dependence of spectral line-width for polyatomic molecule spectra.
$1-CF_2Cl_2$:air; $2-SF_6$:H_2; $3-SF_6$:He; $4-SF_6$:Ar
Solid curves - calculations

To overcome this difficulty, correlation function approach was proposed (7) allowing to obtain some mean value characterized line broadening in polyatomic molecule spectra. Fig.5 shows results of CF_2CL_2 and SF_6 line broadening measurements. At pressures when mean distance between spectral lines becomes comparable with their width slope change in linewidth vs. buffer pressure dependence was observed. This nonlinearity leads to presence in experimental spectra pronounced resonance structure at pressures for which line by line compilation shows disappearance of the structure because of collisional broadening. There is one pair (SF_6:He) for which typical linear dependence occurs. The pair is important

from two points of view. First, it confirms that this slope
change is real effect and isn't result of experimental error.
And second, now we can explain why for this pair linear
dependence should occur.

Slope change on Fig.4 is due to line mixing. Theory
developed (8) predicts such behavior for local interference of
closely spaced line clusters. Thus, to calculate correct
atmosphere transmittance for spectral regions where such
molecules absorb it is necessary to take into account these
effects.

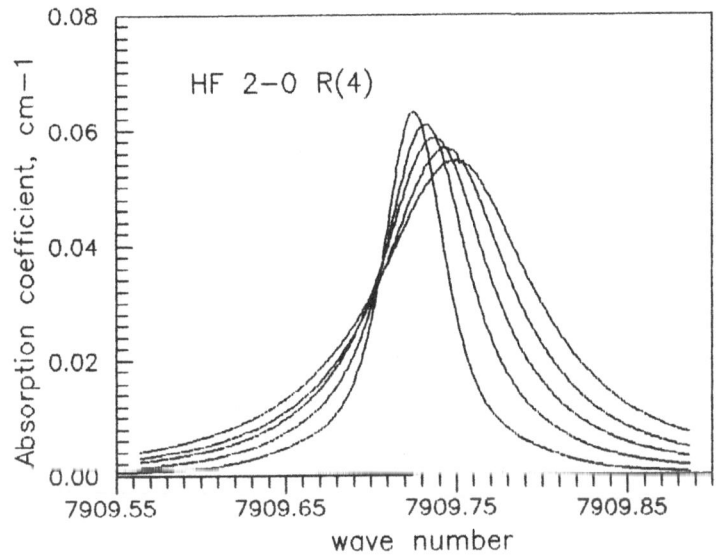

Fig.6 Pressure broadening and shift of HF 2-0 R(4) line

Intensive pressure shift experiments allow one to include
the information into spectral data banks. It was shown that
this information should be taken into account for correct
calculation. As example, lines of HF overtone band demonstrates
pronounced pressure shift (Fig.6).

Another effect to be considered in intermolecular
interaction is connected with collision adiabaticity. Resonance
behavior of collisional broadening cross-section was observed
in v_3 SF_6 band (9). Center of this resonance is located in the
region where adiabatic parameter is close to unity.

4. DLS APPLICATIONS

High sensitivity and selectivity of DLS, ability to work
in real time scale explain why DL spectroscopy is interesting
for diagnostic and analytic applications in environment

protection, high technologies, medicine and so on. This section considers several topics of the DL systems developing and application.

Central Design Bureau for Unique Instrumentation of Ac. Sc. of the USSR developed and manufactured AGL-02 model of DL sensor for CO concentration measuring in open atmosphere. These systems are used now in several regions of our country.

Continuous CO monitoring in Karadag National Park in Crimea was performed by S.Kotlelnikov from 1988 till nowadays (Fig.7). Different correlations have been observed over the system's long term operation between CO concentration and weather, biosphere, and sun activity parameters (Fig.8). For example, increases of carbon monoxide concentration were observed when wind was blowing from various industrial centers.

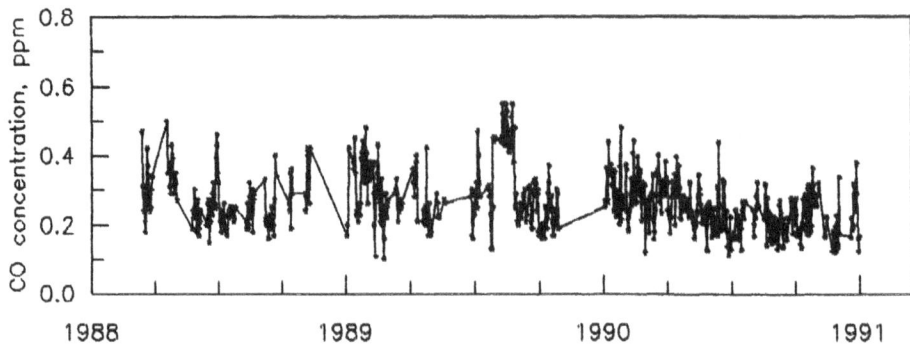

Fig.7 Long-term CO monitoring

As a rule, the lowest concentrations (180-200 ppb) were observed when the wind was from the S, SW, or SE, and the air was coming from the Black Sea. Since last February, a pronounced increase in CO concentration (500-800 ppb) was observed while the wind was blowing from the Persian Gulf.

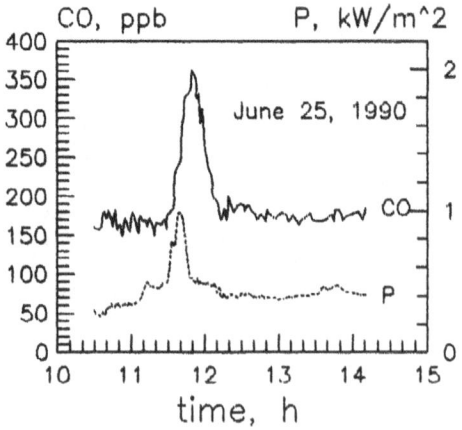

Fig.7 Correlation between CO concentration and incident Sun power P

Developed DL systems correctness was checked by help of calibrated gas mixtures. Some laboratory experiments considering linearity, reproducibility and sources of other systematic errors were done. In all cases experimental error does not exceed 2-5%. Simultaneous operation in field condition of several AGL-02 devices located nearby each other was realized (Institutes of Atmosphere Physics and Applied Geophysics). It was obtained that for closely spaced open atmosphere optical paths CO concentrations measured by these devices differed no more than 5%. Such intercomparison is important because the main systematic error for DLS systems is due to given laser properties. In experiment under consideration different diode lasers and their operation modes, different ro-vibrational analytic lines, different optical alignments were used in different devices.

Problem of small absorption measurement for heavy impurity molecules detection is rather complicated. For solving this problem correlation approach was proposed (11) giving rise of sensitivity increasing. All information of DLS registered spectrum (up to 10000 spectral lines) is used in spectrum under study processing and detected molecule concentration evaluation. This approach was tested during CFC12 detection in atmosphere (12).

In Institute for Chemistry of High pure Materials TDL spectrometer was developed (S.M.Shapin) to measure trace concentration in gaseous substances. System operates now in routine way to certify samples of high purity. One of important applications is connected with trace water detection in samples used in fiber optics technology. Fig.8 presents time dependence of trace water concentration in analytical cell when dry oxygen flow produced by special set was going through it for the first time.

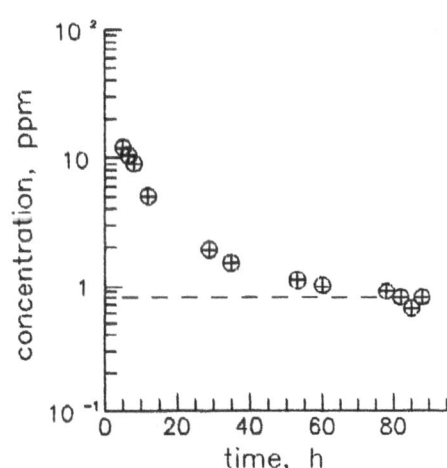

Fig.8 Reduction of water background when dry oxygen was passed through analytical cell for a few days.

To determine minimum water background, new dry oxygen source was developed and analytical signal was reduced under noise level. In this case, test experiments were performed to determine water detection level being 8 ppb for 2 m optical path-length (S/N=3).

Interesting results were obtained when TDL was used as a tool in medicine and biophysics. This area of applications can be considered as very promising in nearest future. Several

experiments were done by E.V.Stepanov and coworkers (General
Physics Institute) to obtain correlation between trace
concentration of different molecules in human breath and
diseases. Fig.9 presents result of simultaneous measurement of
CO and CO_2 concentrations in human expiration as function of
breath holding.

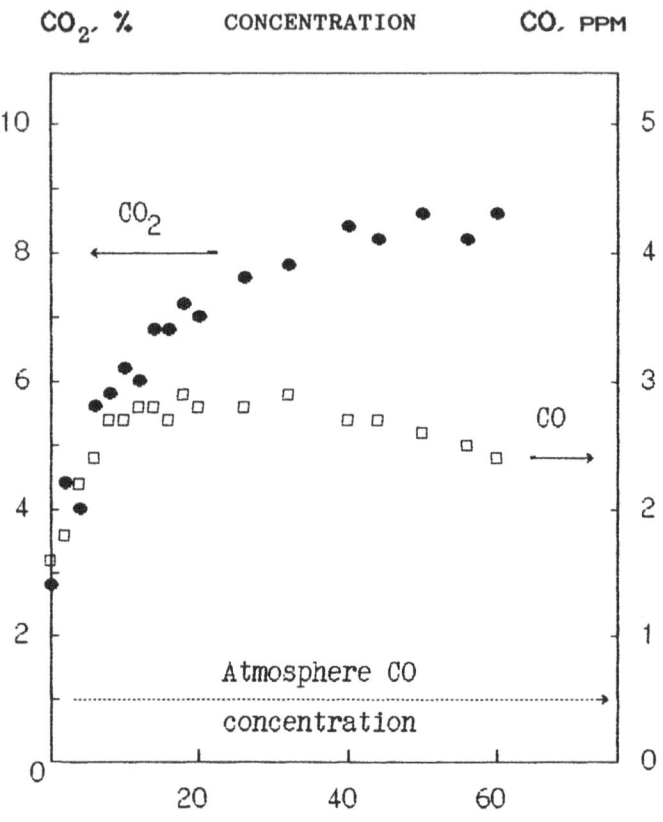

Fig.9 Typical result of simultaneous measurements of carbon
monoxide and carbon dioxide concentrations in human expiration
during breath holding.
Reverse behavior of these two molecules suggests presence of
different regulation mechanisms and demonstrated that TDL
application for trace molecular concentration measurement can
be considered as efficient instrument in biophysical
investigations.

5. CONCLUSION
        Today practically all directions of TDLS are presented in
SU. All-Union Diode Laser Spectroscopy seminar was established
in 1986 and takes place twice per year in Moscow and other

scientific centers of our country. Selected papers of this
seminar was published in Russian last year. New edition in
English "Applications of Tunable Diode lasers" (A.Prokhorov and
A.Nadezhdinskii ed.) is prepared now to be published in SPIE
Technical Publication series in the nearest future.

6. REFERENCES
(1) Avetisov, V.G., et. al. (1990). Pis'ma v Jurnal
     Experimental'noy i Teoreticheskoy Fisiki 51, 6 (in
     Russian).
(2) Merkulov, A.V., et. al. (1990). Jurnal Tehnicheskoy Fisiki
     60, 72 (in Russian).
(3) Artjushenko, V.G., et. al. (1987). SPIE 799, 90.
(4) Shapin, S.M., et. al. (1987). Viysokochistiye Veshestva
     #5, 202 (in Russian).
(5) Guelachvili, G., et. al. (1985). JMS 55, 319; (1987). JMS
     125, 128.
(6) Calculations were performed by O.Uleynikov and coworkers
     for isolated band model
(7) Kosichkin, Yu.V., et. al. (1984). Sov.J.Quantum. Electron.
     14, 1615 (in English); (1990). JQRST 43, 499.
(8) Nadezhdinskii, A.I., et.al. (1987). JQRST 45, 291.
(9) Kosichkin, Yu.V., et. al. (1987). Kratkie Soobsheniya po
     fisike #7, 48 (in English).
(10) Biykov A., Strojnova V. - private communication
(11) Zasavitskii, I.I., et. al. (1983). Kratkie Soobsheniya po
     fisike #9, 11 (in English); (1987). in Monitoring of
     Gaseous Pollutants by Tunable Diode Lasers, ed. R.Grisar
     et.al., D.Reidel Publ.Co., Dordrecht 95.
(12) Zasavitskii, I.I., et. al. (1985). Jurnal Analiticheskoj
     Himii XL, 1903 (in Russian).

Development of a Prototype IR-FM Absorption Spectrometer: Design Criteria and System Performance

Peter Werle, Robert Mücke and Franz Slemr

Fraunhofer-Institut für Atmosphärische Umweltforschung, IFU
Kreuzeckbahnstr. 19, 8100 Garmisch-Partenkirchen, F.R.G.

SUMMARY

This paper reports recent developments of a prototype tunable diode laser absorption spectrometer (TDLAS) based on the high frequency modulation technique (FM). The prototype uses lead-salt diode lasers operating in the mid-IR, where most molecules have strong rovibrational absorption bands.

1. INTRODUCTION

Measurements of atmospheric trace gases impose high demands on instrumentation. Fast, accurate and rugged instruments are needed for aircraft measurements and flux measurements by eddy correlation techniques. Ultrasensitive instruments free of interference by other atmospheric constituents are required to measure free radicals and other reactive species in the atmosphere. High frequency modulation technique (FM) in combination with tunable diode laser absorption spectroscopy (TDLAS) has the potential to satisfy most of the requirements for sensitivity, specifity, detection speed and applicability for many smaller pollutant molecules. The FM technique determines the absorption or dispersion of a narrow spectral feature by detecting the heterodyne beat signal that appears when the FM optical spectrum of the probe wave is distorted by the spectral feature of interest[1]. A potential sensitivity improvement of up to two orders of magnitude in comparison to conventional derivative (2f) spectroscopy can be derived from wideband noise characteristics of lead-salt diode lasers[2]. This sensitivity improvement is achieved by increasing the modulation frequency from the 1/f noise dominated region (10 kHz) into a shot noise limited domain of about 100 MHz [3]. However, to achieve the sensitivity improvement using the FM technique and to build instruments for routine high sensitivity measurements, many practical problems still have to be solved. For optimum performance, FM-TDLAS instruments need high quality lead-salt diode lasers and have to be constructed in a manner to ensure the long term stability which is needed to allow long integration times for digital bandwidth reduction in ultrasensitive measurements. This paper describes our investigations of these prerequisites, a prototype of a newly designed FM-TDLAS instrument and its performance.

2. LASER AND SYSTEM PREREQUISITES

The construction and operation of high sensitivity and/or fast FM-TDLAS instruments requires the fulfillment of several prerequisites. The most important of these is the high quality of the lead-salt diode laser and the system stability. But there are some essentials which dominate the performance of any spectrometer. These two major prerequisites will be discussed below.

High sensitivity and/or fast FM instrument ideally requires a lead-salt diode laser operating in single mode with low noise and high power output and a Gaussian beam profile. The properties of lasers presently available are far from these requirements and have to be substantially improved for use in FM instruments. The single mode operation is required to minimize the mode competition noise and to achieve the shot noise limited performance. Since the detection limit in FM-spectroscopy is proportional to the square root of the laser power impinging upon the detector, the laser output power should be as high as possible. This requirement is further stressed by the use of long absorption paths, multipass cells (White cell or Herriott cell) which strongly attenuate the incoming radiation. The square root dependence of the detection limit on detector incoming power is only valid under quantum limited conditions, which means that the shot noise has to be the dominating noise source in the system. While the shot noise level is dependent on the incident photon flux, the thermal noise is independent of the laser power. This means that the laser power available after passing through the absorption cell should be high enough to generate a shot noise current which is about one order of magnitude above the thermal noise level. With available detectors this requirement means that lasers for FM use should have an output power of at least 1mW. However, calibrated power measurements of most of the lasers presently available show values of about several hundreds of μW. Due to the limited power output, the FM instruments have to be operated at reduced optical absorption path length[4].

An automated test system has been set up to measure several aforementioned properties of lead-salt diode lasers, and the mode characteristics of several lasers has been determined. Fig. 1 shows a mode map of one of the best available LPE-lasers manufactured by Fujitsu, Inc.. The measured single mode power at 6.3 μm was about 400 μW. Due to the high dynamic range of the data acquisition system (10μV..1V), side modes with intensities far below 1% can be detected. The mode characteristics of this predominantly single mode laser still show some spurious side modes. Fig. 2 shows the output laser power measured during the same run. Due to limited space on the plot, the power data are presented only in 30 mA steps. To describe the degree of power contribution by other modes, we also present in Fig. 2 the value of %^2 = 100 Σ (I/ΣI)^2, where I is the intensity of the individual modes detected at a given laser current.

The power iso-plot in Fig. 2 shows the tuning rates in MHz/mA determined from the mode structure, which agree quite well with measurements using a fixed etalon. Typical tuning rates for different lasers vary from several hundreds of MHz/mA to 1.8 GHz/mA. The tuning rate for the Fujitsu laser data was slightly above 1 GHz/mA. These high tuning rates imply that the laser power supply has to be highly stable with a very low noise.

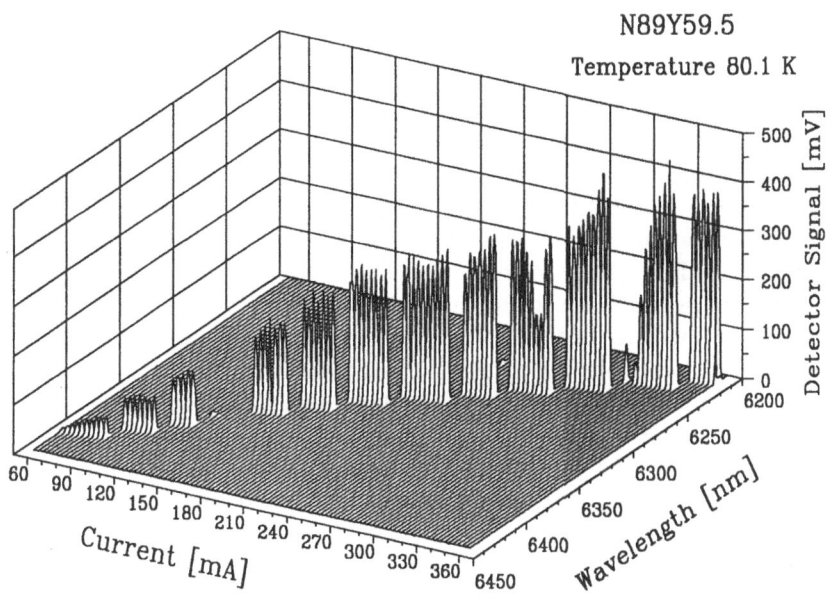

Fig. 1: Modemap of a single mode lead-salt diode laser

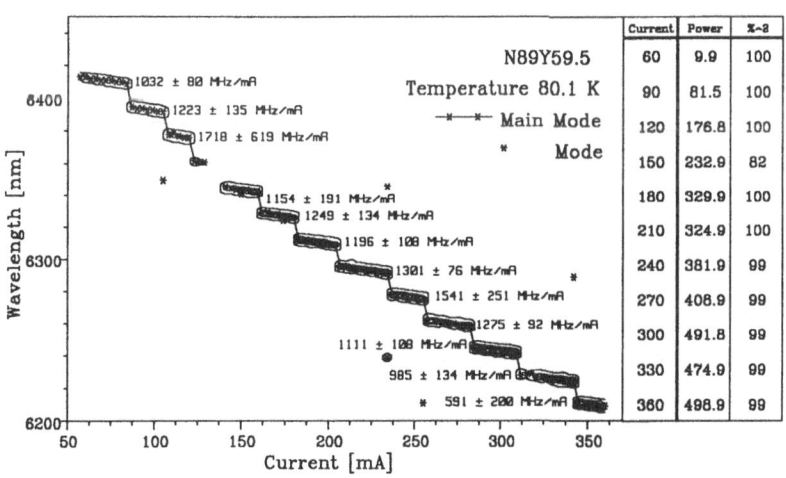

Fig. 2 : Iso-plot of laser N89Y59 with tuning characteristics and power measurements

Another critical factor is the laser beam profile and the emission angle characteristics. Observations made with a high resolution pyroelectric vidicon camera showed that many of the commercially available lasers emit light with a deviation of up to 30° relative to the axis of the laser housing. Such deviation leads to strong feedback, excessive noise, and substancal power losses and thus the performance of any TDLAS instrument deteriorates substantially. The quality of the laser beam profile is also crucial as the beam energy has to be focussed effectively upon a detector which typically has a diameter below 500 μm. As mentioned above, modulation frequencies of 100-200 MHz used in mid-IR FM-spectroscopy usually lead to the use of high bandwidth photovoltaic mercury cadmium telluride detectors. To achieve the required fast response time, the diameter of the detector elements has to be between 200 and 500 μm. We investigated beam profiles of several lasers and found them to be very variable. They varied from a nearly Gaussian beam profile up to a profile consisting of several spatially separated filaments and, in some cases, they changed with changing operating conditions. To optimize the power transfer to the detector, a laser with a nearly Gaussian profile has to be selected and the optics have to be designed to accommodate the divergence of the emitted beam from the axis of the laser housing.

The characteristics of laser noise is essential for the selection of the optimum modulation frequency. Theory predicts that largest FM signals to be found at a modulation frequency corresponding to 1.6 times the HWHM of the molecular absorption line under investigation. Frequency analysis of noise for several lasers showed a laser excess noise up to more than 200 MHz. Consequently, higher modulation frequencies are necessary to achieve near shot noise limited performance of the FM spectrometer. At higher modulation frequencies however, the modulation characteristics of the individual rf components and rf properties of the laser mounts become important. This is very important in multicomponent systems which require a proper impedance matching (BIAS-T) to minimize reflection of rf power and laser cross-talk.

Whether to use a single-tone or two-tone modulation scheme is a frequently discussed question. Both techniques work with modulation frequencies in the flat white noise dominated high frequency region. However, the single-tone FM technique detects the signals at the modulation frequency while the two-tone scheme uses the difference frequency of the two applied discrete rf-signals. If laser noise is the limiting factor inside the system, the two-tone beat frequency has to be choosen high enough to operate outside the 1/f noise region in the shot noise dominated plateau. To enjoy the theoretically predicted sensitivity improvement, the two-tone technique has to use high frequency detectors as well, and the frequently stated advantage of the use of cheaper detector elements is no longer valid.

In addition to laser performance, a second critical factor influencing the high sensitivity measurements is the system stability. From a theoretical point of view, the signal from a perfectly stable system could be averaged infinitely. Infinite averaging should lead to extremely sensitive measurements if limitations of the dynamic range of the system are neglected. Unfortunately, real systems are not stable forever and the fundamental question is how long can signals be averaged to achieve an optimum sensitivity with a given spectrometer? It is obvious that every real unstable system will have an optimum averaging time given by the drifts in the system such as temperature drifts, moving etalon, background changes etc. Therefore, a quality of the spectrometer can be described by the characteristics of the instrument stability. In analogy to other stability investigations, the instrument stability can be described using the Allan variance[5,6], which is defined by the following equation :

$$\sigma_A^2(\tau) = 1/2m \sum_{k=1}^{m} (A_{k+1}(\tau) - A_k(\tau))^2$$

The Allan variance is the time average of the sample variance of two adjacent averages of time series data.

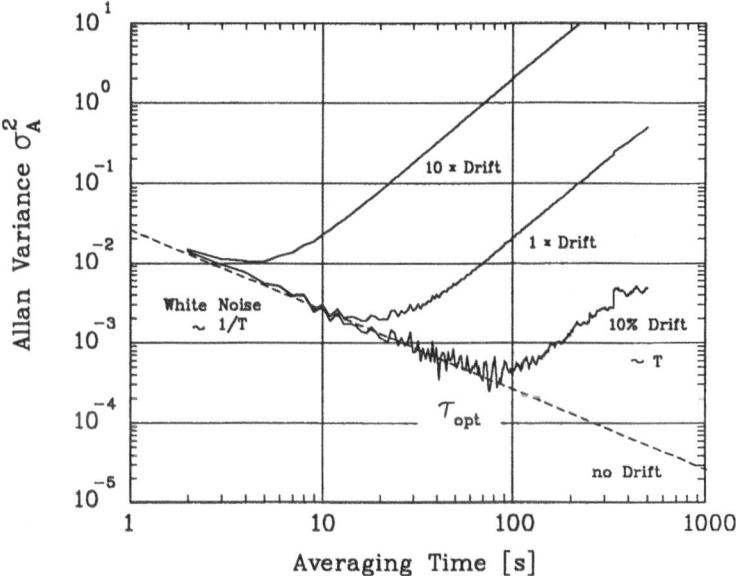

Fig. 3 : Allan plot - Allan variance as a function of the integration time

Noise contributions, S(f), which are encountered in most systems are frequency independent "white noise" and frequency dependent 1/f and $1/f^2$ noise. The last is noise at very low frequencies, which can be considered as drift. The time domain stability[7], $<\sigma_A^2(\tau)>$, can be derived from the frequency domain stability, S(f). As an approximation, the Allan variance can be written in the following form :

$$\sigma_A^2(\tau) = k_{drift}\ \tau + k_{1/f} + k_{white\ noise}\ 1/\tau$$

This means that, depending on the constants, k, at short integration times, τ, within the white noise dominated region, the Allan variance improves with increasing integration time. As the integration time increases, the Allan variance shifts from this region into a drift dominated region where it again starts to increase. In the white noise dominated region the Allan variance is equivalent to the detection limit and can be used to predict the detection limit of a given system as a function of the integration time. Allan variance plotted as a function of the integration time leads to the "Allan plot" such as shown in Fig. 3. The minimum of the Allan variance corresponds to the optimum integration time, τ_{opt}, indicated in Fig. 3. The optimum integration time is a characteristic property for a given instrument; it reflects the system stability (e.g. line locking, etalon drifts, changing background, etc.). We used the Allan variance as a tool to characterise the stability of the whole spectrometer or of its individual parts and to determine the optimum integration time. An example of an Allan variance analysis will be given later.

After the discussion of the two major prerequisites for the development of sensitive FM-TDLAS instrument we will now turn to a description of an instrument developed for sensitive trace gas analysis of air pollutants.

3. SYSTEM DESIGN

A schematic view of the optical system layout is given in Fig. 4. The system is mounted on a 1 m x 0.6 m breadboard which is covered by a box for a) temperature stabilisation and for b) flushing the system with dry nitrogen to reduce changing broadband atmospheric absorptions (e.g. by H_2O), which will limit the system´s stability for signal averaging. To accommodate the possible deviation angle between the laser emission and the laser mount axis, the LN_2 - Dewar is mounted on a $+/\_ 30°$ alignable xyz-stage. Using off-axis parabolic (OAPs) and flat mirrors the beam is collimated and steared through a commercial multipass White cell with a base length of 62.5 cm. For optimum signal-to-noise ratio[4] the absorption path length in the White cell is adjusted to about 25 m. The optical system is prealigned using a 670 nm semiconductor laser which is coupled to the system via a removable pellicle beam splitter.

New electronic components were developed to allow fast and complex data acquisition and software control of the main subunits required for unattended operation. All components were integrated on compact printed circuit boards. The newly developed electronics consist of a rf-detection and demodulation board, a board for real time signal averaging and data reduction, and a board for laser control, i.e. for current and temperature control and laser protection. All modules are designed to be used in a multi component FM system using a separate microprocessor for each channel.

To achieve the best possible sensitivity, the instrument should be made as stable as possible and and the measuring time should be as short as possible. Due to the high modulation frequencies (MHz) and the subsequent demodulation in a "high frequency lock-in amplifier", the laser can be tuned over the spectral feature of interest with frequencies in the kHz range. We prefered higher scanning frequencies to discriminate against low frequency mechanical vibrations. Scanning frequencies of the order of kHz, however, result in a data acquisition rate of the order of several 100 kHz. To cope with the high data acquisition rate, the spectral data are acquired and averaged by a transputer (T4XX) and the real time signal analysis is made by an additional signal processor on the signal processing board. The real time signal analysis is performed using complex data reduction software. The data analysis is shown schematically in Fig. 5. The instrument automatically provides signal, background and calibration spectra consisting typically of 64 channels and averaged over 256 individual spectra. For each averaged data point, the mean and the variance are stored temporarily in the memory. Subsequently, a multiple linear regression is performed to fit the background and the calibration spectra simultaneously to the ambient air spectrum. The simultaneous multiple regression analysis also allows us to apply different filters and to take into account other spectra; for example background spectra preceeding and following the ambient signal can be fitted simultaneously to compensate for the drift effect.

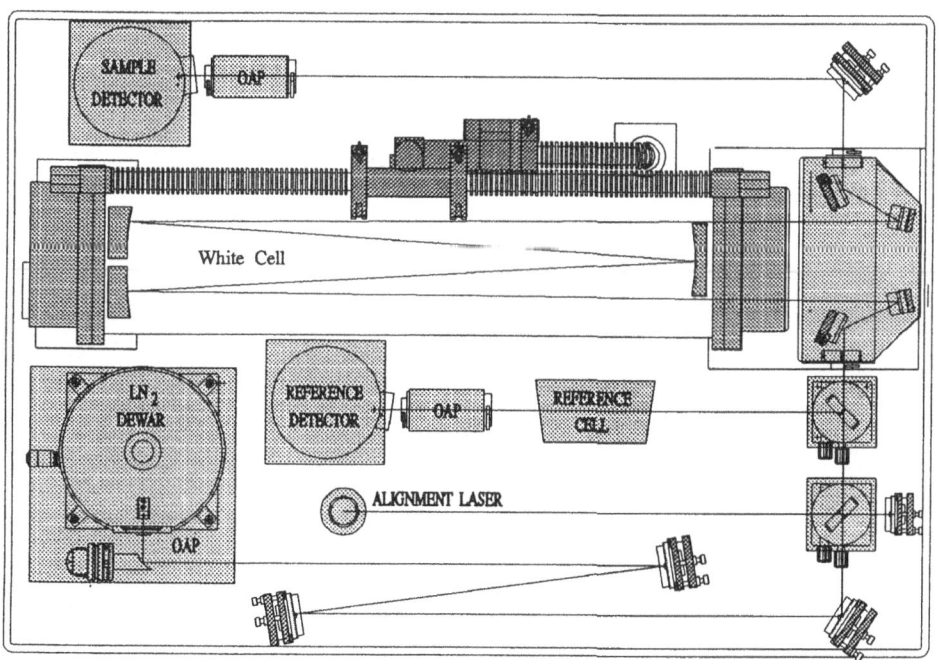

Fig. 4: Optical layout of the FM-spectrometer

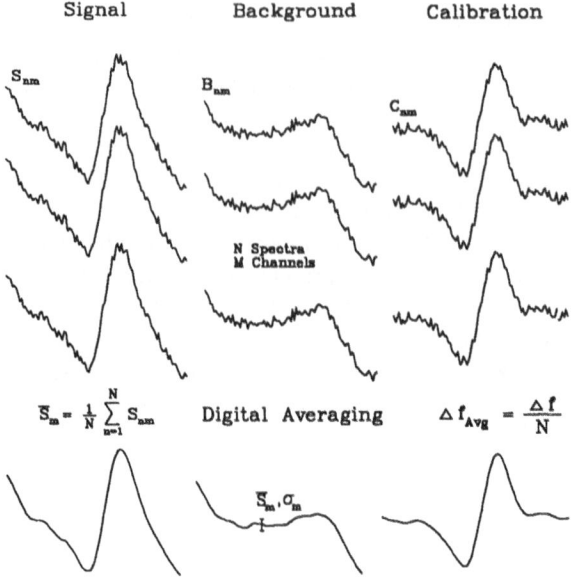

Signal　　　　　Background　　　　Calibration

S_{nm}　　　　　B_{nm}　　　　　C_{nm}

N Spectra
M Channels

$$\bar{S}_m = \frac{1}{N}\sum_{n=1}^{N} S_{nm} \qquad \text{Digital Averaging} \qquad \Delta f_{Avg} = \frac{\Delta f}{N}$$

$\bar{S}_m, \sigma_m$

Multiple Linear Regression

$$\begin{pmatrix} \bar{S}_1 \\ \bar{S}_2 \\ \bar{S}_3 \\ \vdots \\ \bar{S}_M \end{pmatrix} = \begin{pmatrix} 1 & \bar{B}_1 & \bar{C}_1 \\ 1 & \bar{B}_2 & \bar{C}_2 \\ 1 & \bar{B}_3 & \bar{C}_3 \\ \vdots & \vdots & \vdots \\ 1 & \bar{B}_M & \bar{C}_M \end{pmatrix}_{\substack{\text{linear} \\ \text{quadratic} \\ \text{etc.}}} \cdots \times \begin{pmatrix} a \\ b \\ c \\ \vdots \end{pmatrix}$$

$$\vec{S} = \vec{M} \times \vec{C} \qquad\qquad \Sigma = \begin{pmatrix} \sigma_1^2 & & & \\ & \sigma_2^2 & & \\ & & \sigma_3^2 & \\ & & & \sigma_M^2 \end{pmatrix}$$

Solution	$\vec{C} = (\vec{M}^T\vec{M})^{-1}\vec{M}^T\vec{S}$
Error	$\vec{\Delta}_C = (\vec{M}^T \Sigma \vec{M})^{-1}$
Bandwidth	$\Delta f_{eff} = \dfrac{\Delta f_{Avg}}{M} = \dfrac{\Delta f_{Sys}}{M\,N}$

Fig. 5 : Scheme of data reduction and signal analysis

4. SYSTEM PERFORMANCE

The performance of the system, characterized by Allan variance as proposed in Section 2, is shown in Fig. 6. The Allan plot was constructed from a continuous measurement of zero air spiked with 12 ppb of NO_2 from a permeation source for a period of 600 seconds with a time resolution of 1.5 sec. The resulting Allan plot showed a pronounced minimum at about 60 sec. The dashed line shows the theoretically expected behavior of a white noise dominated spectrometer.

The Allan plot predicts the optimum integration time to be about 60 s for this particular instrument. Within this time a complete measurement sequence consisting of the determination of the ambient air, zero air, and calibration gas spectra has to be completed. A substantial part of the time is taken up by gas exchange in the White cell after switching, say, from zero to ambient air. The Allan plot implies that the residuals after digital background subtraction increase if the the whole measurement sequence takes more time than 60 s. Since the linearly decreasing part of the Allan variance (white noise domiated) is equivalent to the statistical variance, the square root of σ_A^2 is an estimate the detection limit which in this particular case was 34 ppt ($^+/_-$ 25%).

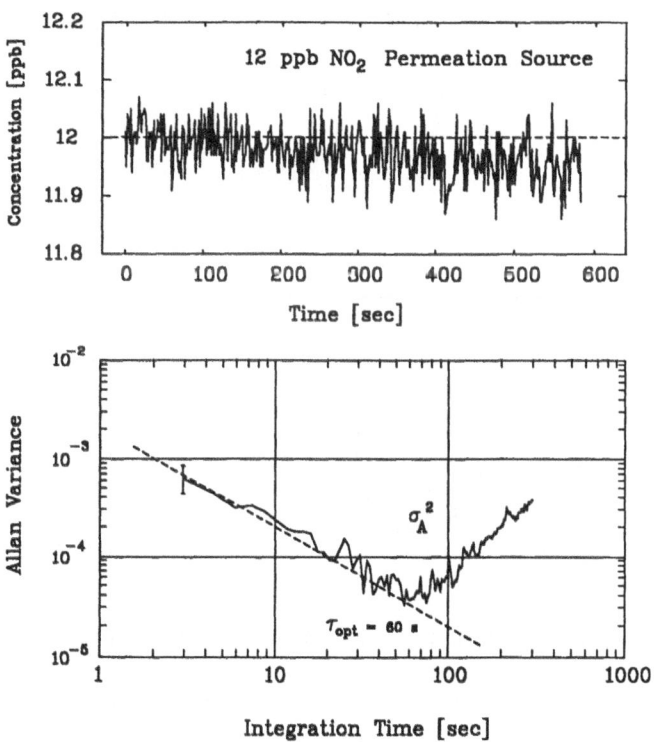

Fig. 6 : Time series data obtained from a stable calibration source and the corresponding Allan plot

The factors controlling system stability were investigated by monitoring many parameters during the measurement. The analysis of these measurements revealed a good correlation between the deviation of line locking from the line center (in MHz) and the confidence range obtained from the fit.

The upper trace of Fig. 7 shows the measured drift of the line locking deviation, the second trace represents the absolute deviation values which correlate well with the calculated confidence range shown in the third trace. The confidence range under "line locked" conditions is about 50 ppt and it may degrade up to a factor of 10 when the deviation is large. This finding illustrates the importance of a highly stable and transient free laser current source.

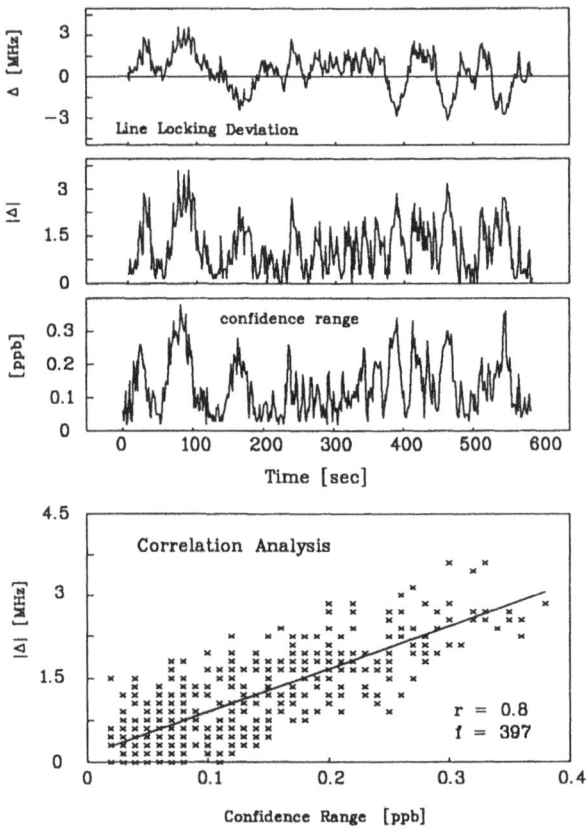

Fig. 7: Correlation analysis of laser stability and confidence range

The most important measure of system performance is the ultimate detection limit that can be achieved. To be able to express the detection limit in terms of optical density, the measured NO_2 line was identified by scanning the laser over 1 cm^{-1} in a wavelength range previously characterized by a mode map (Fig.1). The linearized absorption spectra of NO_2 at high concentration were then compared with a NO_2

spectrum from the HITRAN data base. The comparison is shown in Fig. 8. The upper trace shows the corrected FM spectrum from a 100 ppb NO_2 permeation source while the lower trace shows the corresponding spectrum from the HITRAN database.

The measurements shown in Fig. 6 were made with the NO_2 doublet line at 1600.413 cm^{-1} since its background was almost free of disturbance from pressure broadened H_2O lines nearby. The optical density of the 100 ppbv NO_2 calibration gas in a White cell adjusted to a pathlength of 27.5 m was calculated to be 5.3 $^o/_{oo}$ using a Voigt profile. Consequently, 1 ppb of NO_2 in the following data corresponds to an optical density of about 5×10^{-5} .

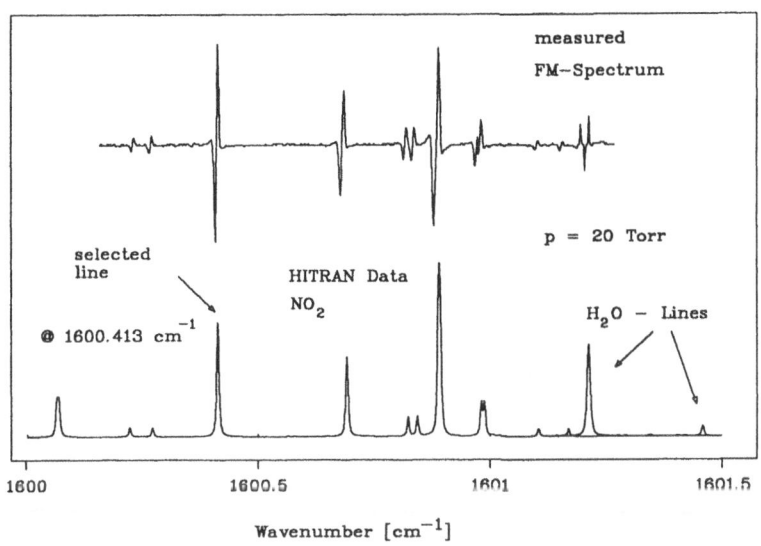

Fig. 8: Line identification : measured FM spectrum of NO_2 HITRAN data

After locking the laser to this line, a series of ambient measurements.were made. A characteristic spectrum, after background subtraction, is shown in Fig. 9. This spectum has been averaged 256 times in 740 ms and is almost noise free at this scale. The electronics bandwidth had been set to 1.5 kHz. The spectra averaging then results in an effective bandwidth of 5.86 Hz. In the subsequent regression analysis 27 channels were used for the fit to the calibration spectrum. The fit reported a concentration of 1.17 ppb with an 1 σ error of 31.5 ppt. This is in a good agreement with the Allan plot in Fig. 6, which predicted 34 ppt from an independent measurement. In our data the confidence range is defined as the 1 σ fit error times the STUDENT-t-factor at 95% statistical significance. Using 25 degrees of freedom in the fit, the t-factor is found to be 2.06 leading to a confidence range of 63 ppt. These data correspond to an optical density of 62 $^+/_-$ 3.34 $\times$ 10^{-6} measured in 740 ms .

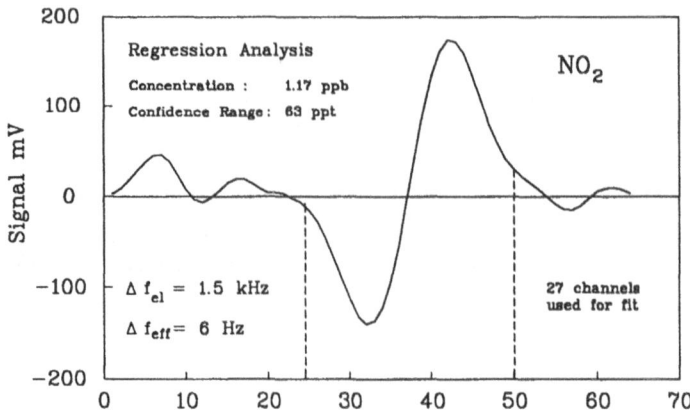

Fig. 9 : 1.17 ppb NO₂-FM-signal from ambient air spectra averaged over 740 ms.

Fig. 10 shows a continuous NO_2 measurement in ambient air. The data were obtained with a time resolution of 1.5 sec during one afternoon near the institute. The observed NO_2 spikes are due to cars coming in and out of a nearby parking lot. The calibration and background measurements have been removed from the plot to avoid confusion.

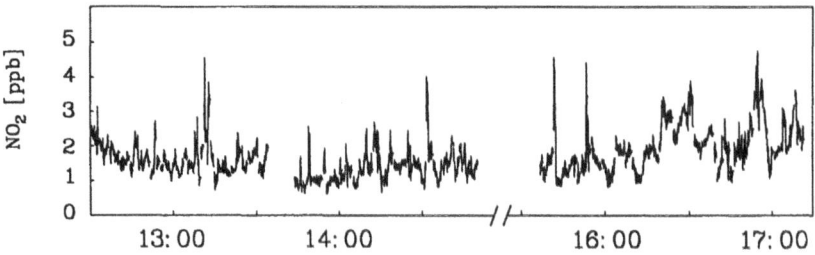

Fig. 10 : A record of NO₂ monitoring in ambient air by the FM-instrument

Fig. 11 shows an expanded part of these data and the record of some control parameters. The presentation of NO_2 concentrations in the upper trace includes background and calibration measurements omitted in Fig. 10. The information from the Allan plot has been used to set the background repetition rate to 60 sec to reduce drift influence on sensitivity. The second trace shows the confidence range for each measured point. For a complete description of the measurement, frequency and amplitude stability of the reference channel are also shown in the lower two traces.

Ambient Air

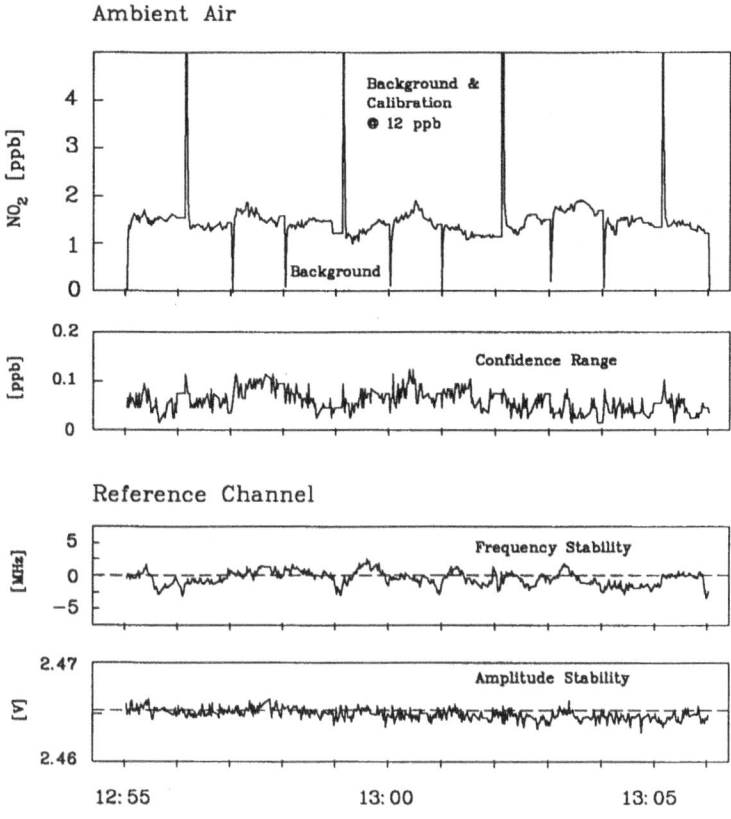

Fig. 11 : Detailed data from the previous plot.

5. CONCLUSIONS

Investigations of the performance of lead-salt diode laser with respect to mode characteristics, tuning rate, far field pattern, laser output power and laser noise indicated that only a small fraction of commercially available lasers is suitable for FM-TDLAS instruments. Some of the laser deficiencies can be corrected by a proper design of the instrument but a widespread use of the FM technique will require a substantial improvement in the laser properties.

An FM-TDLAS instrument was developed and its stability was analysed in terms of an Allan variance. This analysis indicates that the best detection limits can be obtained when the whole measuring sequence consisting of the ambient and zero air spectra is completed within 60 s. For NO_2 measurement in ambient air, a detection limit of 63 pptv with an effective bandwidth of 5.86 Hz was assessed. Further work is needed to find out which factors limit the instrument stability and to reduce their influence.

REFERENCES

[1] Bjorklund, G.C. (1980). Frequency-modulation spectroscopy: a new method for measuring weak absorptions and dispersions. Opt. Lett. 5, 15-17.

[2] Werle, P., Slemr, F., Gehrtz, M. and Bräuchle, Chr. (1989). Wideband noise characteristics of a lead-salt diode laser: possibility of quantum noise limited TDLAS performance. Appl. Opt.28, 1638-1642.

[3] Werle, P., Slemr, F., Gehrtz, M. and Bräuchle, Chr. (1989). Quantum-Limited FM-spectroscopy with a lead-salt diode laser. Appl. Phys. B 49, 99-108.

[4] Werle, P. and Slemr, F. (1991). Signal-to-noise ratio analysis in laser absorption spectrometers using optical multipass cells. Appl. Opt. 30, 430-434.

[5] Werle, P., Josek, K. and Slemr, F. (1991). Application of FM spectroscopy in atmospheric trace gas monitoring: a study of some factors influencing the instrument design. Proc. SPIE Vol. 1433, 128-135.

[6] Allan, D.-W. (1966). Statistics of atomic frequency standards. Proc. IEEE Vol. 54, 221-230.

[7] Barnes, J.A. et. al. (1971). Characterization of frequency stability. IEEE Trans. Instr. Meas., IM-20, 2, 105-120.

A TTFM SPECTROMETER FOR DETECTION
OF TRANSIENT RADICAL SPECIES:
$2\nu_1$ OVERTONE ABSORPTION LINES OF HO_2 AT 1.5 μm

T.J. Johnson, F.G. Wienhold, J.P. Burrows, G.W. Harris*,

Max Planck Institute for Chemistry
Atmospheric Chemistry Department
P.O. Box 3060, W–6500 Mainz
Federal Republic of Germany

and H. Burkhard

DBP Telekom Research Institute
P.O. Box 10 00 03, W–6100 Darmstadt
Federal Republic of Germany

ABSTRACT

A Two-Tone Frequency Modulation (TTFM) Spectrometer has been constructed using a 1.5 μm InGaAsP laser and White cell optics. The spectrometer can consistently achieve detection limits corresponding to <5 x 10^{-6} OD ($\Delta f_{eff} \approx 1Hz$) and has been used to investigate the near–infrared absorptions of the important atmospheric intermediate HO_2. Several lines within the $^2A''(200) \leftarrow {}^2A''(000)$ HO_2 overtone band near 1.509 μm have been recorded, and their cross sections were determined. A description of the spectrometer and of advances in the FM technique are presented. Preliminary experiments investigating the detection limit for the HO_2 radical were undertaken. Results are interpreted in relation to possible atmospheric measurements.

I. INTRODUCTION

Frequency modulation spectroscopy (FMS) in conjunction with diode lasers continues to hold strong promise for enhanced detection sensitivity of gas phase atoms and molecules (1,2). Several authors have recorded improved detection sensitiviy using different variations of the FM technique. This is especially important in fields such as pollution monitoring and chemical kinetics where low concentrations of (weakly absorbing) species makes *in situ* monitoring of their concentrations an ongoing challenge. We have constructed a system using Two–Tone Frequency Modulation Spectroscopy (TTFMS) similar to that described, for example, by Cooper et al. at the previous Freiburg conference (3,4). The goal of our work, however, has been to implement TTFMS in conjunction with long path absorption cells in an effort to achieve maximum sensitivity to low mixing ratios of radical intermediates.

One intermediate of key importance in atmospheric and combustion chemistry is the hydroperoxyl radical, HO_2. It is both an oxidant and closely coupled to OH, the initiator of most photochemical cycles. HO_2 is known to have three fundamental vibrational modes that are infrared active (5), but these have not as yet been exploited for atmospheric monitoring because of lack of sensitivity due to the low mixing ratios and relatively weak line strengths, (6,7) and because of the sampling problems associated with the reduced pressure measurement techniques common for mid–infrared TDL spectroscopy. Nevertheless, HO_2 is known to have near–infrared absorptions (8), the first overtone of the ν_1 O–H stretch centered near 1.504 μm, and a vibronic progression in the low lying $^2A' \leftarrow {}^2A''$ electronic transition beginning at 1.425 μm (9), both of which have previously been recorded only under low resolution. We have chosen to employ our laser spectrometer to investigate the rotational–vibrational $2\nu_1$ overtone lines of the HO_2 radical in the near–infrared using an InGaAsP laser.

II. SPECTROMETER

The TTFM spectrometer used in these experiments has been partially described elsewhere (10). However, some improvements have been made as depicted in Figure 1. For the reference modulation, the 5.5 MHz $= \frac{1}{2}\Omega$ frequency was doubled with a Mini–Circuits MK–3 doubler, and subsequently amplifed and filtered using active LC–filter circuits. After passing the delay–line phase shifter, the Ω signal was input to the local oscillator port of a Mini–Circuits ZFM–3B demodulating mixer. On the laser input side, greater isolation was provided by coupling the $\nu_m = 525.4$ MHz frequency into the LO oscillator port of a Mini–Circuits ZLW–2 mixer at +4.0 dBm. The strength of the $\frac{1}{2}\Omega$ frequency was then empirically adjusted to maximize the amplitude of the two $\nu_m \pm \frac{1}{2}\Omega$ peaks (observed using a Tektronix 2710 spectrum analyzer) relative to the supressed ν_m frequency, as well as to the sidebands at $\nu_m \pm \Omega$, $\nu_m \pm \frac{3}{2}\Omega$,. . . The mixer IF output signal was further amplified and passed through a bias–tee and a 150/550 MHz bandpass filter to the laser diode, the bandpass filter eliminating the higher frequency ν_m harmonics and lower frequency noise including any Ω components. On the other half of the bias–tee, the DC–bias and current ramp were provided by a Spectra–Physics 5820 Laser Current Module (LCM) modified for low current diodes and to reduce 50–Hz line noise.

The laser was a distributed feedback (DFB) InGaAsP/InP III–V diode. This diode provided superior sidemode rejection (>37 dB) and exhibited a linewidth $\Delta\nu =$ 15 MHz. The laser further delivered high output power, over 3 mW, and reasonable temperature tuning within a single mode (11). The collimated laser output traversed a 132 m path length within the White cell and was focused onto a high speed 13PD–100 detector from Telecom Devices. The detector output current was converted to voltage by a high impedance load resistor to reduce the effects of Johnson noise, and the voltage impedance matched to 50 Ω (10). The resulting signal was amplified 60 dB by an SAT electronics IR–500 RF amplifier, low–pass filtered to remove any components at frequencies greater than Ω, and coupled to the RF port of the demodulating mixer. An adjustable low–pass filter after the mixer helped remove high frequency noise and governed the bandwidth of the measurement (4). For both the modulation and detection electronics great effort was expended to avoid "pick up" at frequencies which

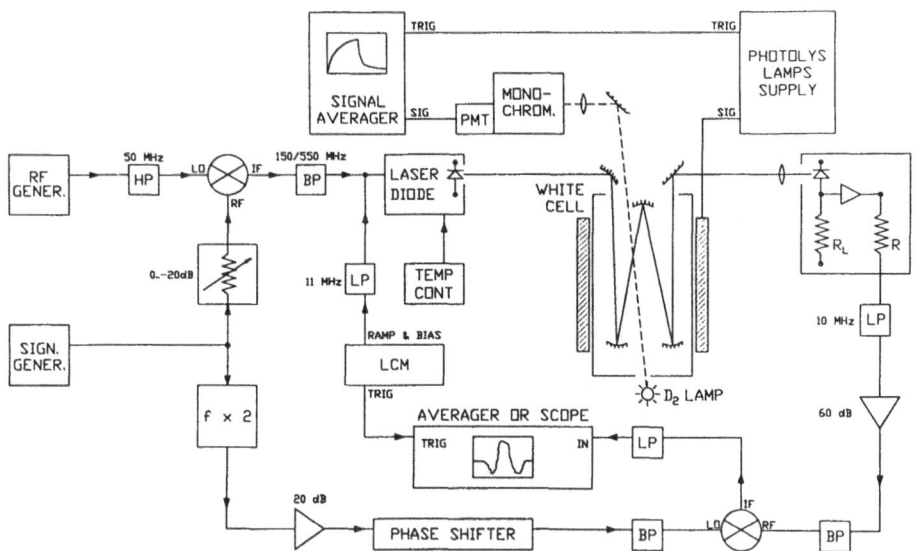

Figure 1. Schematic diagram of TTFM/MMS spectrometer including the InGaAsP laser. BP = band–pass, LP = low–pass, HP = high–pass filters, R_L = load resistor, R = 50 Ω circuit resistor.

interfere with signal recovery. Herein lies one advantage of two–tone versus single–tone FMS: in TTFMS the signal is recovered at an RF frequency (Ω) different from either of the two oscillators ($\frac{1}{2}\Omega$ and ν_m).

The exact frequency of the TDL laser at different HO_2 absorption wavelengths was measured in a second optical channel. The laser was tuned to line centre with the ramp turned off, and the infrared beam was directed into a Bomem DA 2.02 Fourier transform infrared (FTIR) spectrometer equipped with an InSb detector. Full resolution (0.02 cm^{-1}) was used to attain the greatest precision in the wavelength measurement. Since the intensity was high, only four FTIR scans were averaged.

The UV emission from a deuterium lamp traversed the White cell in a single 1.5 m pass and was focused onto a Spex minimate monochromator equipped with a Hamamatsu R–955 photomultiplier tube. The PMT output was amplified and monitored on an EG&G 4203 averager for detection of transient absorptions at 220 nm.

III. ABSORPTION SPECTRA

The HO_2 radical was generated by the modulated photolysis of air/H_2/Cl_2 mixtures using six Philips TLD 36W/08 black lamps:

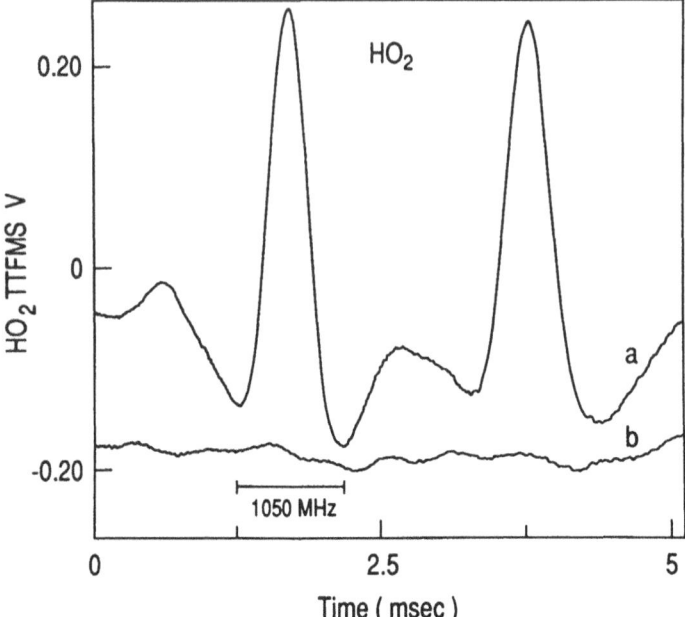

Figure 2. TTFM spectra recorded near 1.509 μm, with six photolysis black lamps on (Trace A) and off (Trace B). The HO_2 optical density of the right peak corresponds to 1.3×10^{-3} OD, and the horizontal bar corresponds to $2 \times \nu_m = 1050$ MHz. Measurement pressure was 61.0 Torr.

$$Cl_2 + h\nu \rightarrow 2Cl \tag{1}$$
$$Cl + H_2 \rightarrow HCl + H \tag{2}$$
$$H + O_2 + M \rightarrow HO_2 + M. \tag{3}$$

HO_2 radicals were removed via the disproportionation reaction:

$$HO_2 + HO_2 \rightarrow H_2O_2 + O_2 \tag{4}$$

where the rate constant (12) for the self reaction is given by $k_{298} = 1.7 \times 10^{-12} + 4.9 \times 10^{-32}[M]$ ($cm^3 molec^{-1} sec^{-1}$). The resulting molecular modulation (5) spectroscopy (MMS) signal at 220 nm (13) was used to monitor the HO_2 concentration, typically in the range 5 to 12×10^{11} molec cm^{-3}.

The TTFM spectra of two the strongest recorded HO_2 lines are shown in Figure 2, with the "photolysis lights off" spectrum (at the same scale) displayed directly underneath. The TTFMS signals were not observed in the absence of Cl_2, and their rise and decay lifetimes were qualitatively the same as that of the UV signal; the signals can therefore not be due to any longer–lived species such as HCl or H_2O_2. Because there were no carbon–containing species in the mixture, the modulated near–IR signal must thus arise from HO_2.

The near–infrared line strengths for the measured lines were calculated using the HO_2 concentrations determined from the MMS ultraviolet measurements. To measure the near–IR optical densities from the TTFM signals, a calibration signal was obtained using a strong water line whose optical density was measured in direct absorption mode. Measuring at lower signal strengths was then accomplished by removing calibrated attenuation steps. TTFM has already been demonstrated to be linear in signal strength over several orders of magnitude (3).

Measurements of the OD of HO_2 lines were corrected for variation in laser power (I_o) at different currents and temperatures. However, because the TTFMS calibration was done using an H_2O absorption, an additional responsivity correction had to be made because of differing HO_2 and H_2O linewidths (14). That is, the TTFM signal strength for a given $\frac{I_o}{I}$ depends on the frequency of the modulation ν_m in comparison to the absorption feature linewidth Γ. For H_2O $\Gamma_{fwhm} = 579$ MHz which, with our $\nu_m = 525$ MHz modulation frequency, results in a 9.0% lower sensitivity than for HO_2 $(\Gamma_{fwhm} = 428$ MHz) based on the analysis of Cooper and Warren (14). A total of more than 20 lines have been measured (15), but in this paper we report only the six strongest lines thus far recorded (Table I). The FTIR wavelength measurements have only recently been obtained and will be used to help assign the lines in a forthcoming publication.

Absorption λ (cm^{-1}) *	Cross section σ (cm^2 molec^{-1})	Absorption λ (cm^{-1}) *	Cross section σ (cm^2 molec^{-1})
6629.011	6.2 x 10^{-20}	6625.636	9.4 x 10^{-20}
6627.993	7.3 x 10^{-20}	6625.263	5.6 x 10^{-20}
6625.777	10.4 x 10^{-20}	6623.569	7.0 x 10^{-20}

* Wavenumbers are accurate to approximately 0.02 cm^{-1}.

Table I. Near–infrared absorption cross sections for some of the strongest measured lines within the $2\nu_1$ band of HO_2.

IV. DETECTION LIMITS

The cross sections reported in Table I are, at first glance, large for overtone transition lines. However, spectroscopic studies of the ν_1 fundamental (6) band strength have shown that the lines in the fundamental are weak, and this might be an instance where the overtone band strength is of the same order of magnitude as the fundamental (15). This is important since it enhances the possibility of using these near–IR lines for atmospheric monitoring.

We chose to investigate the detection limits by employing one of the strongest lines ($\tilde{\nu} = 6625.636$ cm^{-1} in the Table, peak on the right in Figure 2). A relatively small [3.6 x 10^{11} molec cm^{-3}] concentration of HO_2 was generated using a single black lamp for a "reference spectrum", Trace A of Figure 3. A second spectrum

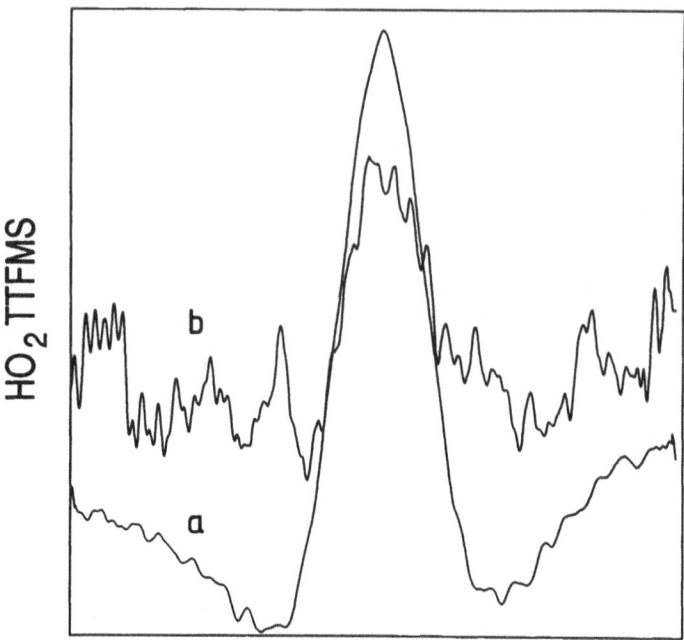

Figure 3. Detection limit test of HO_2 sensitivity. Trace A is due to 3.6×10^{11} HO_2 cm^{-3}, and served as a calibration for Trace B. Trace B is due only to sunlight Cl_2 photolysis. Both spectra recorded at 47 torr (Doppler broadened coniditions).

was then recorded using only late afternoon sunlight for the Cl_2 photolysis, and a dark laboratory spectrum subsequently subtracted. The spectrum thus generated is Trace B of Figure 3. A least–squares fit of the two spectra then determined the concentration in Spectrum B to be $[1.6 \pm 0.3] \times 10^{10}$ molec cm^{-3} observed with a signal–to–noise ratio of $\sim$10 at a 47 torr measurement pressure. The detection limit of our system is therefore estimated to be approximately 1.5×10^9 molec cm^{-3} ($\Delta f \approx 1$ Hz) for these low pressure measurements.

Hydroperoxyl concentrations in the tropics have predicted maxima of 5×10^8 molec cm^{-3} ($\sim$ 20 pptv) (16). This implies that in order to make effective measurements of HO_2, a detection limit of at least 1×10^8 molec cm^{-3} is needed, and furthermore preferably at tropospheric pressures. Hence, in order to utilize the present spectrometer under tropospheric field conditions, two improvements must be realized: 1) approximately one order of magnitude increase in detection sensitivity, and 2) achievement of the same detection limit (in terms of optical density) at higher measurement pressures.

Improvements in sensitivity are clearly possible. The signal–to–noise ratios depicted in Figure 3 were in part limited by an étalon within the White cell; oscillating the mirrors at a dither frequency has proven effective at spoiling such étalon fringes (17,18). Secondly, an increase in the path length (up to a kilometer) should be realizable outside the laboratory by using a longer base path White cell. Finally, for shot–noise limited measurements, increasing the power at the detector by a factor N increases the sensitivity by a factor $\sqrt{N}$ (19); although the laser monomode powers are not likely to radically increase in the short term, more efficient recovery optics and dielectrically enhanced mirror coatings could increase the detected power.

We are presently modifying the spectrometer to measure at pressures closer to those found in the free troposphere. As discussed earlier, Cooper and Warren (14) have shown that if the high frequency modulation ν_m is greater than the linewidth Γ_{fwhm} there is no loss of TTFM responsivity. Thus, so long as the high frequency can be effectively coupled to the diode, the RF frequency can be increased to recover the full signal intensity as the linewidth expands due to pressure broadening. This points to the possible advantage of measuring HO_2 in the near–IR as opposed to the mid–IR: Because there are fewer interferences in the near–infrared, spectra can be recorded at higher pressures before interference from other lines becomes important. This in turn suggests use of open long–path (cell) measurements, which also avoids the handling problems associated with reduced pressure measurements.

V. SUMMARY

We have constructed a TTFM spectrometer using a 1.5 μm InGaAsP laser. The spectrometer employs a White cell for the near–IR path and an ultraviolet system for molecular modulation spectroscopy. Modulated photolysis was used to produce the hydroperoxyl radical, HO_2, and the absorption cross sections σ along with the accurate wavenumbers $\tilde{\nu}$ of several lines in the $^2A''(200) \leftarrow {}^2A''(000)$ band have been determined for the first time. The cross sections are relatively large for overtone transitions, but can be understood in terms of the nature of the fundamental. The detection limits thus far obtained suggest that tropospheric monitoring using these lines remains difficult but may be feasible.

BIBLIOGRAPHY

1. Carlisle, C.B.; Cooper, D.E., Tunable diode laser frequency modulation spectroscopy through an optical fiber: High sensitivity detection of water vapor, *Appl. Phys. Lett.* <u>56</u>, 805, (1990).

2. Stanton, A.C.; Silver, J.A., Measurements in the HCL 3←0 band using a near–IR InGaAsP diode laser, *Appl. Opt.* <u>27</u>, 5009, (1988).

3. Cooper, D.E.; Carlisle, C.B., High–sensitivity frequency modulation spectroscopy with lead–salt diode lasers, in "Monitoring of Gaseous Pollutants by Tunable Diode Lasers, Freiburg Symposium 1988", R. Grisar et al., eds. Kluwer Academic Publishers, Dordrecht, Holland, p. 180, (1989).

4. Carlisle, C.B.; Cooper, D.E.; Preier, H., Quantum noise–limited FM spectroscopy with a lead–salt diode laser, *Appl. Opt.*, <u>28</u>, 2567, (1989).

5. Paukert, T.T. and Johnston, H.S., Spectra and kinetics of the hydroperoxyl free radical in the gas phase, *J. Chem. Phys.*, <u>56</u>, 2824, (1971).

6. Zahniser, M.S.; McCurdy, K.E.; Stanton, A.C., Quantitative spectroscopic studies of the HO_2 radical: band strength measurements in the ν_1 and ν_2 vibrational bands, *J. Phys. Chem.* <u>93</u>, 1066, (1989).

7. Zahniser, M.S.; Stanton, A.C., A measurement of the vibrational band strength for the ν_3 band of the HO_2 radical, *J. Chem. Phys.*, <u>80</u>, 4951, (1984).

8. Hunziker, H.E.; Wendt, H.R., Near infrared absorption spectrum of HO_2, *J. Chem. Phys.* <u>60</u>, 4622, (1974).

9. Hunziker, H.E.; Wendt, H.R., Electronic absorption spectra of organic peroxyl radicals in the near infrared, *J. Chem. Phys.*, <u>64</u>, 3488, (1976).

10. Johnson, T.J.; Wienhold, F.G.; Burrows, J.P.; Harris, G.W., Frequency modulation spectroscopy at 1.3 μm using InGaAsP lasers: a prototype field instrument for atmospheric chemistry research, *Appl. Opt.*, <u>30</u>, 407, (1991).

11. Hillmer, H.; Hansmann, S.; Burkhard, H., Realization of high coupling coefficients in 1.53 μm InGaAsP/InP first–order quarter–wave shifted distributed feedback lasers, *Appl. Phys. Lett.* <u>57</u>, 534, (1990).

12. Demore W.B.; Sander, S.P.; Golden, D.M.; Molina, M.J.; Hampson, R.F.; Kurylo, M.J.; Howard, C.J.; Ravishankara, A.R., "Chemical kinetics and photochemical data for use in stratospheric modeling," JPL Publication 90–1, Jet Propulsion Laboratory, Pasadena, California, USA (1990).

13. Crowley, J.N.; Simon, F.G.; Burrows, J.P.; Moortgat, G.K.; Jenkin, M.E.; Cox, R.A., The HO_2 radical UV absorption spectrum measured by molecular modulation, UV/diode–array spectroscopy, *J. Photoch. A,* <u>60</u>, 1, (1991).

14. Cooper, D.E.; Warren, R.E., Two–tone optical heterodyne spectroscopy with diode lasers: theory of line shapes and experimental results, *J. Opt. Soc. Am.* <u>B 4</u>, 470, (1987).

15. Johnson, T.J.; Wienhold, F.G.; Burrows, J.P.; Harris, G.W.; Burkhard, H., Measurements of line strengths in the HO_2 ν_1 overtone band at 1.5 μm using an InGaAsP laser, *J. Phys. Chem.* <u>95</u>, 6499, (1991).

16. Kanakidou, M.; Singh, H.B.; Valentin, K.M.; Crutzen, P.J., A two–dimensional study of ethane and propane oxidation in the troposphere, *J. Geophys. Res.* <u>D8</u>, 15395, (1991).

17. Webster, C.R., Brewster–plate spoiler: A novel method for reducing the amplitude of interference fringes that limit tunable–laser absorption, *J. Opt. Soc. Am.* <u>B2</u>, 1464, (1985).

18. Silver, J.A.; Stanton, A.C., Optical interference reduction in laser absorption experiments, *Appl. Opt.* <u>27</u>, 1914, (1988).

19. Werle P.; Slemr F., Signal–to–noise ratio analysis in laser absorption spectrometers using optical multipass cells, *Appl. Opt.* <u>30</u>, 430, (1991).

LINE NARROWING AND FREQUENCY CONTROL OF LEAD-SALT DIODE LASERS BY OPTICAL FEEDBACK

M. MÜRTZ, M. SCHAEFER, M. SCHNEIDER, J. S. WELLS[1] and W. URBAN
Institut für Angewandte Physik der Universität Bonn
Wegelerstr. 8, D-5300 Bonn, F.R. Germany

U. SCHIESSL and M. TACKE
Fraunhofer-Institut für Physikalische Meßtechnik
Heidenhofstr. 8, D-7800 Freiburg, F.R. Germany

SUMMARY

We report on a frequency stabilization experiment with lead-salt diode lasers.
An optical feedback technique is used to narrow the diode laser linewidth. A small amount of laser light is retroreflected by an external feedback mirror. In this way narrowing of the linewidth by 2 orders of magnitude (down to 1-2 MHz) is achieved. Frequency stabilization and control of the diode laser is accomplished by frequency-offset locking to a sealed-off CO laser. TDL and CO laser outputs are heterodyned in a HgCdTe detector (1 GHz bandwidth). The resulting beatnote is employed to control the length of the external resonator via an electronic servo loop. The improved spectral properties of the diode laser provide a new tool for accurate frequency measurements and high-resolution sub-Doppler applications in the mid infrared.

1. INTRODUCTION

It is well known that diode lasers are extremely sensitive to small amounts of retroreflected light from external optical elements. Due to the low Q cavity and the strong carrier induced coupling between amplitude and phase of the electrical field, changes in the spectral characteristics of the diode laser occur even for power reflectivities less then 10^{-4}. Uncontrolled optical feedback e. g. from windows or lenses within the beam path can affect both frequency and amplitude stability. Often the sensitivity and resolution of a TDL spectrometer is limited due to optical feedback effects (optical noise).

In the past ten years there was an extensive investigation of effects induced by optical feedback with III-V semiconductor lasers in the near infrared (1)-(11). Stimulated by experiments on diode-to-fiber optical coupling, different groups experimentally investigated and theoretically analysed various feedback setups. Schemes utilizing feedback from mirrors and gratings as well as resonant feedback from a high-finesse cavity were studied. Depending on the feedback conditions (distance of reflector, reflectivity), people found different effects on the spectral properties of the laser radiation: changes of the mode

---

[1]Permanent address: Time and Frequency Division
National Institute of Standards and Technology (NIST)
325 S. Broadway
Boulder, Colorado 80303 USA

characteristics, linewidth broadening, enhanced intensity noise, dynamical instabilities (coherence collapse, self pulsations) and also linewidth reduction. The latter, so-called optical stabilization, was obtained with single-mode GaAlAs type lasers coupled to an external high-finesse resonator. This is a powerful way to reduce amplitude and frequency fluctuations up to very high Fourier frequencies[2].

As much less effort has been put into the investigation of Pb-salt diode lasers with respect to III-V semiconductor lasers, their performance is still much worse. For instance lead-salt lasers mostly exhibit multi-mode operation due to their partly homogeneous and partly inhomogeneous broadend gain profile (spectral and spacial hole burning). The object of our investigation was to find out which behaviour is exhibited by lead chalcogenide homostructure diode lasers with optical feedback and whether it is possible to use controlled optical feedback to narrow the linewidth of these lasers. In the following, the experimental setup and the method of the linewidth measurements are described. The results of a line narrowing experiment with optical feedback from an external mirror are presented. Finally we report on stabilizing the center frequency of the narrowed line by a frequency-offset locking technique.

2. EXPERIMENTAL DETAILS

Figure 1 depicts the laser housing. To ensure that we don't get any unintentional optical feedback due to uncontrolled reflections from external elements, we use some non-standard components. The laser crystal is fixed on a special mounting which allows a

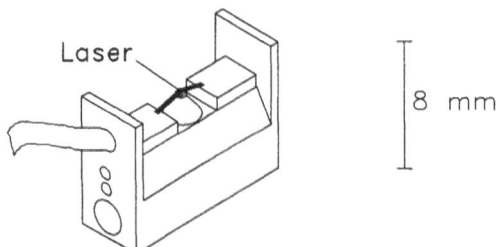

FIG. 1 Laser housing for bidirectional beam output

bidirectional laser beam output. This rules out the possibility of reflections from the backside of the laser as they may occur with the standard mounting which permits beam output in only one direction. Moreover twice the power is available. In our experiments one of the beams was used to examine the linewidth, with the second beam we applied external optical feedback to the laser. This work was done with two homostructure diode lasers which operate near 5 μm at temperatures between 20 K and 40 K. The threshold current is about 180 mA. We usually operated the lasers 40 to 120 mA above threshold

---

[2]The response time of the optical feedback loop is given by the distance of the reflector over the speed of light, i. e. a few nanoseconds.

which turned out to be a region with several single-mode operation regimes[3]. The output power was measured by means of a power meter and was in the range between 1 mW and 5 mW.

Vibration-free cooling is performed by a temperature controlled evaporation cryostat, which we modified for bidirectional beam output. The diode laser is mounted at the bottom of the heat exchanger in such a way that both beams emerge unhindered from the cold station. This cryostat can be operated both with liquid helium or liquid nitrogen. The temperature of the laser station is coarsely adjusted by the flow of the coolant. For accurate temperature stabilization we use an electronic temperature controller, employing electrical resistance heating in combination with a silicon diode as a temperature sensor.

To collimate the laser beams we use two off-axis parabola mirrors (focal length = 10.2 mm) which are located inside the vacuum chamber of the cryostat (figure 2).

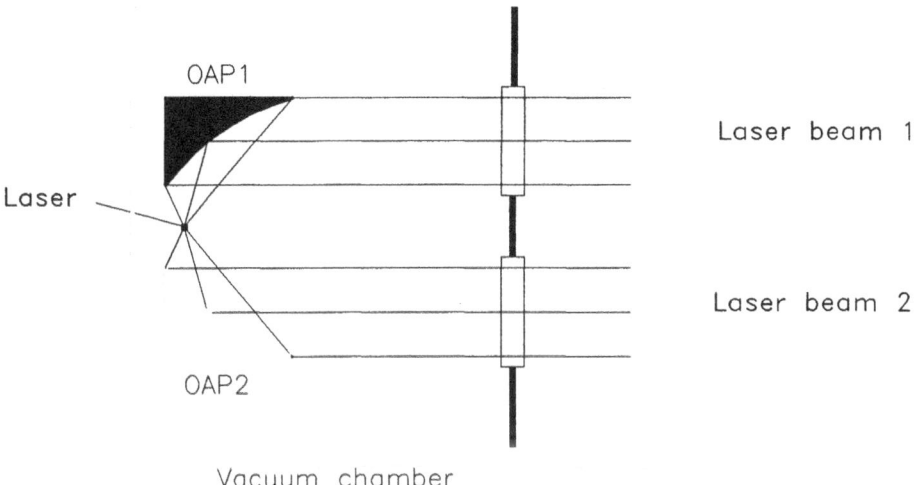

FIG. 2 Collimation of the beams by two Off-Axis-Parabolas

They are adjustable from the outside in the x-, y-, and z-direction. Both collimated beams (diameter: 20 mm) emerge from the cryostat through wedged and tilted CaF_2 windows.

3. LINEWIDTH MEASUREMENTS

We investigated the spectral linewidth of the diode laser by heterodyning the diode laser output with radiation from a sealed-off CO laser (local oscillator). This gas laser provides a few hundred laser lines in the wavelength region from 5 - 8 µm; the output power is 0.01 - 3 W single line and the linewidth is less than 1 MHz. A block diagram of the setup is shown in figure 3. The beams of both lasers were superimposed on a CaF_2

---

[3]The proper current/temperature values in order to run the laser nearly single mode at the desired frequency were found with the help of a modechart (12).

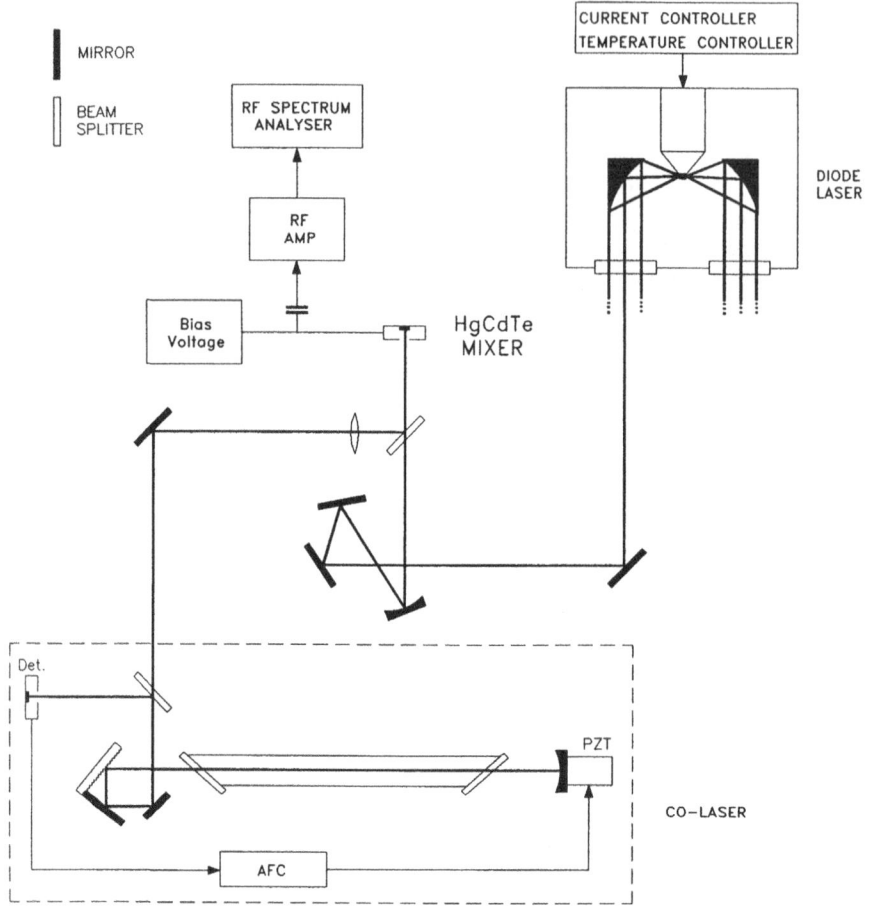

FIG. 3 Schematic setup of the linewidth measurements

beam splitter and focussed on a liquid-nitrogen-cooled photovoltaic HgCdTe detector/mixer. This was done by means of a lens (CO laser beam) and an astigmatically compensated toroidal mirror (diode laser beam) to avoid feedback from a lens to the diode laser. The detector bandwidth is about 1 GHz, under favourable conditions beatnotes up to 5 GHz can be observed. The detector was followed by a microwave amplifier (0.1 - 1.8 GHz, 30 dB). The output of the amplifier was fed to a spectrum analyser, which was mostly operated with 10 ms/div sweeptime.

A typical beatnote, recorded with the diode laser No. 214-B-7, is depicted in figure 4. The center frequency of the beatnote is near 1 GHz. Since the linewidth of the CO laser is less than 1 MHz, the FWHM of the beatnote, which is about 70 MHz, reflects the linewidth of the diode laser. For other temperatures and currents we observed even broader lines, especially in case of multi-mode operation. In earlier experiments with other homostructure diode lasers we found linewidths between 20 MHz and 100 MHz.

In contrast to these measurements the theoretical value for the linewidth, given by the modified Schawlow Townes formula (13), is about 0.1 - 5 MHz (depending on the α-value, output power, etc.). The causes for this huge discrepancy are not well-understood. To ensure that the current supply to the TDL is not responsible for the observed linewidth,

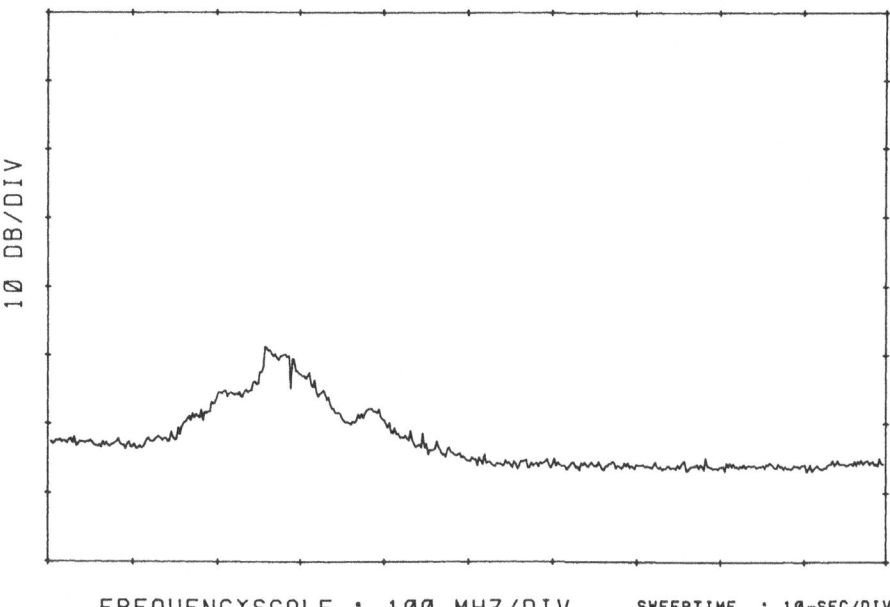

FIG. 4 Typical beatnote between a TDL (214-B-7) and a CO laser

it was carefully tested. It turned out that the ripple of the current supply corresponds to 1 - 2 MHz frequency jitter at most. Furthermore we took very much care in avoiding any possible external feedback to the diode, nonetheless one cannot exclude the possibility of slight optical feedback, e. g. coming from the (tilted) photomixer, which might broaden the line. Other authors report linewidths less than 1 MHz for carefully selected diodes (14). This dramatic variation in the linewidths of diode lasers of the same type may be due to a noise generating mechanism inherent to the TDL which produces the observed linewidth. We assume that mode competition noise and carrier density fluctuations are responsible for the linewidth in these gain-guided homostructure laser diodes. Also laser action in different lasing filaments with slightly different frequencies might contribute to the linewidth. By contrast, linewidth measurements of high performance GaAs laser diodes with better electron and photon confinement are in good agreement with the predictions calculated from the modified Schawlow Townes formula (13).

4. LINE NARROWING

In order to investigate the effect of controlled optical feedback on TDL linewidth we placed mirrors and other simple reflectors such as CaF_2 windows in a distance of about 60 cm from the diode into the second beam, while observing the beatnote on the spectrum analyser. The first noticable effects were different kinds of instabilities. The applied

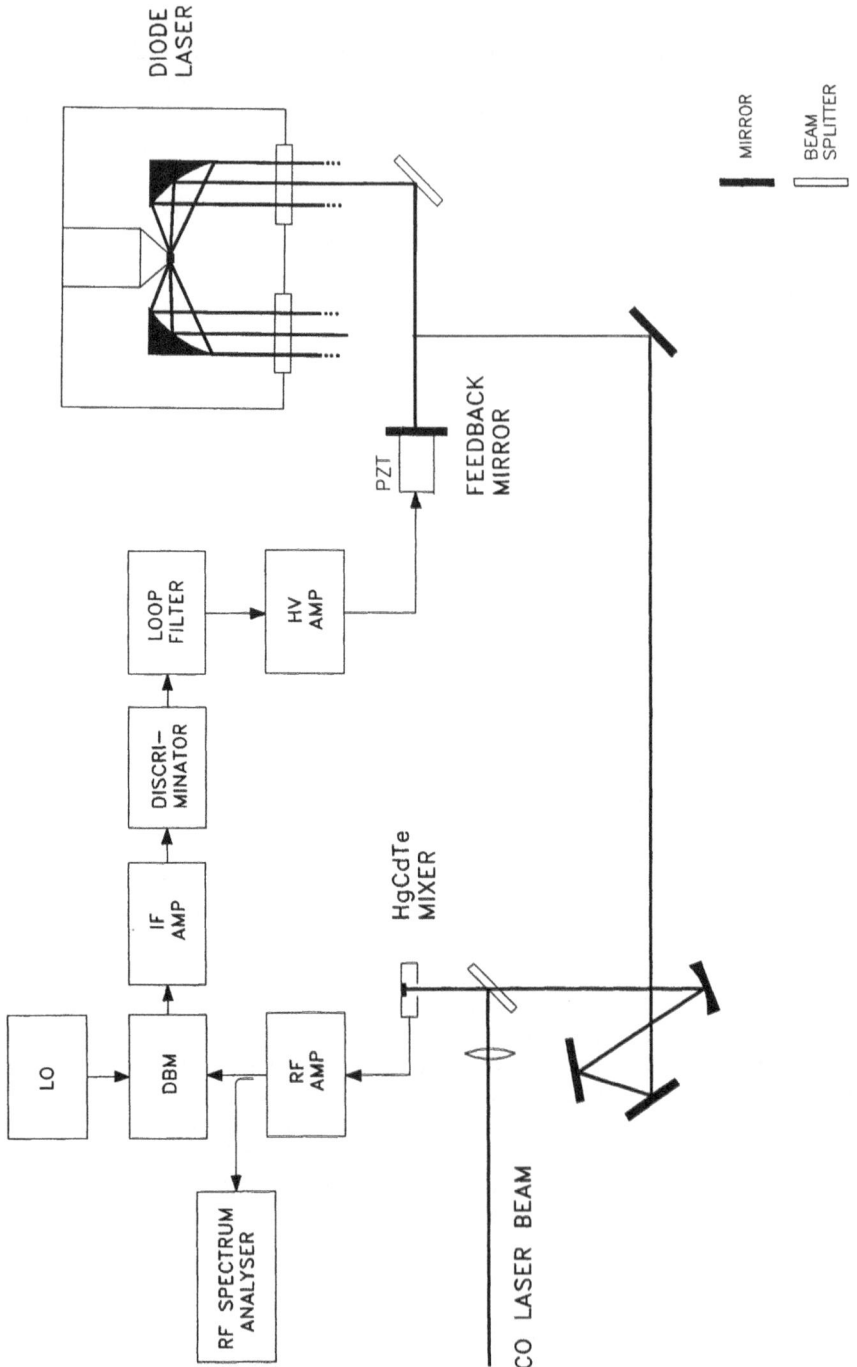

FIG. 5 Diagram of the frequency-offset locking scheme

feedback induced line broadening and in the worst case mode jumps which meant loss of the beatnote. In order to reduce the feedback level we employed a mirror in combination with a beam splitter as external reflector[4] (figure 5). With this setup we found dramatic linewidth reduction by two orders of magnitude down to 1-2 MHz. Figure 6 shows the narrowed beatnote. This linewidth reduction can be explained by the coupling of the diode laser cavity to the external cavity formed by the mirror and one facet of the laser crystal.

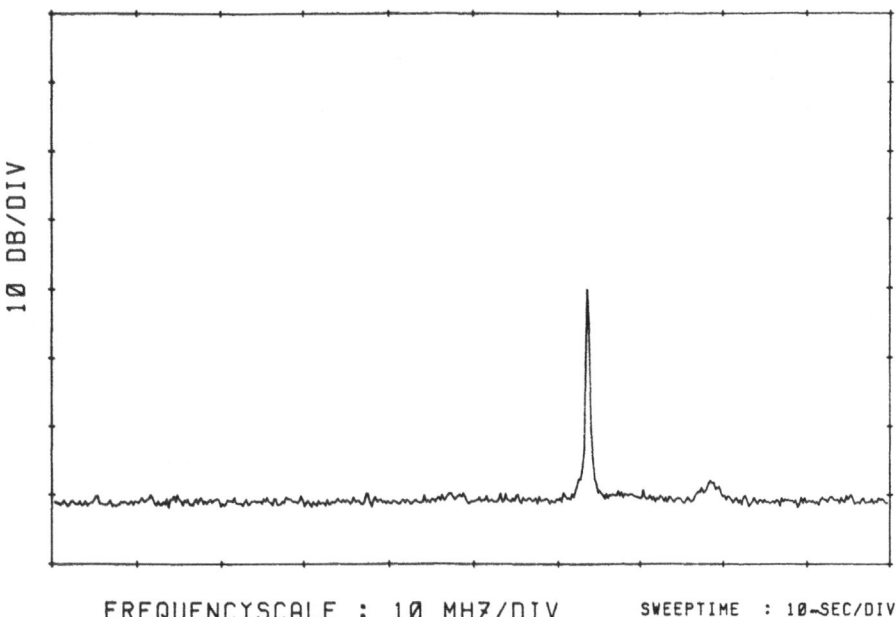

FREQUENCYSCALE : 1Ø MHZ/DIV   SWEEPTIME : 1Ø-SEC/DIV

FIG. 6 Beatnote after line narrowing by optical feedback

The line narrowing was accompanied with the appearance of external cavity modes: depending on the feedback strength we observed one to three or even more external cavity modes simultaneously with a FSR of about 200 MHz (corresponding to the 60 cm cavity length). It turned out that a defined coupling to the external resonator is not possible if the laser is working multimode, i.e. if there are two or more modes with nearly the same output power. However in the case of near single mode operation[5] the compound cavity laser behaved exactly as predicted by a rate equation model for extended cavity diode lasers (4,5,8,9).

An estimate of the feedback strength was obtained in the following manner. The theoretical analysis of the phase and gain conditions of a compound cavity shows that the tuning rate of the diode laser with optical feedback is decreased by a factor of

---

[4]This results in a feedback level below 10^{-3}.

[5]I. e. the power level of any other mode was less than 10 % of the dominant mode power level.

$$1 + X\sqrt{1 + \alpha^2}$$

compared to the solitary laser, where X = feedback strength parameter and α = linewidth broadening factor (5). We found tuning rates of 700 MHz/mA for the solitary laser and 160 MHz/mA for the laser with optical feedback. From these measurements we can deduce the feedback parameter X to be about 1.5 to 2.5 (depending on the α-factor). This corresponds to an external reflectivity of about 10^{-4}. These values are in good agreement with the results of a stability analysis of the rate equations, which predict a linewidth reduction in the case of weak optical feedback ($X \approx 1$) and the onset of instabilities, when a critical value for the feedback strength is exceeded ($X \gg 1$).

Tuning of the extended cavity diode laser can be performed in two different ways. Either one tunes the diode laser cavity by changing the current or one tunes the external cavity by changing its length. As discussed above, the current tuning rate of the extended cavity diode laser is reduced. Moreover we observed discontinuous tuning behaviour. The frequency of a mode can be changed - depending on the feedback strength - up to 50 MHz. When the current is tuned further, a mode hop occurs: the laser emission jumps to the next external mode (1 FSR to higher frequency with increasing current). On the other hand we can tune the laser frequency by changing the length of the external resonator. The feedback mirror can be moved a few microns with a piezo-ceramic transducer (PZT). This results in a continuous tuning range of about 200 MHz (= 1 FSR), followed by a mode hop into the opposite direction. To achieve longer continuous tuning ranges one can either decrease the length of the external resonator in order to increase the FSR or one can tune the external cavity with the PZT and change the current synchronously.

5. FREQUENCY STABILIZATION AND CONTROL

The acoustical noise in the laboratory introduces vibrations of the feedback mirror and the other optical components which are elements of the external cavity. This results in a jitter of the center frequency of the narrowed line, which is visible on the left side of figure 7, where the sweeptime is increased to 1 sec/div[6]. The center frequency jitters over a bandwidth of about 30 MHz. By a detailed experimental analysis we found that this frequency instability is produced by low frequency vibrations of the mirrors below 1 kHz. For the purpose of stabilization we installed a servo loop which offset-locks the TDL frequency to the reference laser frequency. A schematic diagram of the setup is shown in figure 5. This frequency-offset locking technique was used in a different approach by Freed at al. (14).

The beatnote is down-converted by means of a double-balanced mixer and a frequency synthesizer as local oscillator. The frequency of the synthesizer v_{syn} is adjusted such that the frequency of the down-converted beatnote is roughly 160 MHz. The intermediate frequency (IF) signal is amplified by 70 dB and fed to a discriminator. The difference between the IF and the center frequency of the discriminator (160 MHz) is converted to an error signal with a sensitivity of 100 mV/MHz. The loop filter contains an amplifier with low pass characteristics and an integrator. Moreover we can introduce a

---

[6]The peak on the right side of figure 6 is due to the local oscillator described below.

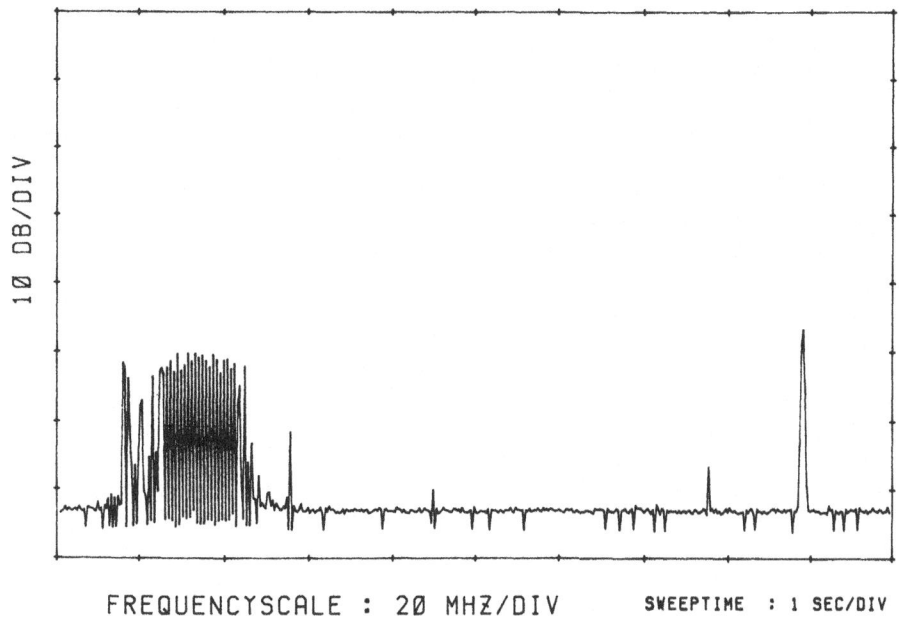

FREQUENCYSCALE : 20 MHZ/DIV   SWEEPTIME : 1 SEC/DIV

FIG. 7 Narrowed beatnote displayed with 1 sec/div sweeptime

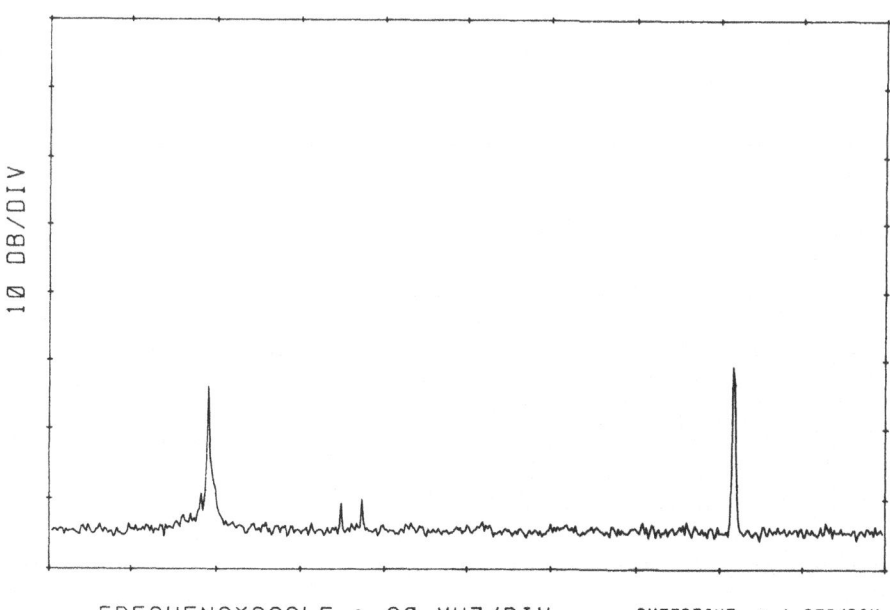

FREQUENCYSCALE : 20 MHZ/DIV   SWEEPTIME : 1 SEC/DIV

FIG. 8 Narrowed beatnote of the offset-locked diode laser

variable offset voltage to the lock loop in order to sweep the frequency of the TDL within the bandwidth of the discriminator (65 MHz). The signal is finally amplified to a high voltage level and fed to the PZT. When the servo loop is closed, the TDL frequency is locked to the reference CO laser frequency[7] with a variable offset of $v_{syn} \pm 160$ MHz, provided that there is no additional offset in the loop. The sign depends on whether we choose v_{syn} either above or below the beatnote frequency. A preliminary result is depicted in figure 8, where the same sweeptime (1 sec/div) was used as in figure 7. The former frequency jitter has disappeared. Optimizing the lock features is still in progress, some improvements concerning the stability of the lock have to be made.

6. CONCLUSION AND PLANS

Our experiments pointed out that it is possible to achieve linewidth reduction for a lead-salt diode laser by applying controlled optical feedback. We combined this method with frequency stabilization and control by offset-locking the TDL to a CO laser. This tunable narrow-linewidth light source in the mid infrared will be a new tool for high resolution spectroscopy with sub-Doppler accuracy. By introducing one or two Lamb dip stabilized CO_2 lasers, we gain the capability to make very accurate frequency measurements (15). We will extend our investigations to shorter wavelengths (3 µm) and to longer wavelengths (10 µm) and also to different diode laser structures (DH-, BH-laser, etc.). Further experiments with improved optical feedback elements like a reflection grating or an external high-finesse cavity are in preparation.

ACKNOWLEDGEMENTS

One of us (JSW) would like to thank the Alexander von Humboldt Foundation for an Award leading to this further IAP/NIST collaboration. We are also indebted to the Upper Atmospheric Research Office of NASA for continued support of heterodyne techniques with tunable diode lasers.

REFERENCES

(1) R. LANG AND K. KOBAYASHI
"External Optical Feedback Effects on Semiconductor Injection Laser Properties"
IEEE J. Quantum Electron. QE-**16**, 347 (1980)
(2) L. GOLDBERG, H.F. TAYLOR, A. DANDRIDGE, J.F. WELLER AND R.O MILES
"Spectral Characteristics of Semiconductor Lasers with Optical Feedback"
IEEE J. Quantum Electron. QE-**18**, 555 (1982)
(3) S. SAITO, O. NILSSON AND Y. YAMAMOTO
"Oscillation Center Frequency Tuning, Quantum FM Noise, and Direct Frequency
    Modulation Characteristics in External Grating Loaded Semiconductor Lasers"
IEEE J. Quantum Electron. QE-**18**, 961 (1982)
(4) C.H. HENRY AND R.F. KAZARINOV
"Instability of Semiconductor Lasers due to Optical Feedback from Distant Reflectors"

---

[7]The CO laser was line-center locked by means of a standard frequency modulation technique

IEEE J. Quantum Electron. QE-**22**, 294 (1986)

(5) H. OLESEN, J.H. OSMUNDSEN AND B. TROMBORG
"Nonlinear Dynamics and Spectral Behaviour for an External Cavity Laser"
IEEE J. Quantum Electron. QE-**22**, 762 (1986)

(6) B. TROMBORG, J.H. OSMUNDSEN AND H. OLESEN
"Stability Analysis for a Semiconductor Laser in an External Cavity"
IEEE J. Quantum Electron. QE-**20**, 1023 (1984)

(7) B. DAHMANI, L. HOLLBERG AND R. DRULLINGER
"Frequency Stabilization of Semiconductor Lasers by Resonant Optical Feedback"
Opt. Lett. **12**, 876 (1987)

(8) PH. LAURENT, A. CLAIRON AND CH. BREANT
"Frequency Noise Analysis of Optically Self-locked Diode Lasers"
IEEE J. Quantum Electron. QE-**25**,1131 (1989)

(9) H. LI AND N.B. ABRAHAM
"Analysis of the Noise Spectra of a Laser Diode with Optical Feedback from a High-Finesse Resonator"
IEEE J. Quantum Electron. QE-**25**, 1782 (1989)

(10) H. LI AND H.R. TELLE
"Efficient Frequency Noise Reduction of GaAlAs Semiconductor Lasers by Optical Feedback from an External High-Finesse Resonator"
IEEE J. Quantum Electron. QE-**25**, 257 (1989)

(11) J.S. COHEN, F. WITTGREFE, M.D. HOOGERLAND AND J.P. WOERDMAN
"Optical Spectra of Semiconductor Laser with Incoherent Optical Feedback"
IEEE J. Quantum Electron. QE-**26**, 982 (1990)

(12) M. PETRI, T. FINK AND W. URBAN
"New Development in Computer-Controlled Diode Laser Spectroscopy"
This volume

(13) C.H. HENRY
"Theory of the Linewidth of Semiconductor Lasers"
IEEE J. Quantum Electron. QE-**18**, 259 (1982)

(14) C. FREED A, J.W. BIELINSKI AND W. LO
"Programmable, secondary frequency standard based infrared synthesizer using lead-salt diode lasers"
Proc. SPIE **438**, 119 (1984)

(15) A.G. MAKI AND J.S. WELLS
"Wavelength Calibration Standards from Heterodyne Frequency Measurements"
NIST Special Publication 821 (approx 700 pages), Nov. 1991

FIBER OPTIC ACCESSORIES FOR MOLECULAR SPECTROSCOPY AND GAS
ANALYSIS WITH TUNABLE DIODE LASERS IN THE MIDDLE INFRARED

E.V.STEPANOV, A.I.KUZNETSOV,
K.L.MOSKALENKO, A.I.NADEZHDINSKII
Institute of General Physics of the Academy
of Science of the USSR
38 Vavilova Street, 117942 Moscow,USSR

Summary

The results on middle infrared fiber optic component usage
for spectroscopy and analysis with A4B6 tunable diode
lasers are presented. Spectral parameters of the fibers
available for middle IR are described. The units for
integration of the lasers with fibers and their cooling
were developed as well as various fiber optic analytical
parts of the spectrometers intended to solve different
analytical problems. As a result complete spectroscopic
systems on the base of middle IR fibers and Pb-salt diode
lasers was created. Further development of presented
systems with laser radiation transfer by fibers is
promising for applications in remote monitoring,
multicomponent gas mixtures analysis and evanescent wave
spectroscopy of gases in nearest future

1.INTRODUCTION
    A current success in development of middle infrared
fibers (MIRF) with low optical losses is inducing now a new
tendencies in design of devices based on high resolution
IR-spectroscopy. A new generation of chemical sensors,
analyzers and spectrometers using new fiber-design approaches
becomes real when fibers are coupled with wide spectral range
tunable sources in middle IR. Incorporation of the fibers could
joins advantages of high resolution spectroscopy and fiber
optics and originates to additional functional possibilities of
such apparatus, e.g. their use in remote control, evanescent
wave spectroscopy and multicomponent analysis. This also could
give appreciable opportunity to transfer to some extent
technologies and principals developed previously for silicon
glass fibers in visible and near IR-regions to spectroscopy of
middle IR range. So due to progress in this unification the
generation of sophisticated spectral apparatus with new
consumption advantages could appear in several years.
    Our efforts in this field were concentrated on integration
of MIRF with Pb-salts TDL. The last ones are used now very
intensively for high resolution molecular spectroscopy as well

as for high sensitive analysis of gas mixtures (1,2). Our aim consisted in development of complete spectroscopic systems on the base of TDL and MIRF. It suggested creation of a special source-receiver unit for generation and detection of tuned IR radiation as well as an analytical units both oriented on the use of MIRF for radiation transmission.

2.MIDDLE INFRARED FIBERS

Practically the whole of middle IR region in which this TDL are usually used (from 3 up to 20 microns) is fully covered by two kinds of fibers having now good consumer parameters. The first ones are chalcogenide glass ($As_x S_{x-1}$) fibers with minimal optical losses region from 1 to 6 microns. Such fibers are produced now in the USSR by the Institute of Chemistry of High-Purity Substances, Nizshnii-Novgorod, in collaboration with the General Physics Institute (GPI), the Laboratory of Dr. V.G.Plotnichenko (3). Within the above mentioned spectral region optical losses of the $As_x S_{x-1}$ fibers are down to 200

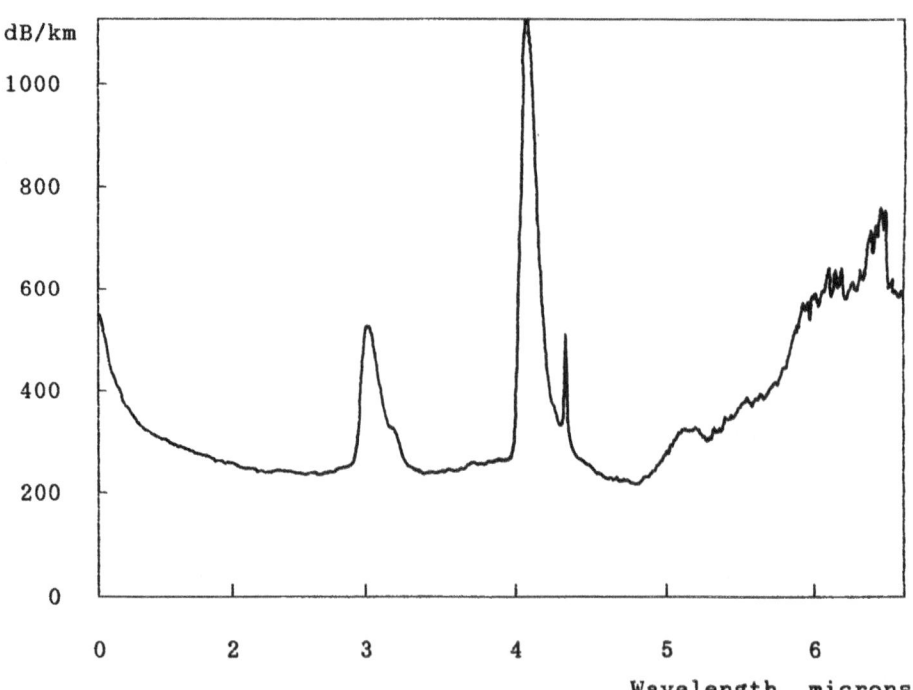

Fig.1.The spectrum of optical losses in chalcogenide glass fibers.

dB/km (Fig.1). Fibers with cores of diameters from 150 to 500 microns are now available. Usually the fibers have waveguiding cladding and plastic coating to improve elastic properties and to prevent their mechanical damage.

The second type of IR-fibers - so called MIR-fiber - are

polycrystalline fibers, extruded from crystals of solid
solutions of silver halides (AgCl-AgBr) (4). These fibers are
developed in the GPI, the Laboratory of Dr. V.G.Artjushenko,
and now they are commercially available from
Soviet-German-American joint venture "Ceram-Optec Systems",
Moscow, with rather small optical losses - 200-500 dB/km in the
region of spectra from 6 to 16 microns, Fig.2. These non-toxic,
non-hygroscopical and non-brittle MIR-fibers could be

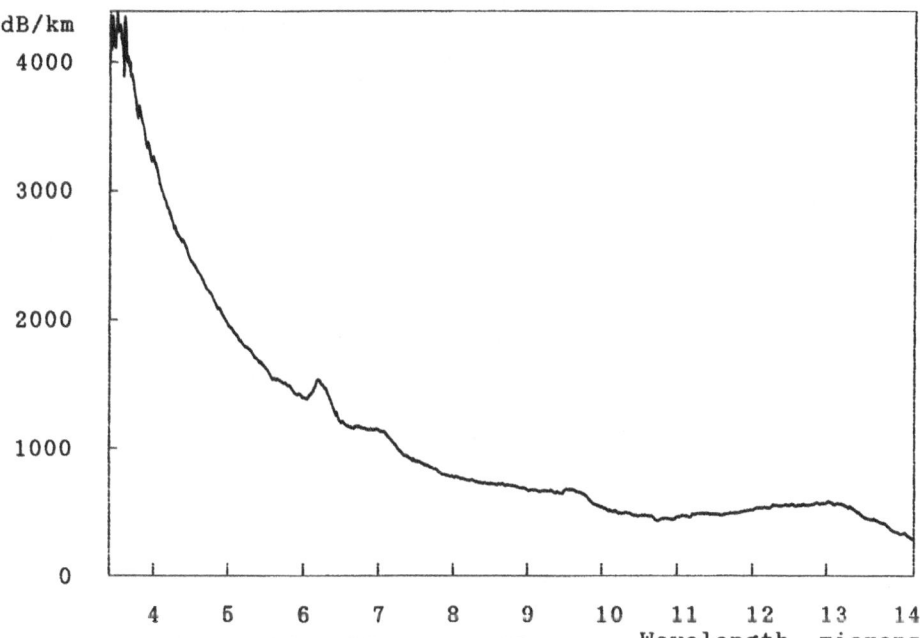

Fig.2.The spectrum of optical losses in halide crystal fibers.

fabricated either as bare core fiber in loose polymer tube, or
core-clad structure in polymer or metal coating (5). In spite
of little bit higher optical losses in core/clad fibers (600
dB/km at 10.6 microns) in comparison with bare core fibers they
possess by such advantages as twice smaller numerical aperture,
better protection from ambient environment leading to longer
life of fiber-optic zonde (6).

3.TUNABLE DIODE LASERS AND DETECTORS
      Pb-salts based TDLs were used as sources of transmitted IR
radiation in this work. Part of such lasers have been
fabricated by diffusion techniques in the Laboratory of Narrow
Gap Semiconductors, the Lebedev Physics Institute, Moscow,
headed by Prof.A.P.Shotov and others have been produced in the
Tbilisi State University, Georgia, by O.I Davarashvilli. It
should be noted that the second ones have been created by
liquid phase epitaxy with control of carriers concentration
profile ($^{+}$PbSe-nPbSe-p^{+}PbSe) for 7 micron spectral region and
with double heterojunction PbSe/PbSnSeTe for 8.5 micron region.

These technologies have provided slow wavelength tuning of about $10^3 \mathrm{cm}^{-1}$/s and radiation power of about some hundreds micro watts in single modes when pulse generation regime was used.

Of cause wavelength tuning the diode lasers were pumped by current pulses in our experiments. As our experience has shown the pulse generation regime is very convenient for spectroscopy and gas analysis with TDL. It's due to both fast rate of laser frequency tuning which permits more efficient noise selection and soft criteria on laser threshold parameters at higher operational temperatures. Amplitude and duration of the laser pulses were usually varied from 0.5 to 8 Amperes and from 5 microseconds to 10 milliseconds, respectively. Their repetition rate was usually about 100-200 Hz. At such regimes of generation an emitted wavelength is tuned during the pulse due to laser crystal heating by current. Starting wavelength usually depends on semiconductor composition and temperature of laser. A tuning range covered by a laser in a pulse depends on duration of the pumping current pulse and value of laser threshold current.

To detect IR laser radiation returned back from investigated object also by fibers, different photodiodes based on InSb or HgCdTe compounds were used. The first ones were used to cover spectral region from visible to 5.4 microns, the second for detection in the region up to 12 microns. Time constants of the used detectors were usually less than 50 ns. Detectors with facets size of more than 1 mm were generally

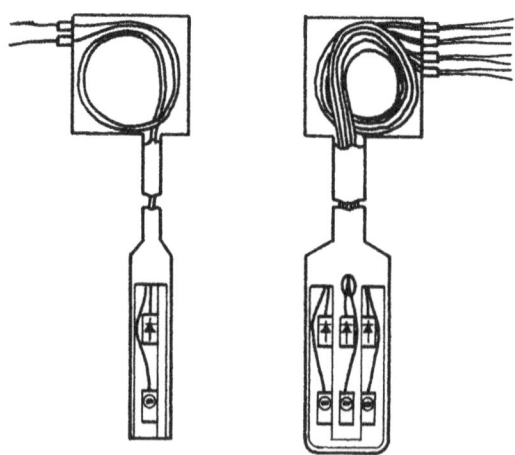

Fig.3.The systems for coupling of tunable diode lasers and IR detectors with IR fibers providing their cooling down to cryogenic temperatures.

used to provide their easy efficient coupling with wide core
fibers.

Mentioned TDLs and IR detectors operate mainly at
cryogenic temperatures. So a special units for cooling of TDLs
and detectors coupled with fibers were developed (7) (Fig.3).
They provide both exact positioning of all the elements to
obtain efficient input and output of radiation and lasers
temperature control. Several pairs (up to four) of TDLs on
special holders and detectors coupled with fibers could be
mounted inside the unit shown on Fig.3, right. Such
configuration gives possibility to develop a multiwavelength
source-receiver laser system on this base convenient for
multicomponent gas analysis. A cooled part of the unit is
immersed into the Dewar vessel intended for transportation of
cryogenic liquids. It ensures long time (several weeks)
continuous operation of the system without handling, long time
stability of temperature and convenience in system usage.

Electrical signals obtained with the detectors are
preamplified by proper electronics at the warm top of the unit.
Operation of this IR source-receiver system is supported by
special hardware driven by the IBM compatible computer (8). It
provides laser temperature control, pulse current pumping of
lasers, selection of parameters of detected electric signals
(control of radio-frequency band and amplification
coefficient), fast one-pulse registration of analog signals
containing information of transmission spectra, their storage
and development. Electronic unit consists of modules joint by
an eight bit data bus. These modules include (see Fig.4):
-interface card allowing direct computer access to the unit
data bus, and the corresponding card for IBM computer;
-temperature controller to stabilize diode laser temperature
with the accuracy up to 1 mK under computer control;
-laser drive unit providing complete control of the laser

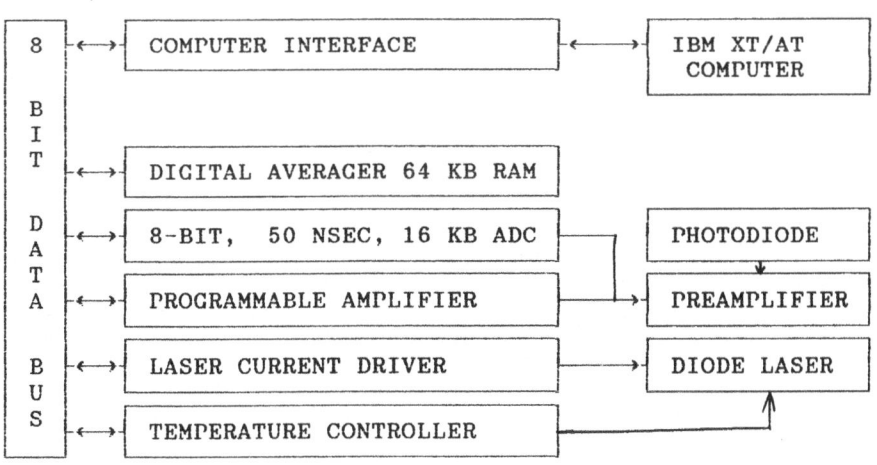

Fig.4. Laser control and data registration system.

current parameters;
-programmable high bandwidth amplifier for analog processing;
-8-bit, 50 nsec ADC with 16 KB memory;
-high speed digital signal averager with 32 bit word length and 16 K word memory.
So the above described unit gives opportunity to generate simultaneously laser radiation of several different wavelengths with tunable frequencies, to transmit them to investigated object for it's remote spectral analysis and to return the transmitted radiation back for detection and development. On Fig. 5 the general scheme of alone spectral channel diode laser spectrometer using above described design principals is shown.

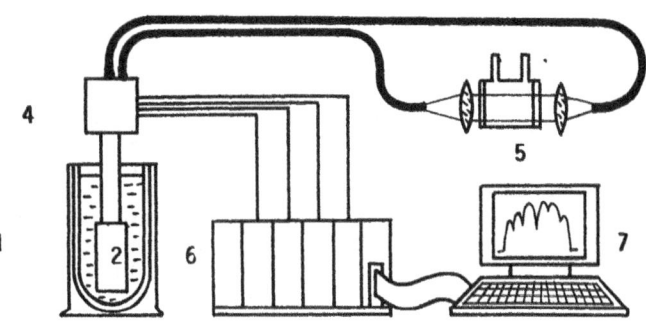

Fig.5.The general scheme of the diode laser based fiber optic spectrometer. 1-nitrogen Dewar vessel; 2-fiber optic cryostate with diode laser and photodetectors; 4-preamplifier; 5-analytical gas cell; 6-electronic unit; 7-IBM computer.

4.ANALYTICAL GAS CELLS
    In our experiments several modifications of analytical part for fiber-optical spectrometers or analyzers using described source-receiver systems with TDLs have been verified (see Fig.6).There was an alone-pass gas cell with fluoride used as lenses collimating optics (9,10). A short hollow core waveguiding cell made from stainless steel and containing focon element for collimation of light on a facet of receiving fiber have also been tested. A TDL-based fiber-optical analyzer with open atmosphere optical path and retroreflection mirror returning radiation back to the system for detection was tested too (11).
    For further discussion in details a spectrometer with

multipass gas cell used as an analytical part is the most
interesting version of such sensors. Special accessories have
been developed to realize operation of multipass gas cell using

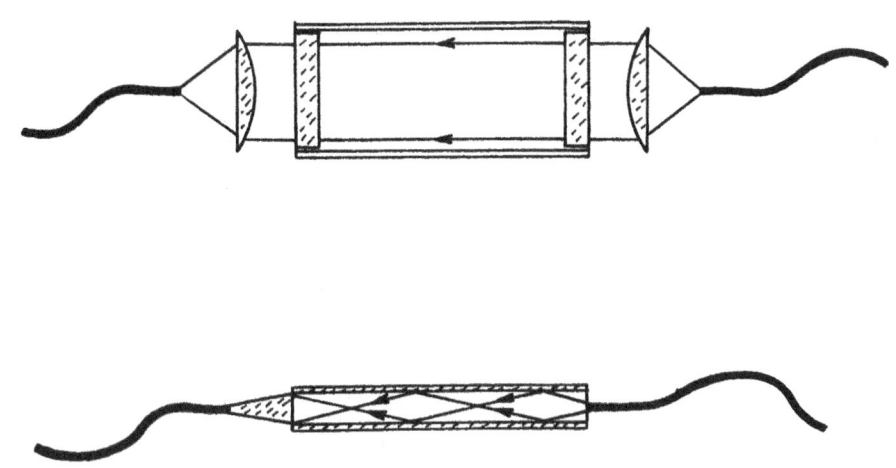

Fig.6.Tested optical schemes for gas analysis in which fibers
      were used for transmission of tunable diode laser
      radiation.

fiber-optic input and output of IR-radiation (Fig.7). For this
purpose a multipass cell based on generally used White-cell

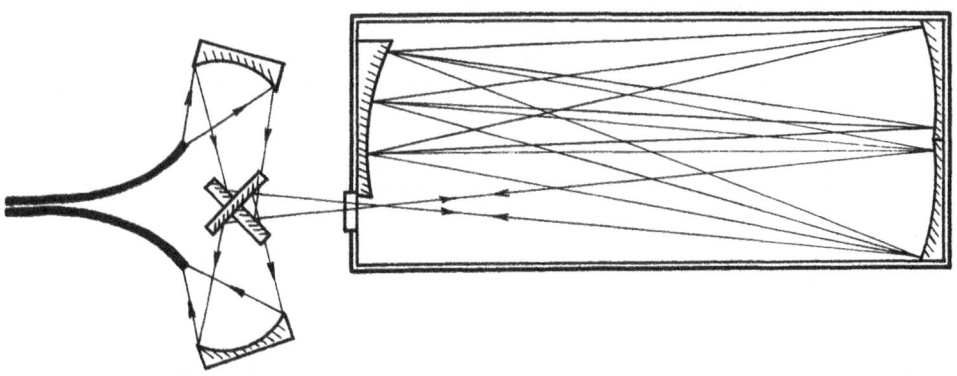

Fig.7.The multipass gas cell with fibers input and output of IR
      radiation used as analytical part of the chemical sensor
      on the base of tunable diode lasers.

scheme with base length of 25 cm, 35 millimeter objectives of F:8 and of 1,5 liter volume has been modified to Barskaia (12) scheme. In this modification of White-cell the same confocal mirrors optics is used, but incoming and outcoming laser beams are located at the same side of collimating mirror of the cell, see Fig.7. In the Figure they differ in height. Such way of alignment as known provides lower off axis aberrations due to more strict their compensation when beams pass the cell in opposite directions.

In the both output and input channels two identical spherical aluminum mirrors were used to focus radiation. Diameter and focal distances of the mirrors have been selected accordingly to fit divergency of beam coming from the source fiber to the geometry of the cell. Unfortunately it's very difficult to use laser radiation completely without strong off axis aberration in such scheme. So we've used only a half of radiation available with divergency of not more than 30 degrees. This solution ensures also not so large image of fiber facet on a plane of the collective mirror of the cell. This dimension becomes an important parameter of cell alignment because it determines the maximum number of passes through the multipass cell. For the same reason fibers of different core dimensions were used at outlet and inlet of the cell. The narrowest fiber of about 130 microns in core diameter have been used in source channel. It has resulted to the dimension of laser beam spots on the collective mirror of about 520 microns. At 30 millimeters across length of the mirror it ensured about 118 passes of laser beam without spots interference. At the receiving channel a fiber with 450 microns in core diameter was used to reduce the influence of laser beam distortion due to spherical and off axis aberrations and to collect as much radiation as possible.

One important advantage of multipass cell with fiber-optic output and input should be noted here. The cell is very convenient in alignment when only mirror optics is used and silica multimode fiber with He-Ne laser visible radiation is applied for this purpose. In this case it appears to be very easy to make a chose of right focal positions of all mirrors of the cell and to count the number of passes through the cell because of adequate modeling of IR-radiation source in visible light.

5.OPTICAL LOSSES AND SIGNAL TO NOISE RATIO

As a sensitivity of any sensor based on spectral measurements depends on signal to noise ratio of obtained spectra, the reduction of this value in our apparatus should be considered. There are several main reasons responsible for additional S/N ratio reduction of TDL spectra when fiber optics is used. The first group is connected with radiation intensity decrease due to different losses when light is transmitted by the fiber. One of them are Fresnel reflections of light on inlet and outlet facets of the fiber. Used IR fibers have relatively high refractive indices. For AgCl n is close to 2.2 at 10 microns (4), for $As_x S_{x-1}$ n is close to 2.4 (3). So the losses of optical signal just after one pass through the fiber due to the Fresnel reflection are about 17-20% at one surface.

But fibers with anti reflecting coating of fiber end surfaces are still unavailable.

Optical losses for scattering and absorption in fiber itself leads to drop in transmission for several percents at each meter additionally to Fresnel losses. If fiber optical line is used only for convenient transmission of IR radiation inside a spectral apparatus they could be neglected. The length of a two-way fiber line should not be more then 2 meters. And thus the total optical losses due to fiber loss and Fresnel reflection at all ends of two fiber pieces in such schemes could provide more, than 50% of transmission.

It should be noted here that the use of fibers to transfer IR-radiation from diode laser to investigated object gives an additional advantage consisting in more efficient utilization of laser radiation. In fact, higher numerical apertures become available in this case in comparison with commonly used lenses lines or mirrors optical schemes. Unfortunately the back side of the benefit is the more strong divergency of IR radiation leaving the wide aperture fibers compared with the divergency of light emitted by diode laser. Even at the lowest fiber diameter used in our experiments of about 150 microns multimode waveguiding regimes are realized. The value of the beam divergency solid angle may amounts to 60 degrees. It causes aperture limitation and strong losses due to spherical mirror aberrations. This demands careful fitting of used optic elements to diminish losses when laser radiation is passed through any analytical part of spectrometer and then back to the receiver fiber. Specific solution depends on used scheme of analytical part of spectrometer. As mentioned above in this work simple spherical mirror optics were used. In more sophisticated schemes aperture limitation could be reduced using carefully fitted wide aperture elliptical optics, which can lead to more than one order magnitude increase of signal.

In the described fiber-optical systems S/N ratio is reduced also by additional noisy modulations of signal amplitude due to interference of TDL radiation spread in different fiber modes. As mentioned above pulse regime generation of diode laser leads to laser wavelength tuning due to heating of laser crystal by current. Radiation running in different modes of fiber has different time delay. If fast tuning of laser frequency is carried out it causes essential frequency differences in modes reaching the exit end of the fiber simultaneously. Thus effective laser amplitude curve will be disturbed by interference modulations. Depth of this modulations depends on the number of waveguiding modes which exist in the fiber and efficiency of their collection on a detector or on a facet of receiving fiber. Period of the interference modulations in wavenumber scale is determined by the rate of frequency tuning, delay times between modes and thus depends on length of used fiber and highest order of waveguided modes. The last one is in turn connected with the width of waveguiding core. A total efficient interference perturbation of the laser intensity curve is complicated by overlapping of many individual interference patterns, which leads to averaging and reduction of this modulations. Residual hindrance modulation curve as our experiments shown is rather

stable in time and varies with changes of fibers space
configuration. Additional noise level due to this interference
hindrances is usually from 10^{-2}% up to 1% of signal value. But
these hindrances seem could be reduced radically by
simultaneous compulsory storage of signal and mechanical
modulations of fiber position to cause effective mixing of
waveguiding modes. Use of wide aperture optics leads to
decrease of interference hindrance too because it reduces the
depth of amplitude signal modulations due to variations of
individual modes intensity. These notes seems to be general for
all types of optics used in analytical part of spectral
apparatus with fiber optic transmission of IR radiation.

6.EXPERIMENTAL RESULTS
    Spectroscopic and analytical possibilities of the diode
laser fiber-optic apparatus have been demonstrated on
measurements of gaseous ammonium (NH_3) at 10 microns, water
vapor (H_2O) at 7,3 microns and carbon monoxide (CO) at 4,7
micron region. PbSe and PbSeSn diode lasers slightly varied in
chemical composition and operated at liquid nitrogen
temperatures have been used for this purpose. The both above
mentioned types of fibers have been tested for different
wavelength regions to transmit radiation inside the
spectrometer.
    On Fig.8 a transmission spectrum of CO natural isotope
mixture at 4.7 microns is presented. It has been detected when
the optical scheme shown on Fig.5 was applied. A single pass
cell of 85 cm length with 27 Torr of investigated gas mixture
was used. Laser radiation was transfered to the analytical gas
cell by chalcogenide fibers. The absorption lines of three
different isotope modifications of carbon monoxide are seen on
the spectrum.

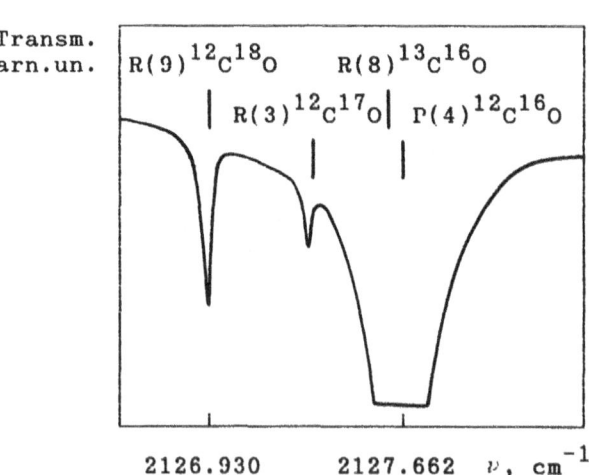

Fig.8.An example of transmission spectrum of CO natural isotope
    mixture in the 4.7 microns region obtained with fiber
    optic spectrometer based on diode laser.

When multipass gas cell is used in analytical channel of spectrometer an optimum number of laser beam passes through the cell should be previously chosen to obtain the most high efficient regime of trace gas detection. As usually the number is connected with competition between increase of resonance absorption in detected gas lines and additional reflection losses rising with optical length in the cell. As known at

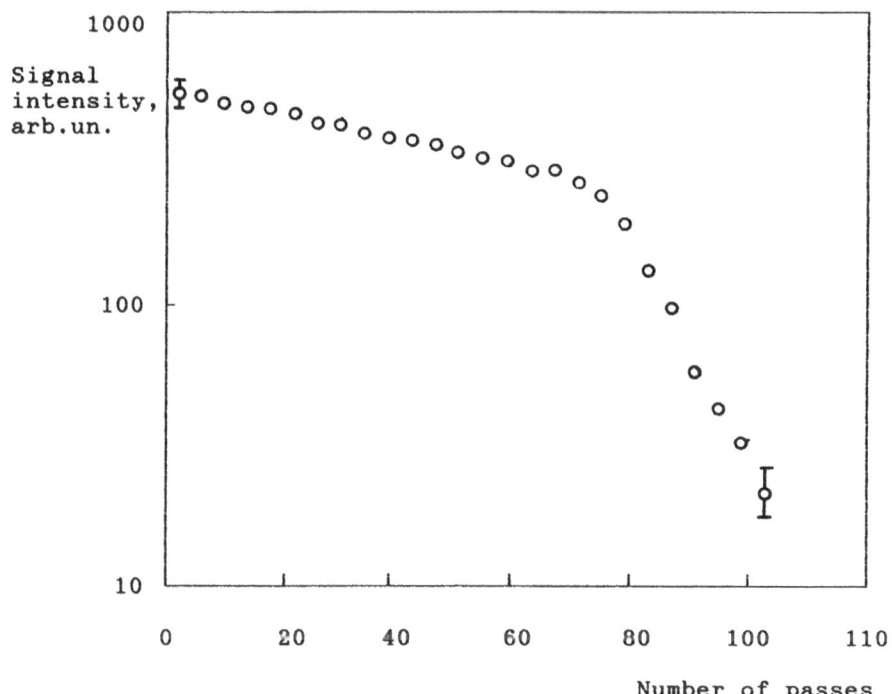

Fig.9.The dependence of output signal on numbers of laser beam
       passes through the multipass cell.

refection coefficient of mirror of about 0.985 (aluminum mirror coating for 10 microns) the optimum number of passes through the multipass cell is equal to about 62. On Fig.9 a dependence of received signal on number of beam passes is shown for developed fiber system. A slow slope of the curve at small pass numbers is determined by only signal losses on the mirror reflections. A more sharp decrease at higher values of N seems to be caused by enlarged laser beam spots on the collective mirror of the cell due to not so strict alignment of the system. Fortunately the break of the curve is located at slightly large numbers of beam passes (of about 76) than that correspondent to above mentioned optimum on reflection losses. So the used alignment was sufficient to obtain optimum signal to noise ratio for spectra detection.

On Fig.10 an absorption spectrum of natural isotope mixture of ammonium obtained by the spectrometer at the spectral region of about 10 microns is shown. Very weak lines belonging to both the hot rotational-vibration band $a2\nu_2-s\nu_2$ of main isotope modification of ammonium $^{14}NH_3$ (the lines ## 2 and 3) and to the rotational-vibration band ν_2 of $^{15}NH_3$ (the lines ## 1, 4 and 5) were registered in this spectral region. A spectral resolution here is practically determined by TDL linewidth and seems not to be disturbed by used fiber lines. This value could be estimated as better than $0.5*10^{-3}cm^{-1}$. An assignment of the lines and their exact wavenumbers are presented at the caption of the Figure. The assignment was provided on the base of previously reported data in the papers by L.Fuzina and K.Singh (13,14). The spectrum have been obtained at partial pressure of natural ammonia of about 0.075 Torr, which was mixed previously with dry nitrogen in proportion 42:720. It corresponds to about 0.25 mTorr of $^{15}NH_3$ partial pressure. Optimum optical pass length through the cell

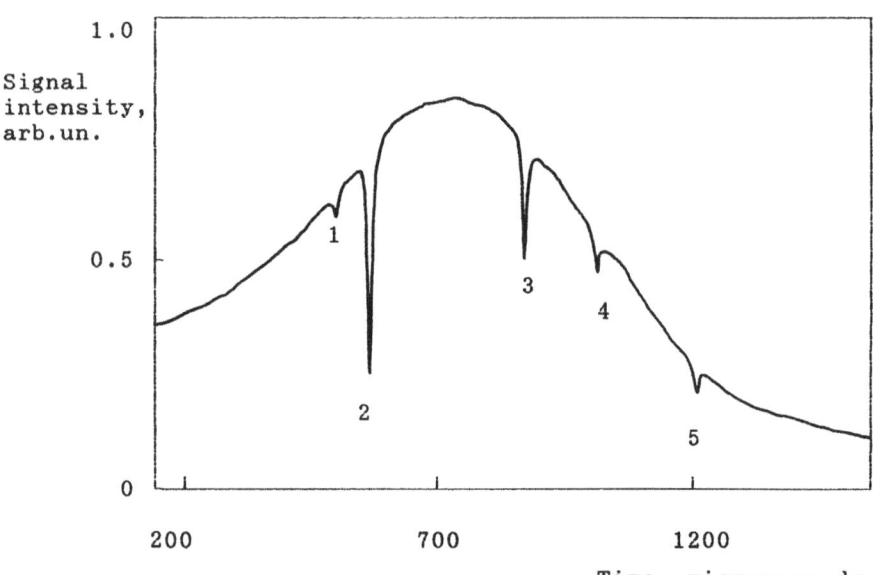

Fig.10.Tunable diode laser pulse intensity with transmission spectrum of natural isotope mixture of ammonium near 990 cm^{-1} region. An assignment of the absorption lines:

1.aR(2,2) of ν_2 $^{15}NH_3$ -987.632 cm^{-1} (13),
2. R(1.0) of $a2n_2-s\nu_2$ $^{14}NH_3$ -987.741 cm^{-1} (14),
3. R(1.1) of $a2n_2-s\nu_2$ $^{14}NH_3$ -988.200 cm^{-1} (14),
4.aR(2,2) of ν_2 $^{15}NH_3$ -988.397 cm^{-1} (13),
5.aR(2,2) of ν_2 $^{15}NH_3$ -988.648 cm^{-1} (13).

of about 15 meters have been used. S/N ratio of this spectrum was better than 100 and was limited by interference hindrances. Presented spectrum have been obtained without any storage and additional filtering of detected radio signal. Obtained data allow to hope on reaching sensitivity of the order of tenths ppb ore better in detection of some gases. Such sensitivity is enough for monitoring of some pollutions in atmosphere, for some medicine analytical problems as well as for combustion processes control.

7.CONCLUSION

Synergy of diode laser advantages with development of MIR-Fiber optic zondes opens the new generation of devices for remote IR-spectroscopy, highly necessary for wide application.

The our first experience in development of complete TDL-based fiber-optical system for chemical gas analysis has given promising results. Such systems could operate over wide spectral region and thus could be applied for high sensitive detection of large variety of molecules having absorption spectra in the middle IR-region. Near IR range (from 0.8 to 2.5 microns) could also be easy incorporated in fiber-design analyzers. Due to use of multimode fibers for radiation transmission a highest expected sensitivity of the IR-sensors would be of only two orders less than one for general TDL spectrometers. The additional interference hindrances described above seems to be a main reason for this limitation. Nevertheless the obtained sensitivity is high enough to solve majority of problems touching atmosphere pollutants monitoring, medicine breath diagnostics, and industrial technologies control. In turn the use of fibers is bringing several new advantages in TDL-based spectral apparatus like easy alignment of their optical schemes, possibility for remote spectral measurements, and compactness of the devices.

Presented device could be already used for some measurements, and its parameters could be further improved by use of core/clad MIR-fibers with smaller losses and numerical apertures, by AR-coating of fiber ends and design of optimal optics cells. Described IR fiber systems with tunable coherent light sources like TDLs appear to be perspective as a base for development of evanescent wave gas sensors possessing of high sensitivity and selectivity.

8.ACKNOWLEDGEMENTS

This research was performed during internship with the Laboratory of Narrow Gap Semiconductors, the Physical Lebedev Institute of the USSR Academy of Science, Moscow, with the Institute of Chemistry of High-Purity Substances of the USSR Academy of Sciences, Nizshnii-Novgorod, and with the Tbilisi State University, Georgia. The authors wish to thank Prof.A.P.Shotov, Dr.Yu.G.Selivanov for tunable diode lasers presented for this work. We thank also Prof.M.F.Churbanov and Dr.V.G.Plotnichenko for chalcogenide fibers specially fabricated for this work, Drs. V.G.Artjushenko and L.N.Butvina for presented halide fibers and all stuffs of the Special Design Bureau of the GPI, Troitzk, participating in development of the described system.

REFERENCES

(1) "Monitoring of Gaseous Pollutants by Tunable Diode Lasers", Proceedings of the International Symposium held in Freiburg, FRG, 13-14 November 1986, Edited by GRISAR, R., PREIER,H., SHMIDTKE,G., RESTELLI,G., (1987), 174 p., D.Reidel Publishing Company, Dordrecht, Holland.

(2) "Monitoring of Gaseous Pollutants by Tunable Diode Lasers", Proceedings of the International Symposium held in Freiburg, FRG, 17-18 October 1988, Edited by GRISAR,R., SHMIDTKE,G.,TACKE,M.,RESTELLI,G.,(1989) 305 p., Kluwer Academic Publishers, Dordrecht, The Netherlands,

(3) DEVYATYKH,G.G., CHURBANOV,M.F., et.al., (1989),.,"The role of impurities in the optical losses of chalcogenide glass fibers", Proc.SPIE "Infrared Fiber Optics", Vol.1048, pp.80-84,.

(4) ARTJUSHENKO,V.G., (1990) "Infrared crystalline fibers", Proc. SPIE "Infrared Fiber Optics II", Vol.1228, pp.12-25.

(5) BUTVINA,L.N., DIANOV,E.M., KOLESNIKOV,Yu.G., PROKASHEV,V.A., (1990),"Properties of core-clad IR-crystalline fibers", Proc.SPIE "Infrared Fiber Optics II", Vol. 1228, pp 155-165.

(6) ARTJUSHENKO,V.G., KONOV,V.I., KRYUKOV,A.P., et al., (1991) "Mechanisms of optical losses in silver halide fibers", Proc. SPIE "Infrared Fiber Optics", Vol.1591, #11.

(7) KOSICHKIN,YU.V.,NADEZHDINSKII,A.I.,STEPANOV,E.V.,SHIRSHOV, V.M., (1985) "System for obtaining tunable laser radiation". Patent of the USSR, #1454197, December, 30.

(8) KUZNETSOV,A.I., SHVETZ,E.V., (1989) "System for tunable diode laser driving", Patent of the USSR, #1653509, March,31, 1989.

(9) KUZNETSOV,A.I., NADEZHDINSKII,A.I., STEPANOV,E.V., (1990), "Computerized infrared fiber-optic system for gas analysis based on diode lasers", Proc. SPIE "Infrared Fiber Optics II", Vol.1228, pp.262-265.

(10) KOSICHKIN,YU.V., LEMEKHOV,N.V., NADEZHDINSKII,A.I., PENCHEV,S., STEPANOV,E.V., RADIONOV,A.R., (1986) "A Diode Laser High Resolution Fiberoptical Infrared Spectrometer", Soviet Physical-Lebedev Institute Reports, No.12, pp. 36-38.

(11) ARTJUSHENKO,V.G., DIANOV,E.M., KOLESNIKOV,J.G., LEMEKHOV,N.M., NADEZHDINSKII,A.I., RADIONOV,A.R., STEPANOV,E.V.,(1987) "Remote Diode Laser Spectroscopy Using IR Fibers", Proc SPIE "New Materials for Optical Waveguides", Vol.799, pp.90-92.

(12) BARSKAIA,E.G., (1968) "Multipass optical cell", Patent of the USSR, #206857, Published in 1968.

(13) DI.LONARDO,G., FUZINA,L., TROMBETTI,A., MILLS,I.M., (1982) "The ν_2, $2\nu_2$, $3\nu_2$, ν_4 and $\nu_2+\nu_4$ Bands of $^{15}NH_3$", Journal of Molecular Spectroscopy, vol.92, pp.298-325.

(14) SINGH,K., D'CUNHA,R., KARTHA,V.B., (1988) "Transition Dipole Moment and Intensity Measurements of the $a2_2\nu-s\nu_2$ Hot Band of $^{14}NH_3$," Journal of Molecular Spectroscopy, vol.129, pp.307-313.

NEW DEVELOPMENTS IN COMPUTER-CONTROLLED DIODE LASER SPECTROSCOPY

M. PETRI, T. FINK, W. URBAN

Institut für Angewandte Physik der Universität Bonn
Wegelerstr. 8
5300 Bonn 1

SUMMARY

By the use of mode charts and automatic calibration techniques spectroscopic throughput and ease of use of a diode laser spectrometer are greatly enhanced. Automatic on-line linearization and calibration allow unlimited averaging times over large frequency intervals, independent of frequency drifts of the laser. Furthermore background subtraction can be incorporated easily.

1. INTRODUCTION

One of the main disadvantages of lead-salt diode lasers are their chaotic spectral properties, consisting of multi-mode emission accompanied by substantial mode competition between different modes, small continuous tuning intervals separated by mode jumps, gaps in the frequency coverage and so forth. Diode lasers exhibit dramatic changes in mode output power and frequency on scales down to 1 cm^{-1}, thus resulting in virtually unpredictable emission properties over intervals longer than 1 cm^{-1}. Even the quite predictable current and temperature tuning rates often vary substantially from diode to diode. This makes working with a diode laser spectrometer difficult and time consuming.

In an attempt to increase the ease of use of the spectrometer, several groups use a digital computer for the three tasks of data acquisition, data analysis and the control of the spectrometer, i.e. setting of temperature, current and monochromator transmission frequency (1, 2, 4, 5, 6, 9, 10, 11, 13, 14, 15, 17). Usually these three tasks are implemented as separate entities. Sometimes not all features are implemented fully. For example it is quite common to perform the analysis of the data long after data acquisition, often on a different computer. This separate approach is easy to implement, but doesn't carry very far in tunable diode laser spectroscopy, where one has to deal with the chaotic and everchanging properties of the diode lasers. Use of a computer gets interesting, as soon as one starts to couple the data analysis task with the control of the spectrometer, through an extensive on-line analysis of the recorded spectra. This alliance provides very fruitful and opens up a broad range of new possibilities, a few of which will be discussed in this paper. Not only can the spectroscopic throughput of the system be increased tremendously by the use of mode charts and automatic calibration techniques, but also novel spectroscopic techniques such as digital averaging over unlimited time intervals become possible.

2. MODE CHARTS AS A SPECTROSCOPIC TOOL

A very time consuming procedure when working with diode lasers is the search for the optimum operating conditions at a given laser frequency or frequency interval. Because of mode jumps and gaps in the frequency coverage there might be no temperature/current setting at all, which allows operation at the desired laser frequency. Even if the frequency coverage is 100%, there might be several choices of operating conditions of which only very few are optimal.

Usually the optimum operating settings are selected by trial and error, starting with a mode close to the desired frequency and changing temperature and/or current in small steps until laser emission at the desired frequency is attained. This procedure is very much dependant on the experimental prowess of the operator and doesn't guarantee an optimum setting. If the laser has a gap at the desired frequency, under trial and error conditions it might take a long time until this fact is discovered. When the laser frequency has to be changed often, as is the case in spectroscopic applications, the time spent in searching for an optimum mode, sums up to a large percentage of idle time lost to the experiment.

Searching for the best possible mode can be greatly facilitated by the use of mode charts. A mode chart contains the frequency and intensity of all lasing modes of the diode in tabulated form, at a large number of different current/temperature sampling points. A mode chart is recorded fully automatically by the computer. Prior to recording a temperature/current grid has to be specified by selecting an interval and an appropriate stepwidth for both temperature and current. Then the mode structure of the diode at each point of the temperature/current grid is sampled, by scanning the monochromator over the gain profile of the laser and recording the transmitted intensity with a detector. The frequency positions and relative intensities of the modes are determined automatically and stored in the chart after each monochromator scan. The mode positions at the previous sampling points are used to predict the frequency range of the gain profile - and thus the monochromator scan interval - at the next sampling point. This allows a very rapid evaluation of the mode structure. As soon as the grid is completely sampled, all mode positions belonging to one distinct mode, either via current or temperature tuning, are sorted together. This is also done automatically, by starting out with reasonable values for the temperature and current tuning rates and sorting the chart in several passes, recalculating the tunings rates in each pass. The sorting process allows us to determine the statistics of each mode (continuous frequency coverage, tuning rates, etc.) and delivers interesting global diode characteristics such as temperature and current tuning rates, threshold currents, overall frequency coverage, single mode frequency coverage etc.

A raster of 1000 points is sampled in about one hour. This is quite sufficient for determining the optimum operating conditions at a specific frequency interval. A mode chart over the full temperature and current range of a typical diode requires approximately 5000 to 10000 sampling points and is preferentially recorded over night, where the automatic recording process doesn't interfere with the operation of the spectrometer.

The mode chart, or selected parts of the chart, can be viewed in various different ways, which allow a quick graphic determination of the best temperature/current settings. Fig. 1 shows a current plot of the diode 181-6-11, a view of the current tuning characteristics at a fixed temperature of 32 K. The various mode positions are indicated by vertical lines. The frequency and current coordinates of a mode are given by the lower starting point of the line, the modes intensity is proportional to the length of the line. Mode positions belonging together via current tuning are connected. The mutual distance between two connected positions is proportional to the current tuning rate.

The current plot allows the determination of the optimum current value at a fixed temperature. It also enables the operator to determine those current settings, at which the diode emits radiation in a single mode. In order to simultaneously view temperature and current tuning behavior, a temperature plot is constructed by vertically arranging several compressed current plots belonging to adjacent temperatures. Fig. 2 shows a temperature plot of the diode 181-6-11, which contains the current plot of the previous diagram. In order to facilitate the determination of the optimum operating conditions, a freely moveable graphical crosshair can be projected on the screen, which shows the temperature/current setting corresponding to its vertical and the frequency corresponding to its horizontal coordinate.

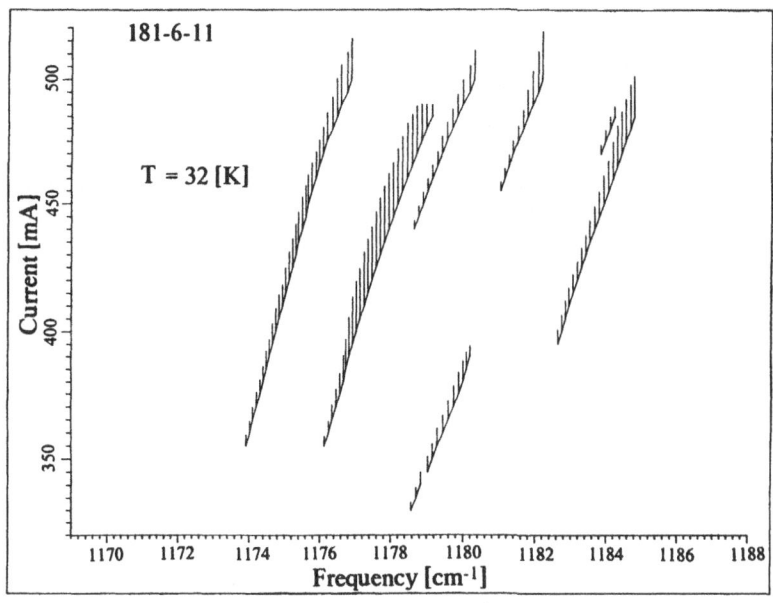

Fig. 1: Current plot of the diode 181-6-11 at a temperature of 32 K

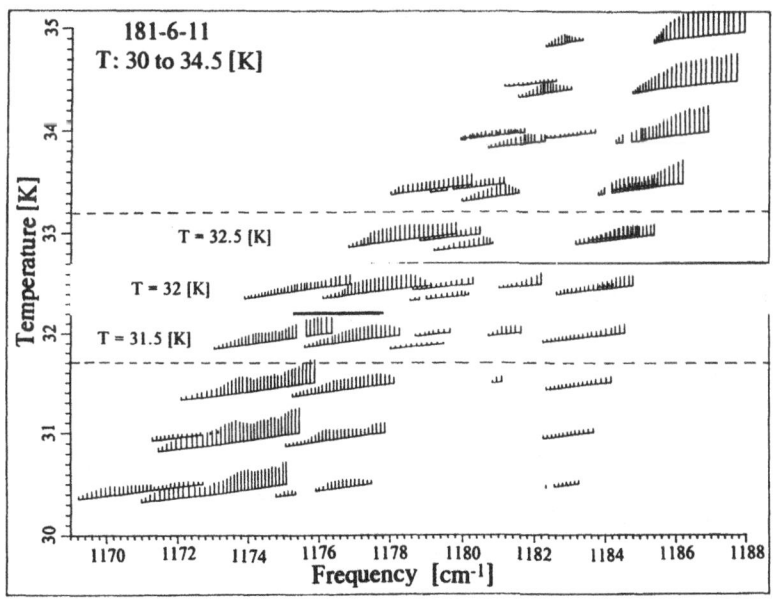

Fig. 2: Temperature plot of the diode 181-6-11

When a temperature plot is comprised of many current plots, such as Fig. 3, the strong vertical compression of the current plots makes the determination of the single mode regions of the diode - or other interesting spectral features - quite impossible. Therefore several restrictions can be placed on the mode positions shown in a plot. For example a minimum mode separation can be specified, which allows one to view only sufficiently isolated modes, that is modes which don't have any neighboring modes within the specified separation interval. By using a combination of restrictions, the spectroscopically most promising regions of the laser can be singled out, which turns out to be very useful in determining the optimum operating conditions. Fig. 4 shows a temperature plot of the diode 181-6-11, where all but the single mode regions of the diode are eliminated from the plot.

By the use of mode charts it is not only possible to quite accurately determine the optimum temperature/current settings of the diode at a given laser frequency, but also mode search time can be cut almost completely. A high resolution mode chart for a specific frequency interval is recorded in about half an hour, less the time one usually spends in manually searching the specified frequency region. When the mode chart is recorded over night, as is normally the case, there is virtually no time penalty involved in finding the best available mode.

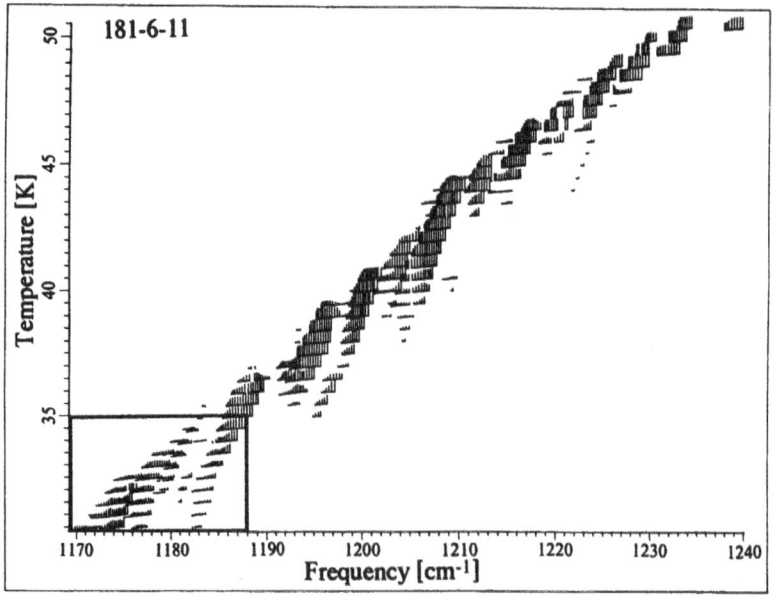

Fig. 3: Temperature plot of the diode 181-6-11 over the full temperature range

As diode characteristics change with time, mode chart have to be updated once in a while. The timescale of aging very much depends on the diode. We found that for our system, as long as the diode remains in the cryostat and isn't stored on the shelf for periods longer than a month, the calibration drifts of the temperature control unit (up to one Kelvin over a few days) were much more a problem concerning the reproducibility of the chart than aging. In all cases where we compared mode charts taken several days (up to four weeks) apart, and where the temperature scale was adjusted to correct for the drift of our temperature controller, we couldn't find any significant differences in spectral behavior attributable to aging. In fact, the reproducibility of the mode charts is so good, that we were able to operate a diode (324-5-19) for over a month without using the monochromator at all, relying completely on the mode chart taken at the beginning of the one month period. When comparing mode charts taken several months apart (with the diodes having been stored at room temperature for a significant amount of time), changes in spectral behavior can be detected which cannot solely be attributed to temperature drifts. Although diode characteristics don't change dramatically even over large time intervals - distinct features of a diode are still clearly recognizable and most of the modes are still present - , there seems to be an overall increase in threshold current accompanied by an reduction of the continous tuning ranges of most modes.

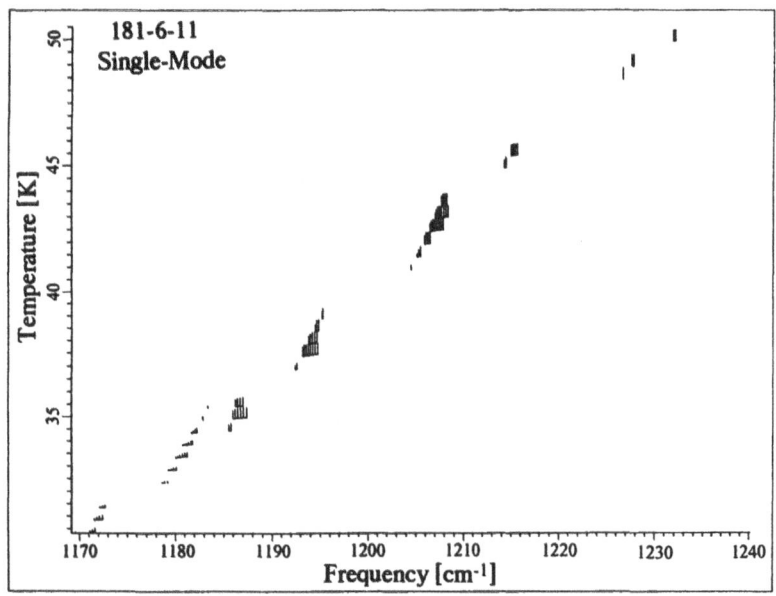

Fig. 4: Single mode temperature plot of the diode 181-6-11

3. DETERMINING LASER PROPERTIES AND QUALITY WITH MODE CHARTS

Mode charts are not only extremely useful for increasing spectral throughput of the spectrometer. The wealth of information immediately available within the chart allows one to easily and systematically determine some interesting laser parameters and allows a quite reliable determination of a lasers spectral quality. Furthermore, as each mode chart serves as a fingerprint for the corresponding diode, it is for the first time possible to conduct systematic studies concerning long term changes in a diodes spectral behavior due to aging.

Throughout the sorting process the tuning rates are immediately available. Figures 5a and 5b show the temperature and current tuning rates of the diode 160-1-13, plotted versus temperature. As can be seen from these plots the tuning rates vary substantially with temperature and exhibit discountinous jumps at certain temperatures. From both temperature and current tuning rates quite an interesting laser parameter, the heating coefficient $\partial T/\partial I$, can be determined:

$$\partial T/\partial I = \partial f/\partial I / \partial f/\partial T$$

This coefficient describes how much the internal temperature in the active zone of the diode (T_{act}) is increased with respect to the temperature of the mount (external temperature T_{ext}), when a current is applied to the diode. Fig. 5c shows the temperature dependance of the heating coefficient of the diode 160-1-13. Values for $\partial T/\partial I$ of 20 K/A are quite typical, although we found a high variability in the values of $\partial T/\partial I$, ranging between 5 and 150 K/A. The heating coefficient allows an estimate of the thermal contact of the diode, it is directly proportional thermal resistance between active zone and the laser mount.

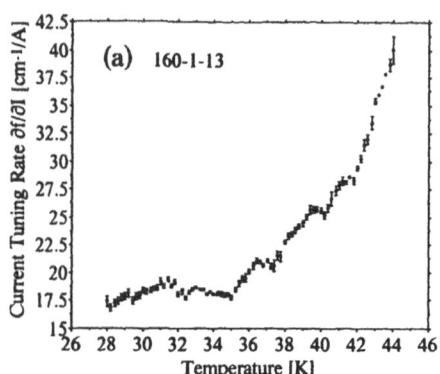 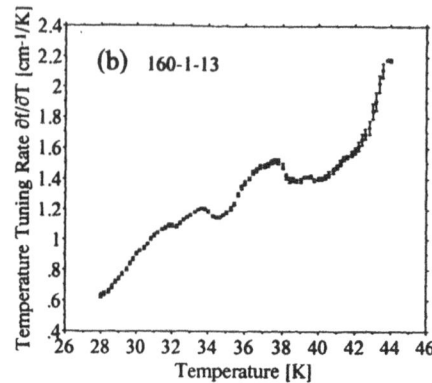

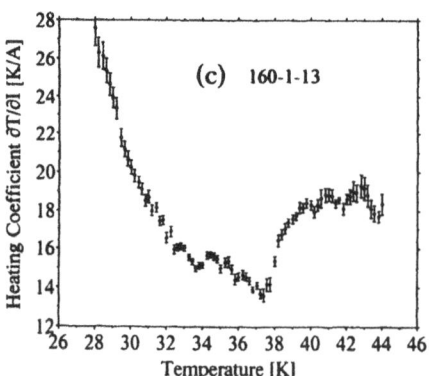

Fig. 5: Temperature dependance of the current tuning rate (a), the temperature tuning rate (b) and the heating coefficient (c) of the diode 160-1-13

Knowledge of the heating coefficient allows one to quite accurately calculate the internal temperature of the laser, as it depends on the external temperature and the current:

$$T_{act} \quad = \quad T_{ext} + \partial T/\partial I * I$$

This allows one to construct a physically very interesting version of the mode chart, where the mode frequencies are plotted versus internal temperature. Fig. 6 shows such a physical mode plot of the diode 324-5-19. The quite rapid temperature tuning of the gain profile and the lower tuning rates of the modes are immediately apparent, also the narrowing of the gain profile as the maximum temperature is approached. This diagram allows us to immediately determine some important aspects of a diodes spectral quality, such as occurrence of lateral modes, gaps in the continuous frequency coverage of a mode, overall frequency gaps of a diode etc. The diode 324-5-19 is an extremely good laser, lasing only on longitudinal modes with no lateral modes present.

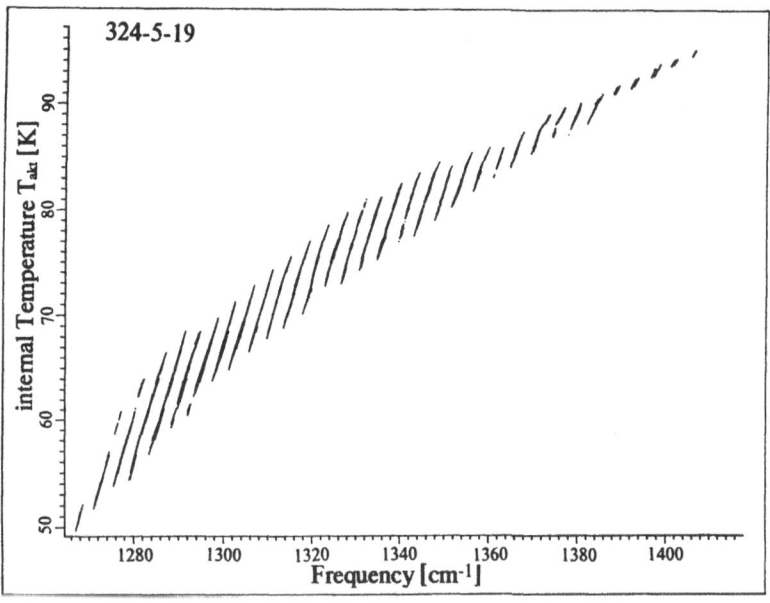

Fig. 6: Mode frequencies of the diode 324-5-19 plotted versus internal temperature

4. AUTOMATIC CALIBRATION

Another time-consuming process in tunable diode laser spectroscopy is the exact determination of the laser frequency. As highly precise and reliable wavemeters operating 3 to 30 μm range aren't readily available (3) and the monochromator is not accurate enough, commonly reference gas lines are used to calibrate the recorded spectra. A standard technique is to first linearize the frequency axis of the data by using the equidistant fringes of a highly stable marker etalon and then establish an absolute frequency scale with the help of reference gas lines (8).

We highly automated the calibration procedure in order to further increase the throughput of the spectrometer. A sophisticated peak finding routine calculates the precise positions of the reference gas lines and the marker transmission minima. Then a linearization of the frequency axis is carried through, using the marker transmission minima to establish an accurate linear frequency scale. After the linearization procedure only two parameters are needed for the absolute calibration: the exact marker spacing and the absolute frequency of the start of the scan. These two parameters are determined with the help of a pattern recognition routine, which compares the characteristic absorption pattern defined by the calculated positions and intensities of the reference gas lines with the complete reference gas spectrum available in a catalogue on disk. For these purposes parts of the GEISA spectroscopic line parameter catalogue (7) were ported to our diode laser system. Prior to the precise calibration, the absolute frequency of the laser is usually known to an accuracy of 1 cm^{-1}, either through the the the monochromator transmission frequency or through a mode chart. This allows us to use just a small part of the catalogue (10-20 cm^{-1} centered around the expected frequency) for the pattern recognition process, saving time and reducing wrong by-chance-assignments.

The pattern recognition process is extremely reliable. It works especially well with dense reference gas spectra with limited structure, in which case a manual assignment of the reference gas lines is extremely time consuming. Even with complex reference gas spectra we are able to carry through a complete calibration of the spectra within a few seconds after completion of the scan.

5. DIGITAL AVERAGING

Averaging techniques are commonly used in order to increase the measurement sensitivity. According to theory, the signal to noise ratio (S/N) increases with the square root of the averaging time, resulting in a steady increase in S/N as the measurement is prolonged. Albeit this is true in principle, long term frequency drifts of the diode laser set a limit to the useful averaging time, resulting in an optimum averaging time which depends on the stability of the system and the linewidth of the spectral features observed. This optimum averaging time is reached, when the long term frequency drifts of the laser become comparable to the linewidth of the spectral features one wishes to observe. Increasing the averaging time over this optimal amount tends to blur the spectral features and results in a decrease of S/N. Typical values for the optimum averaging time are on the order of a few minutes.

One approach to overcome these limitations is to increase the stability of the system, either by increasing the stability of the current/temperature controller or by an active stabilization scheme (12, 16). Another approach taken in this paper is the technique of digital averaging with on-line frequency correction of the laser drifts. The last technique has the advantage, that no expensive and complex hardware is necessary, furthermore truly unlimited averaging times can be reached. Averaging is done by scanning the diode laser successively over the desired frequency interval and digitally averaging successive scans. The frequency drifts of the laser are corrected for immediately by carrying through an extensive on-line analysis of each single scan. This technique is not alltogether new. It has been already successfully employed in trace gas measurements, where the shift of single reference line position during successive scans was detected and corrected for, either by shifting the scan or by correcting the laser temperature or current (4, 13). We report on an extension of this technique for spectroscopic purposes, which can be applied to arbitrary large scanning intervals containing several reference gas lines.

At the basic level of our technique, the data from the marker etalon are used to keep the laser frequency in check. At the beginning of an averaging scan the frequency interval, over which averaging shall take place, has to be specified. The number of marker fringes contained in this interval is calculated. The scan is started and, as the laser scans along, the marker fringes passing by are counted until the specified count is reached. At this point the tuning direction of the laser is reversed and the laser is scanned back over the same frequency interval. During the reverse scan the data of the forward scan are linearized automatically in the background, using the marker fringes to establish a precise relative frequency scale for all scans. The linearized data are then summed up. Linearization prior to summing up is done in order to eliminate short term frequency drifts of the diode laser during a single scan, which become especially important when averaging is done over long frequency intervals. The linearization procedure also serves to eliminate nonlinearities in the turning points of the scan. As soon as the specified marker fringe count of the backward scan is reached, scanning direction is again reversed and a second cycle is started. The backward data are inverted and then processed in the same way as the forward data.

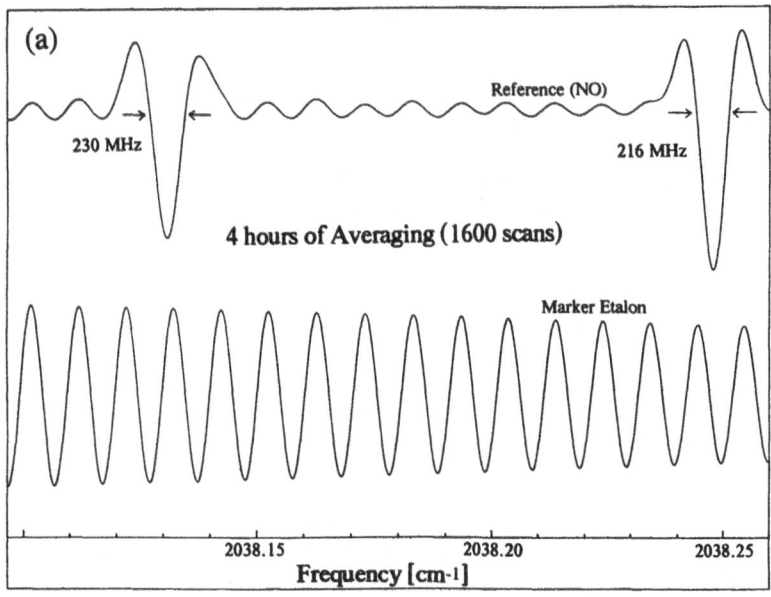

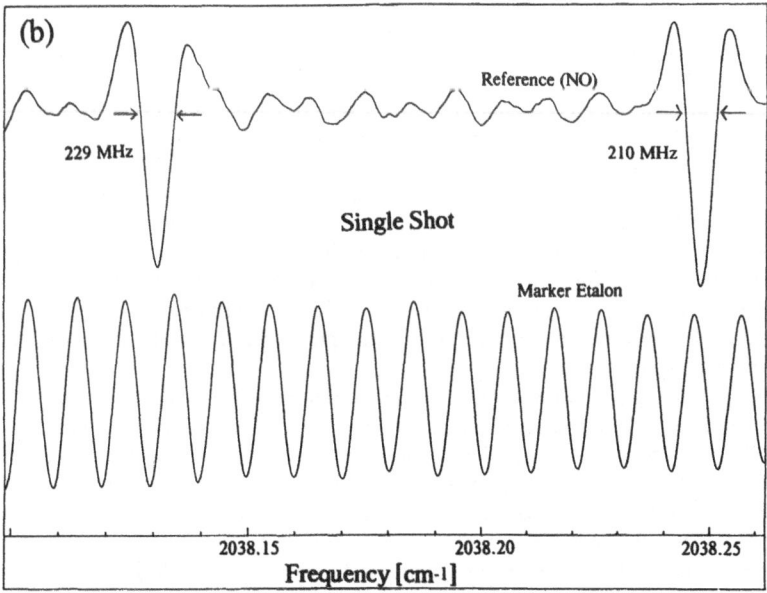

Fig. 7: Digital averaging scan. (a) 1600 averaged scans accumulated in 4 hours,
(b) single shot scan taken in 10 s

By scanning the laser back and forth between two orders of the marker etalon and lineari-zing the single scans prior to averaging, all long and short term frequency drifts of the diode laser can be eliminated, as long as the etalon FSR remains stable. We employ a 300 MHz air spaced etalon mounted on a cerodur casing, which, though not stabilized, exhibits very low frequency drifts. Typical drifts of the fringes (in 10^5 order, corresponding to a diode laser frequency of 1000 cm^{-1}) are less than 10 MHz/hour. When applying our technique to Doppler limited spectroscopy with linewidths around 100 MHz this sums up to a useful averaging time of a few hours. This is demonstrated by Figures 7a and 7b. Fig. 7a shows an averaging scan, where the laser was scanned repeatedly over a frequency interval of 0.2 cm^{-1} including two NO-lines. Within a time interval of 4 hours 1600 scans, each of a duration of approximately 10 sec, were averaged. No significant broadening of the two NO-lines due to drifts of the marker fringes could be observed, as can be seen by comparing the linewidths of the NO-lines in Fig. 7a and in Fig. 7b, a single scan taken over the same frequency interval. The typical drifts of the free running laser during the averaging process were 10 MHz/s, which proves that our technique works even with severe drifts of the laser frequency. Although the S/N ratio in Fig. 7a is high, the sensitivity with respect to the NO-absorption lines is limited by an etalon structure. All not highly stable etalons have been ave-raged out, there is just one etalon remaining with a FSR exactly matching the marker FSR. As this etalon is produced by the marker cavity itself it cannot be completely eliminated, though by slight-ly misaligning the marker beam it can be greatly reduced. For the purpose of demonstration this wasn't done in the shown scan. If the etalon signal is interpreted as noise a minimum detectable absorption of $2 \cdot 10^{-5}$ can be calculated from Fig. 7. Because of the high regularity and reproducibi-lity of the remaining etalon structures, the sensitivity of the digitally averaged spectra can be in-creased either after data acquisition through digital filtering or during data acquisition by the incor-poration of background subtraction techniques. As digital filtering can be performed by any sophi-sticated data analysis package - and usually yields only limited improvement - , we chose to incor-porate background subtraction into our averaging technique.

Only a slight modification of the averaging process was necessary. The laser is scanned back and forth (one full cycle) with the signal present. At the end of the cycle the computer locks the laser to the etalon fringe at the beginning of the scan. The signal is turned off - either per hand or through a switch operated by the computer - and one full cycle is scanned with just the back-ground signal present, which is subtracted from the data of the previous cycle. This procedure can be repeated as often as necessary in order to increase the S/N. We applied this technique to the spectroscopy of the radical GeD, which is produced in a gas discharge. The background signal is generated by turning the discharge off, in which case no GeD should be present. Fig. 8a shows a spectrum of GeD recorded with a single signal-background cycle, whereas Fig. 8b shows the background signal corresponding 8a. Although switching off the discharge is quite a crude back-ground subtraction technique, the background could easily be reduced by a factor of 3, which allowed us to identify two GeD lines with a S/N of 5. For the future more sophisticated back-ground subtraction techniques are planned, for example switching on and off a magnetic field in the case of paramagnetic species.

The averaging technique as presented so far allows us to significantly increase the opti-mum averaging time of our spectrometer. Still there is need for enhancements. Our technique is quite sensitive to counting errors of the etalon fringes. Although these errors occur seldom, one single false count can invalidate the whole so far accumulated data. Rudimentary checks are built into the program which detect and correct counting errors, but these checks don't work perfectly. Another problem is the drift of the marker etalon, which limits the useful averaging time, espe-cially when the linewidth of the spectral features to be observed is small. Both of these problems can be tackled by including the reference data into the analysis of the individual scans. The pattern recognition program described above, which is presently used for the the calibration of the scans, is able to detect shifts of the laser frequency during successive scans - relative to the reference line pattern of the first scan - and to correct for these shifts. When several reference gas lines are present, counting errors of the marker fringes will also be detected easily by analysis of the relative spacing between reference gas lines, resulting in a more reliable operation of our marker counting technique. These enhancements should be incorporated into the program by the end of this year, resulting in truly unlimited averaging times and a even higher reliability.

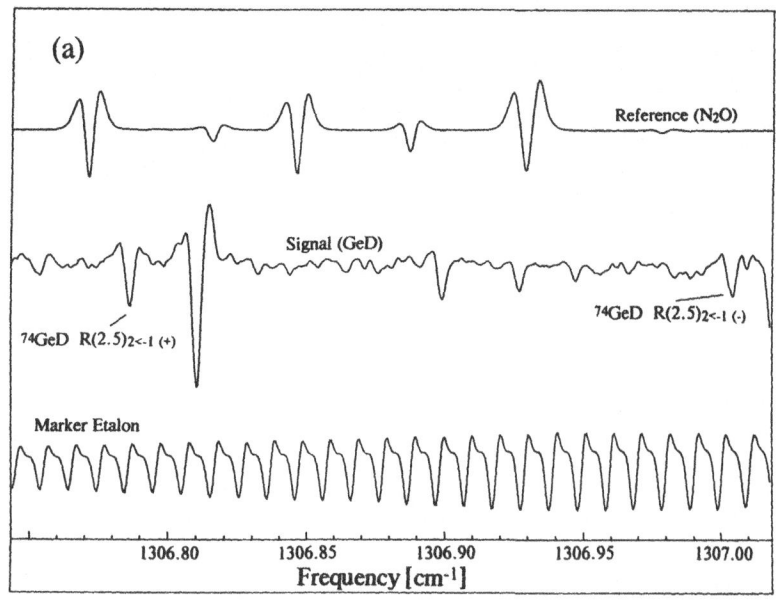

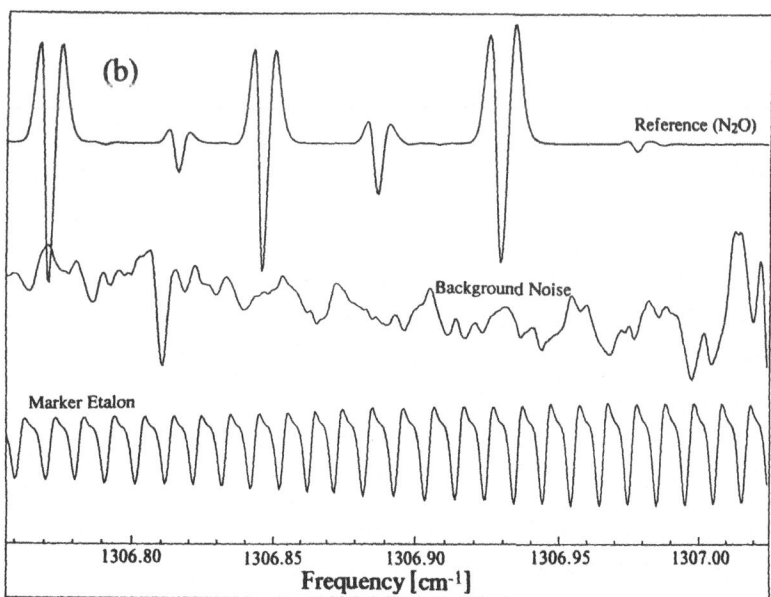

Fig. 8: Background subtraction scan. (a) result of one single signal-background cycle
(b) background of the scan in (a)

6. CONCLUSION

By the use of a digital computer ease of use and spectroscopic throughput of our diode laser system could be greatly increased. Furthermore, by correcting the laser frequency drifts on-line by help of the automatic calibration technique, we reach a very high frequency stability - and thus sensitivity - without having to resort to complex and expensive hardware. In the future we will use the automatic calibration technique to continuously update our mode charts. This will greatly increase the accuracy and the predictive power of the charts. With these tools at hand the dream of just having to specify a frequency interval and letting the system set the temperature-/current accordingly shouldn't be too far away. Thus by combining the advantages of a computer-controlled system with the fabrication of better laser diodes, it might be possible to develop diode laser spectrometers in the near future, which will truly meet industrial standards.

REFERENCES

(1)   BRÜGGEMANN, R.et al. (1989). Computer Controlled Diode Laser Spectrometer with a Helium Evaporation Cryostat and Spectroscopy of the v_3 Vibration of NCO. *Appl. Phys. B*, **48**, 105-110 .

(2)   CLINE, D. S., VARGHESE, P. L. (1988). High resolution spectral measurements in the v_5 band of formaldehyde using a tunable IR diode laser.
Appl. Opt. , **27**, 3219-3224.

(3)   EVANS, W. J., LAMBERT, D. K. (1986). Wavemeter for lead-salt diode laser calibration. *Appl. Opt.* , **25**, 2867-2868.

(4)   FRIED, A., DRUMMOND, J. R., HENRY, B., FOX, J. (1991). Versatile integrated tunable diode laser system for high precision: application for ambient measurements of OCS. *App. Opt.* , **30**, 1916-1932.

(5)   FUKUNISHI, H., OKANO, S., TAGUCHI, M., OHNUMA, T. (1990). Laser hetero-dyne spectrometer using a liquid nitrogen cooled tunable diode laser for remote measurements of atmospheric O_3 and N_2O. *Appl. Opt.* , **29**, 2722-2728.

(6)   GIESEN, T. et al. (1988). High resolution spectroscopy using a stabilized diode laser: the $2v_9$ band of HNO_3. *Z. Naturforsch.* , **43a**, 402.

(7)   HUSSON, N. et al. (1986). The Geisa Spectroscopic Line Parameters Data Bank in 1984. *Annales Geophysicae, Fasc. 2, Series A*

(8)   JENNINGS, D. E. (1984). Calibration of diode laser spectra using a confocal etalon. *Appl. Opt.* , **23**, 1299-1301.

(9)   NADLER, S., DAUNT, S. J., REUTER, D. C. (1987). Tunable diode laser measure-
        ments of the formaldehyde foreign-gas broadening parameters and line strengths in the 9-
        11 μm region. *Appl. Opt.* , **26**, 1641-1646.

(10)   NICOLAS, C., MANTZ, A. W. (1989). Infrared tunable diode laser control: frequency
        stabilization and digitization of spectra leading to high sensitivity and accurate frequency
        scale. *Appl. Opt.* , **28**, 4525-4532.

(11)   NICOLAS, C., SPROUL, J. C. (1990). Sampling of diode laser spectra by a confocal
        etalon fringe pattern. *Appl. Opt.* , **29**, 798-801.

(12)   REICH, M., SCHIEDER, R., CLAR, H. J., WINNEWISSER, G. (1986). Internally
        coupled Fabry-Perot interferometer for high precision wavelength control of tunable diode
        lasers. *Appl. Opt.* , **25**, 130-135.

(13)   SCHIFF, H. I., KARECKI, D. R., HARRIS, G. W. (1990). A Tunable Diode Laser
        System for Aircraft Measurements of Trace Gases. *J. Geophys. Research*, **95**, 147-153.

(14)   SCHMIDTKE, G.et al. (1989). Diode laser spectrometer for monitoring up to five
        atmospheric trace gases in unattended operation. *Appl. Opt.* , **28**, 3665-3670.

(15)   SILVER, J. A., BOMSE, D. S., STANTON, A. C. (1991). Diode laser measurements of
        trace concentrations of ammonia in an entrained-flow coal reactor. *Appl. Opt.* , **30**, 1505-
        1511.

(16)   VALENTIN, A., NICOLAS, C., HENRY, L., MANTZ, A. W. (1987). Tunable diode
        laser control by a stepping Michelson interferometer. *Appl. Opt.* **26**, 41-46.

(17)   WANG, Z., ELIADES, M., CARRON, K., BEVAN, J. W. (1990). A cw planar jet
        computer-controlled tunable IR diode laser spectrometer for the investigation of hydrogen-
        bounded complexes. *Rev. Sci. Instrum.* , **62**, 21-26.

MEASUREMENT OF PRESSURE-EFFECTS WITH A STABILIZED IR-DIODE LASER SPECTROMETER

N. Anselm, Th. Giesen, M. Harter, R. Schieder, and G. Winnewisser

I. Physikalisches Institut, Universität zu Köln,
Zülpicher Str. 77, 5000 Köln 41,
Germany

SUMMARY

Using a precisely stabilized high-resolution IR-Tunable Diode Laser (TDL) Spectrometer, pressure effects of weak rovibrational transitions of CO and H_2O with high angular momentum have been examined with high accuracy. In order to measure these lines, a 50 m multiple-reflection cell of Herriott-type was developed. With an improved shock isolation of the closed-cycle cold head the mechanically induced noise of the laser diodes has been reduced considerably. A sensitivity for lineshift measurements as good as 0.01 MHz/mbar is obtained.

1. INTRODUCTION

Knowledge about phenomena like pressure broadening and pressure shift of molecular lines is very important for a correct interpretation of the structure of molecular absorption lines, in particular when line profiles from different altitudes of the atmosphere are considered. Reliable data in this field can help to achieve a more realistic picture of the atmospheric structure and dynamical processes in the atmosphere [1].

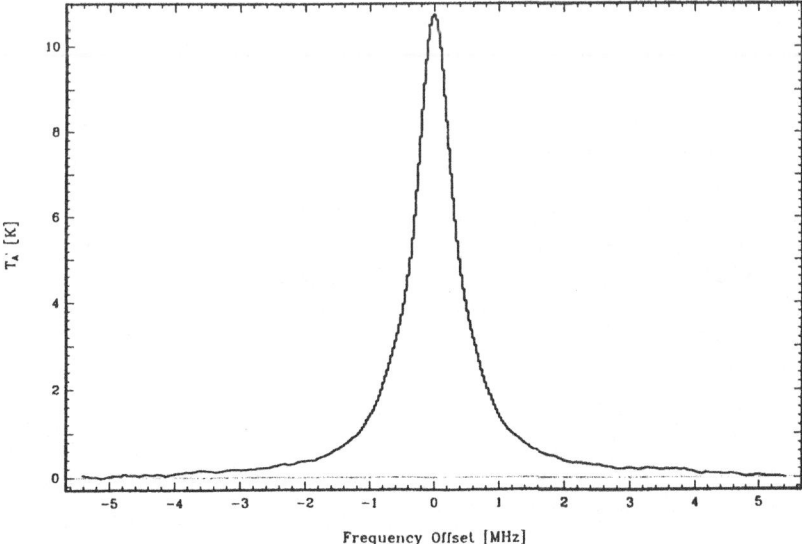

Figure 1: Measurement of an atmospheric CO 2 $\longrightarrow$ 1-transition with the 3m-radiotelescope at Gornergrat, Switzerland

Measurements of spectral lines of an atmospheric molecule supply information about the distribution of this molecule in the atmosphere. Figure 1 shows a typical measurement of a microwave rotational transition of $^{12}C^{16}O$ ($J = 2 \longrightarrow 1$; 230 GHz) with the 3-m radiotelescope at Gornergrat, Switzerland. This spectrum contains the whole information about the altitude distribution of CO. The narrow peak represents the distribution in the upper atmosphere at low pressure, whereas the wide line wings give information about the distribution in the lower part of the atmosphere.

Not all molecules can be detected in the microwave region (e.g. non-polar molecules like CH_4, C_2H_2, ...), thus IR remote sensing is a complementary method for studies of the higher atmosphere. For this we started to develop a tunable IR heterodyne spectrometer using a stabilized TDL as local oscillator. Presently a TDL spectrometer is used for measurements of pressure effects. An interesting result obtained in this field is e.g. the observation of Dicke narrowing of H_2O lines.

In order to derive spectroscopic parameters as accurate as possible, our TDL spectrometer has been optimized. Two striking improvements have been the development of a shock-isolated closed-cycle cold head and a multiple-reflection cell of the Herriott-type.

2. SPECTROMETER

The pressure effects have been determined by using a stabilized IR-TDL spectrometer (Fig. 1) operating in the infrared region where vibration-rotation transitions of molecules are observed. Since a diode-laser oscillates in multi-mode, a grating is used to select the desired main mode. After that the beam diameter is reduced by a telescope consisting of two off-axis parabolic mirrors in order to fit it to a multiple reflection cell of the Herriott-type. This absorption cell has an optical path length of 50 m and a fundamental length of about 1 m. Before entering the cell, a portion of the beam is coupled out by a beam-splitter. This permits to detect signal and background simultaneously. By dividing these two signals the baseline ripple due to frequency dependent transmission through the setup is removed and atmospheric absorptions are corrected.

A frequency stabilization of the spectrometer is achieved by using an 80 cm internally coupled Fabry-Perot Interferometer (icFPI) with a free spectral range of 94 MHz [2]. Therefore, an additional portion of the laser beam is coupled out by a second beam-splitter. A control-loop stabilizes the frequency of the laser to an interference maximum of the icFPI with the help of source frequency modulation and phase sensitive detectors. To calibrate the measurements, a iodine-stabilized HeNe-laser is also coupled into the icFPI [3]. The resulting interference signal and the divided IR-signal are recorded by a personal computer via A/D-converters.

The main problem for high precision measurements by a TDL spectrometer is the mechanically induced frequency noise caused by the closed-cycle cold head. This jitter limits the frequency resolution. Apart from that the mechanical shocks of the cooler cause intensity fluctuations of the laser and vibrations of all optical components. This reduces the sensitivity of the spectrometer.

Two possibilities exist to eliminate the disturbances caused by the cooling process:

1. Cooling the laser diodes by liquid helium or

2. Development of a new cold head which is able to protect the diodes from the mechanical shocks

The second point has been adopted because the development of a new cold-head seemed to be more convenient for every day usage.

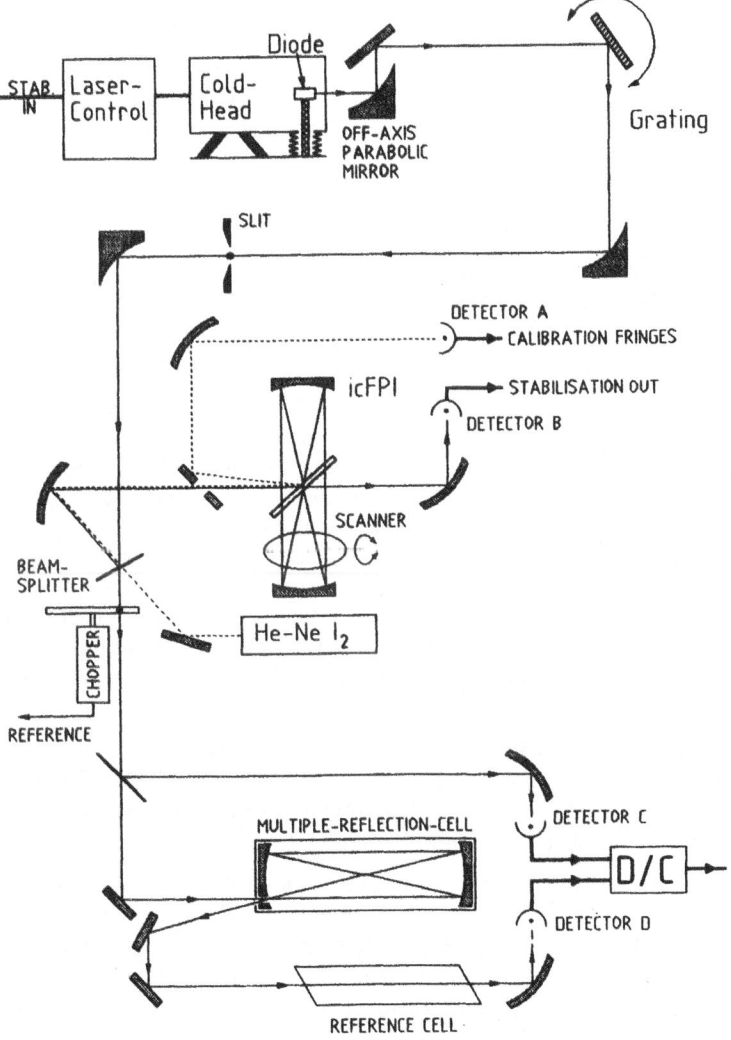

Figure 2: Setup of the stabilized high-resolution IR-TDL spectrometer

3. COLD-HEAD

The mechanical shocks are produced in the cooling machine. The aim was to decouple the diode laser mechanically as good as possible from the cooler. Therefore, the diode is mounted on a rod which is fixed on the optical bench (Fig. 3).

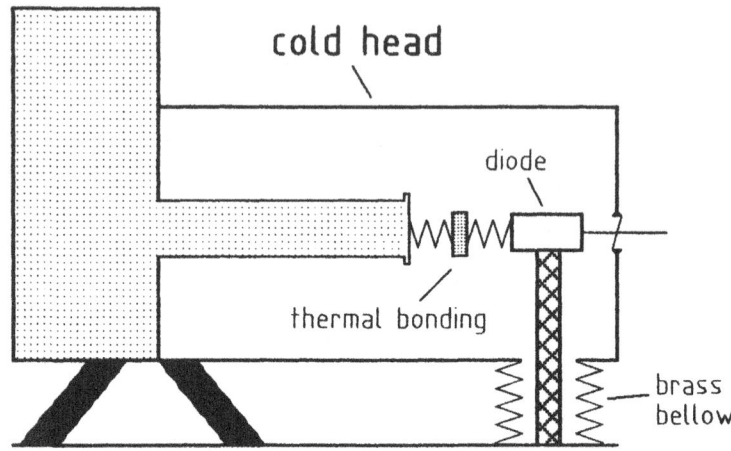

Figure 3: Schematic diagramm of the cold head

There are two different paths of transfering these perturbations to the laser diode. One is the thermal bonding between diode and cold-finger. This direct path could be interrupted by an "impedance mismatch" realized by two "springs" with a weight in the middle connecting them. One "spring" is made out of a soft bundle of annealed copper. The second is a tighter one which is able to carry the copper weight. Another possibility of transfering perturbations is the way over the optical bench. This was prevented by putting the whole cooler on soft rubber buffers and connecting the vacuum dewar with the optical bench by a brass bellow surrounding the rod on which the diodes are mounted. These two innovations were realized in our new shock-isolated cold head which turned out to be a very effective improvement.

A frequency analysis (Fig. 4 a) and b)) of the acoustic noise at the laser diodes shows that this concept of decoupling the laser diode has been very efficient. The low frequency acoustic noise (0-50 Hz) could be nearly eliminated. Also high frequency acoustic noise here analysed up to 25000 Hz is almost completely suppressed.

Another very impressive picture of the efficiency of the improved new cold head is shown in Fig. 4 c) where the frequency jitter of the laser diodes is presented by a measurement of interference maxima of an 80 cm icFPI. It is obvious that the shock isolation leads to a enormous reduction of the frequency jitter. This improves the frequency resolution of the whole spectrometer. Intensity instabilities of the laser are also reduced. Moreover all optical components become free from vibrations. As a consequence, intensity fluctuations at the detectors are reduced. This gives a better signal/noise-ratio, increasing the sensitivity of the spectrometer.

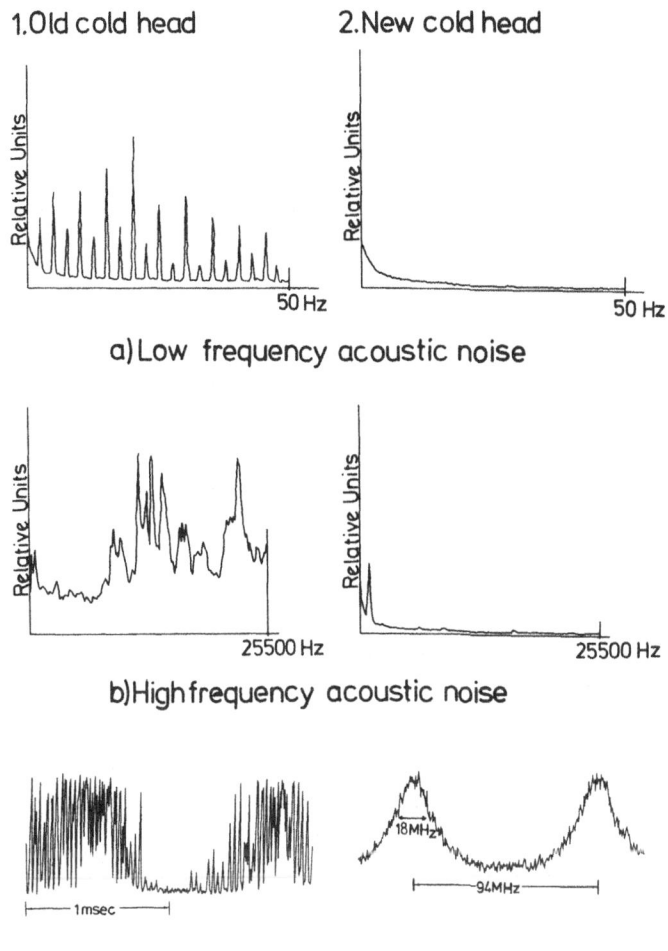

1. Old cold head   2. New cold head

Relative Units — 50 Hz   Relative Units — 50 Hz

a) Low frequency acoustic noise

Relative Units — 25500 Hz   Relative Units — 25500 Hz

b) High frequency acoustic noise

1 msec   18 MHz   94 MHz

c) Frequency jitter of the laser

Figure 4: a) and b) show a frequency analysis of the acoustic noise
c) shows two interference maxima of the 80 cm icFPI

4. MULTIPLE-REFLECTION CELL

A further important improvement is the use of a multiple-reflection cell. This cell
was developed according to the design of Herriott and has proved to be a very good
mean for detecting weak absorption signals. Great advantage of this type of cell is the
very simple adjustment once the two mirrors are brought into their correct positions.
Moreover the optical imaging through the cell is free of any astigmatism, which is
advantageous compared to White cells.

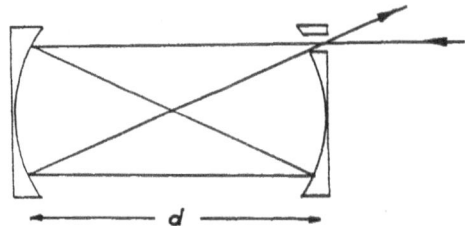

Figure 5: Schematic setup of a Herriott-cell

A Herriott cells [4],[5] consists of two spherical mirrors having a focal length f with a distance d, so that they form a nearly confocal resonator. The setup of a Herriott cell is shown in Fig. 5 schematically. The spots have a circular pattern on the surface of both mirrors. The number of the reflections is a function of distance d and focal length f:

$$d = 2f \left[1 - cos \left(\frac{\pi n}{k} \right) \right] \qquad k = 1, 2, 3, \cdots \tag{1}$$

It is of great importance that the beam waist of the laser is matched to that of the cell. Fig. 6 shows the reflections for both, the matched and the mismatched case. Cavity mismatching causes losses of intensity and is the main reason for fringes. Two cells of fixed path length, a 50 and a 30 meter, are available at present. A 134 m cell is being developed. All cells have a fundamental length of about 1 m.

matched 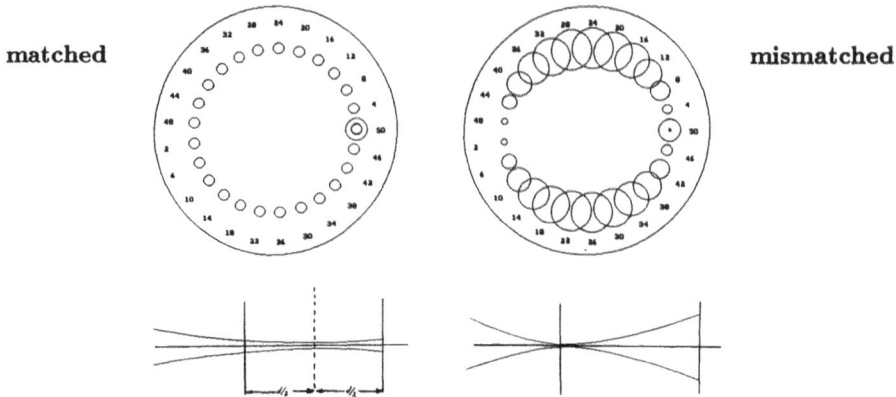 mismatched

Figure 6: Reflection pattern on the entrance mirror in the case of cavity matching and mismatching

5. PRESSURE EFFECTS

There are two different kinds of pressure effects: pressure shift and pressure broadening. In Fig. 7 both effects are demonstrated at a measured absorption line of water. The atmospheric absorption line is obviously much broader than the same at low pressure (14 mbar). Also the position of both lines does not coincide due to pressure shift.

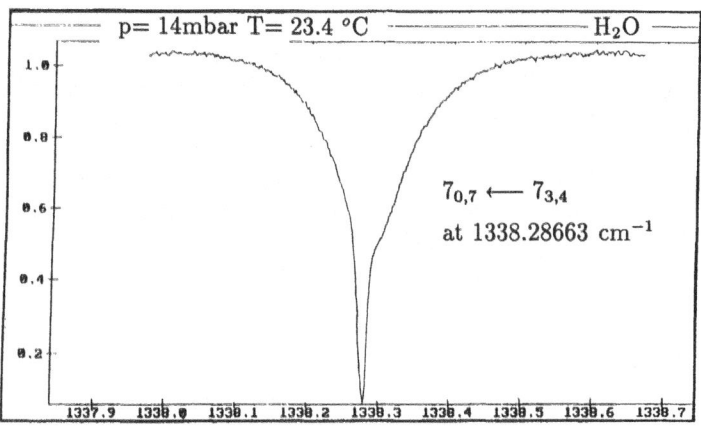

Figure 7: Absorption line of water at high (in atmosphere) and low pressure (14 mbar)

Both effects are very strong. Therefore, it is necessary to determine shift and broadening very accurately to be able to deconvolute measured atmospheric lines which give information about the altitude distribution of the molecules [6]. These data can help to understand the dynamic processes in the atmosphere. Besides, such measurements can be used to prove theoretical models describing pressure effects. In particular, accurate data about shifts are needed.

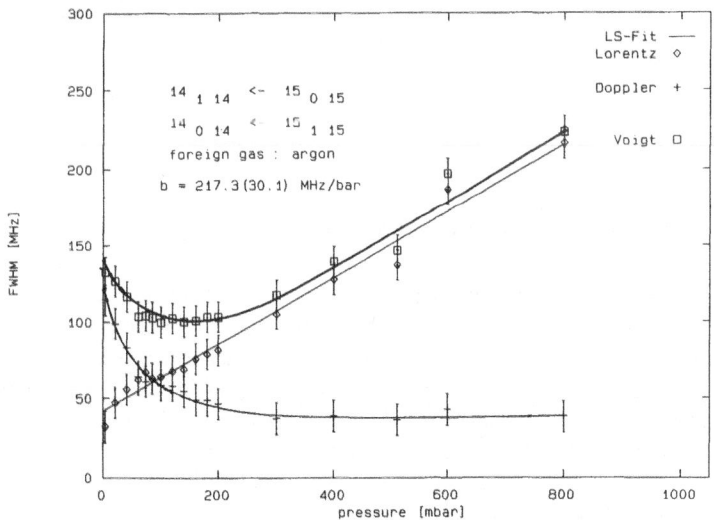

Figure 8: Dicke narrowing of the water $14_{1,14} \longleftarrow 15_{0,15}$ and $14_{0,14} \longleftarrow 15_{1,15}$ absorption doublet. In this analysis the Lorentz widths for the two lines are assumed to be equal.

Pressure effects are difficult to understand since they depend on many different parameters. Measurements at high rotational quantum numbers have shown Dicke narrowing [7] at some molecules like water. Absorption line profiles can be described by a Voigt profile, a convolution of a Lorentz and a Gauss profile. In some cases the Doppler width decreases with higher pressure so much that in spite of a linear increasing Lorentz half width the resulting Voigt half width has a minimum at a certain pressure. Fig. 8 shows Dicke narrowing at the doublet $J_{K_a,K_b} = 14_{1,14} \longleftarrow 15_{0,15}$ and $14_{0,14} \longleftarrow 15_{1,15}$ of water with the foreign gas argon.

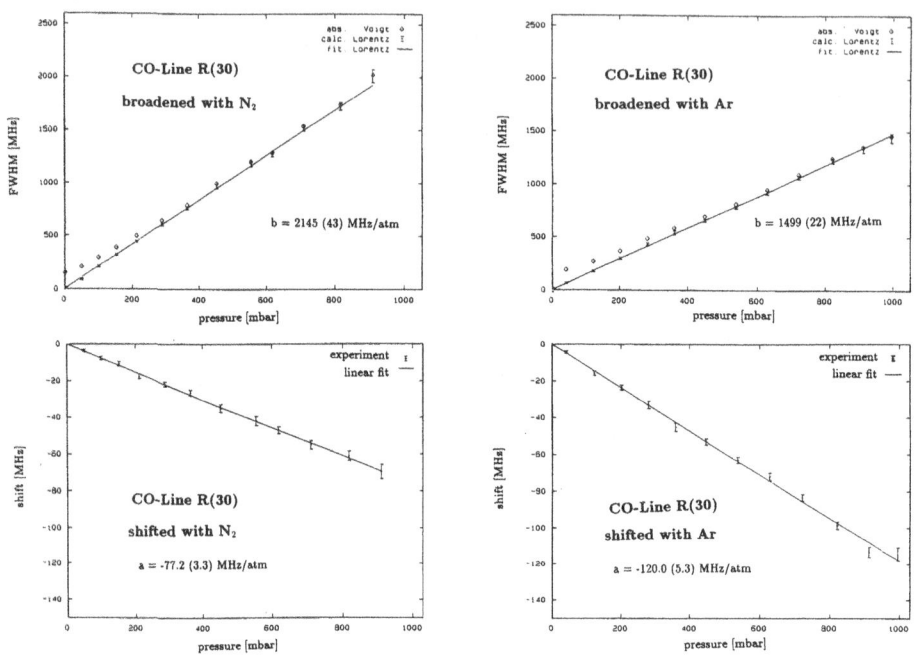

Figure 9: Different behavior of pressure effects by changing the foreign gas

The pressure effects of weak rovibrational transitions of CO with high rotational quantum number in the fundamental band were also measured. The results show that shift and pressure broadening do not behave in the same way when using different foreign gases (Fig. 9). Whereas nitrogen influences the broadening of a CO absorption line R(30) much more than argon, the pressure shift develops in the opposite way. All measurements were carried out at room temperature.

6. THE HETERODYNE EXPERIMENT

A heterodyne spectrometer has been developed for future atmospheric studies. First experiments with a fixed frequency laser (CO_2-laser) as local oscillator (LO) demonstrate that pressure effects can be measured. Figure 10 demonstrates two Ammonia lines, $2sQ(6,3)$ and $2sQ(9,7)$, detected by the heterodyne spectrometer.

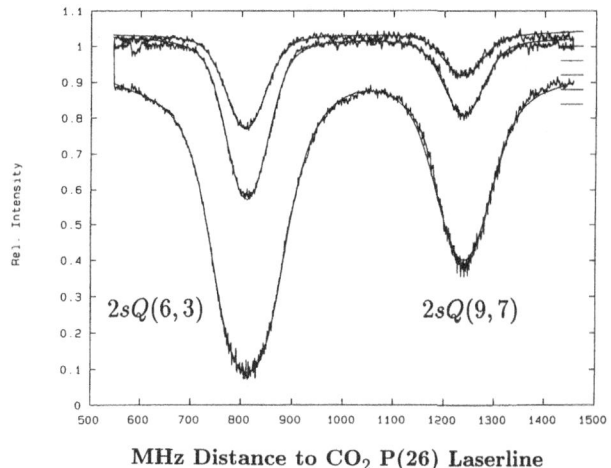

$2sQ(6,3)$ $2sQ(9,7)$

MHz Distance to CO_2 P(26) Laserline

Figure 10: Heterodyne detection of two Ammonia lines at various gas pressures

The setup of the spectrometer is shown in Fig. 11. The superposition of the LO and the radiation of the black body generates a heterodyne-signal on the HgCdTe detector. This signal is preamplified (17 dB) by a HEMT-amplifier. After further amplification the signal is finally analyzed by an acousto-optical spectrometer [8] in 1500 channels. These data are stored on disk by a personal computer.

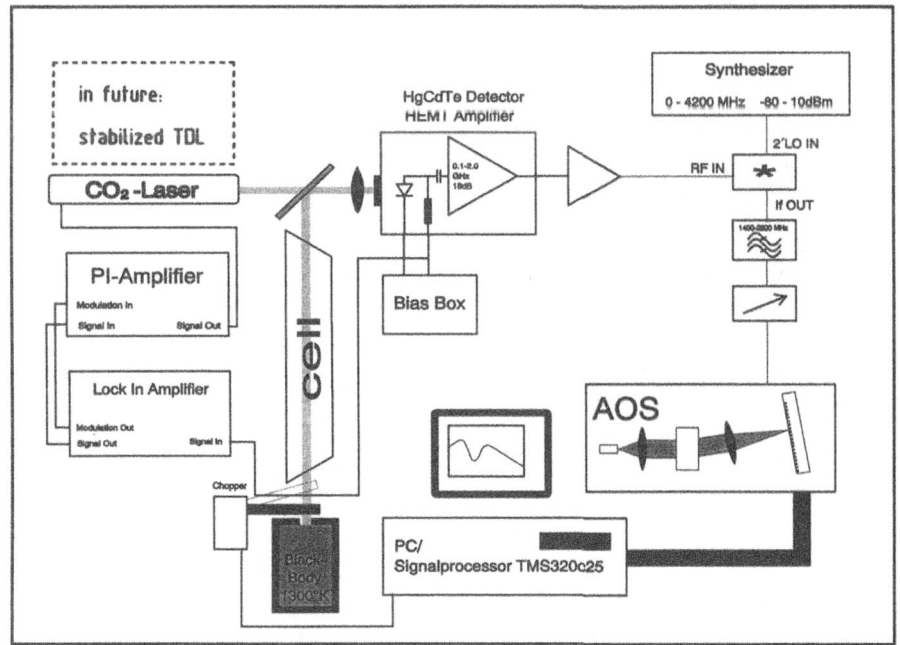

Figure 11: Setup of the heterodyne spectrometer

If the CO_2-laser can be replaced by a stabilized TDL as LO, broad frequency ranges will become accessible by this spectrometer to remote sensing in the upper atmosphere. The necessary stabilization technique can be applied as previously described. At present the only obstacle to realize such an IR-TDL heterodyne spectrometer is the lack of tunable high power (> 1 mW) single mode diode lasers. Hopefully this will be solved in near future.

REFERENCES

[1] S. Solomon, "Photochemistry and Transport of Carbon Monoxide in the Middle Atmosphere", Journal or the Atmospheric Sciences, Vol. 42, No. 10 (1985)

[2] M. Reich, R. Schieder, H.-J. Clar und G. Winnewisser, "Internally coupled Fabry-Perot interferometer for high precision wavelenth control of tunable diode lasers", Appl. Opt., Vol. 25, 130 (1986)

[3] H. J. Clar, R. Schieder, M. Reich, and G. Winnewisser, "High precision frequency calibration of tunable diode lasers stabilized on an internally coupled Fabry-Perot interferometer", Appl. Opt., Vol. 28, No. 9, 1648-1656 (1989)

[4] Donald R. Herriott and Harry J. Schulte, "Folded Optical Delay Lines", Applied Optics, Vol. 4, No. 8, 883–889 (1965)

[5] D. Herriott, H. Kogelnik, and R. Kompfner, "Off-Axis Paths in Spherical Mirror Interferometers", Applied Optics, Vol. 3, No. 4, 523–526 (1964)

[6] H. J. Clar, R. Schieder, G. Winnewisser, K. M. T. Yamada, "Pressure Broadening and Lineshifts in the ν_2 Band of NH_3", Journal of Molecular Structure, Vol. 190, 447-456 (1988)

[7] R.H. Dicke, "The Effect of Collisions upon the Doppler Width of Spectral Lines", Phys. Rev., Vol. 89, No. 2, 472–473 (1953)

[8] V. Tolls, R. Schieder, G. Winnewisser, The Cologne Acousto-Optical Spectrometer, Experimental Astronomy Vol. 1, 101 (1989)

PRESSURE BROADENING PARAMETER OF NO$_x$ USING TDLAS

V.V. PUSTOGOV, Ka. HERRMANN, F. KÜHNEMANN,
J. ORPHAL, B. SUMPF,
J. BERGER*$^{)}$ and H.-D. KRONFELDT*$^{)}$

Abteilung für Molekülphysik/Photobiophysik der Humboldt-Universität zu Berlin,
Fachbereich Physik, Invalidenstr.110, O-1040 Berlin;
*$^{)}$ Optisches Institut der Technischen Universität Berlin,
Straße des 17. Juni 135, D-1000 Berlin 12

SUMMARY

Linewidths measurements provide usefull data for both molecular parameters
and the monitoring of gaseous pollutants in the atmosphere. Line broadening
of NO$_2$ was studied using a double path TDL spectrometer with pulsed
operated lead chalcogenide lasers and equiped with single-pass cells and
multi-pass Herriott cells. Self- and foreign-gas broadening was measured for
individual lines in the fundamental band of NO around 1900 cm^{-1}, in the ν_3
band of NO$_2$ at 1603 cm^{-1} and in the 2ν_2 band of N$_2$O at 1160 cm^{-1}. The
experimental results are compared with literature data. Their application for
the numerical modelling of the NO$_x$ absorption spectra under atmospheric
conditions is discussed.

1. INTRODUCTION

The nitrogen oxides (NO, NO$_2$, N$_2$O) are important atmospheric pollutants
involved in the chemistry of both the troposphere (formation of photochemical
smog) and the stratosphere (ozone destruction). The accuracy of the trace gas
concentration measurements depends on the spectroscopic line parameters: positions,
strengths, broadening coefficients.
The aim of this work is to present measurements of self- and nitrogen-broadening
coefficients for some selected individual lines within the:

1 − 0 band of nitric oxide (NO) at 1935 cm^{-1};
2ν_2 band of nitrous oxide (N$_2$O) at 1163 cm^{-1};
ν_3 band of nitrogen dioxide (NO$_2$) at 1603 cm^{-1}.

2. THE SPECTROMETER

Our tunable diode laser spectrometer (Fig. 1) works in pulsed operation mode.
The lead chalcogenide diode lasers (DL), used in this study of NO$_x$, were made at
Humboldt-University (Berlin), at Fraunhofer-Institut für Physikalische Meßtechnik
(Freiburg) and at Lebedev Physical Institute (Moscow).

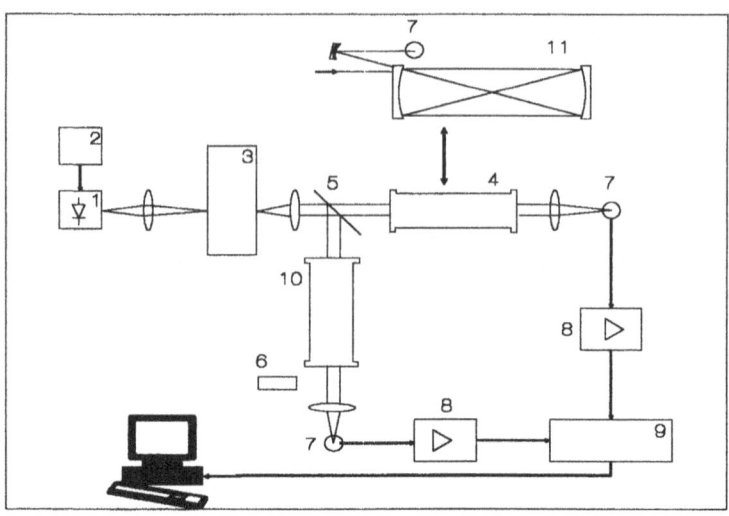

Figure 1: Scheme of the spectrometer:

(1) TDL in LHe cryostat, temp. stability 1mK
(2) laser control unit
(3) monochromator
(4) single pass cell
(5) beam splitter (KRS-5)
(6) Ge etalon
(7) HgCdTe photodetector
(8) amplifier
(9) 2 channel digital oscilloscope
(10) reference gas cell
(11) Herriott multipass cell

The signal registration was carried out by means of a 10 bit two-channel digital oscilloscope (Philips PM 3320A).

For the spectroscopy with diode lasers in pulsed operation mode the response time of the detector-amplifier system is important. For its determination we have used a diode laser mode jump (Fig. 2).

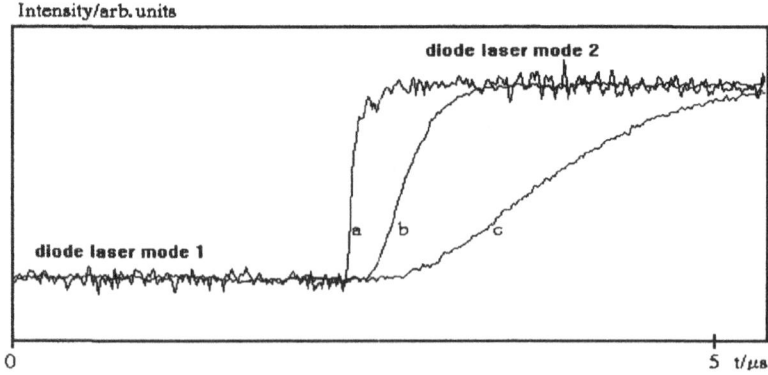

Figure 2: Determination of response time of the Detector-Amplifier-System using a diode laser mode jump.

It is interesting to mention, that with the help of this procedure it is possible to measure the response time of any photodetector at any desirable wavenumber. For example, for our HgCdTe detectors "1" and "2" the response times τ_r at 1040 cm^{-1} are 15 ns and 90 ns, respectively.

To analyse the shape of Doppler-broadened lines we must minimize the effect of the detector's response time. Therefore our spectral resolution should be at least one order of magnitude better as the width (FWHM) of the Doppler-broadened lines (in our case typical values of about $\Delta\tilde{\nu}_D \cong 3 * 10^{-3}$ cm^{-1}). Hence the diode laser tuning rate $D = d\tilde{\nu}/dt$ should be slower as $0.1\Delta\tilde{\nu}_D/\tau_r$ (for example, for our detectors "1" and "2" slower as 3 and 20 cm^{-1}/ms, respectively). All the measurements presented here were carried out with DL tuning rates $D < 1$cm^{-1}/ms. A DL pulse was typically several milliseconds long.

The spectral resolution of TDLAS is finally limited by the total diode laser emission linewidth. To estimate its influence on the registrated absorption linewidth we have compared the experimentally determined Doppler width of the NH$_3$ line at 1040.957 cm^{-1} (T=298 K)

$$\Delta\tilde{\nu}_{exp} = (3.16 \pm 0.04) * 10^{-3} \text{cm}^{-1}$$

with its theoretical value at the same temperature

$$\Delta\tilde{\nu}_D = 3.122 * 10^{-3} \text{ cm}^{-1}.$$

The absorption line was registrated simultaniously with Germanium etalon fringes in a single-shot mode of a digital oscilloscope (without sampling) within a single DL pulse (Fig. 3). The DL tuning rate was in this case $D \cong 0.2$ cm^{-1}/ms, so the influence of the response time of the detector-amplifier system on the original lineshape could be neglected. The only reason for the difference between observed and theoretical values was the linewidth of the DL itself. The observed NH$_3$ line in Fig.3 represents the convolution of the Doppler-broadened molecular line with the emission line of

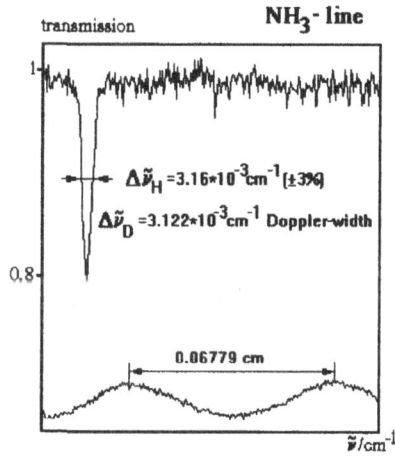

Figure 3. Estimation of the diode laser emission linewidth.

the diode laser. Assuming a Gaussian DL lineprofile, one gets the DL emission line width about 9.8 MHz, what is equivalent to wavenumber width $\Delta\tilde{\nu} \leq 3.3 * 10^{-4}$cm^{-1}.

3. NO$_2$

The nitrogen dioxide molecule contains an unpaired electron and therefore the spectrum of this stable free radical is complicated due to spin-doubling (as in the case of NO) and due to the strong Coriolis resonance between (001), (100) and (020) states. Each rotational level designated by the quantum number N is splitted into two levels with total angular momentum J=N+1/2 (F+ level) and J=N-1/2 (F- level). It is very important to know the individual linewidths $\Delta \tilde{\nu}$ (FWHM) and broadening coefficients γ_{xy} ("x" and "y" mean here NO_2 and a collision partner, respectively) for each NO_2 line, especially for those of atmospheric interest. Only then one can adequate interprete and calculate the NO_2 spectra in any gas mixture at given pressure and temperature. The spectroscopic data banks (HITRAN 1991, GEISA 1984) operate with one broadening coefficient γ_{xy}=0.067 cm^{-1}/atm ("y" here is air), equal for all lines in the whole band. There is only very few information on high resolution measurements of pressure broadening parameters for individual lines from the ν_3 band of NO_2 in literature. The paper of Malathy Devi et al.[1] reports the results on the self- and N_2-broadening coefficients for 20 lines from the ν_3 band of NO_2 in the wavenumber region between 1567.9 and 1586.3 cm^{-1}, covering the N-values between 24 and 46. We have investigated the nitrogen-broadening in wavenumber region from 1602.7 to 1604.2 cm^{-1} and report here about nitrogen-broadening coefficients of 10 ν_3-band lines with N between 8 and 14 near 1600 cm^{-1}. Each experimental absorption spectrum was registrated simultaniously with the interference fringes of the Ge etalon. Each curve in Fig.4 was registrated averaging over 8 DL pulses at DL pulse repetition rate of 100 Hz.

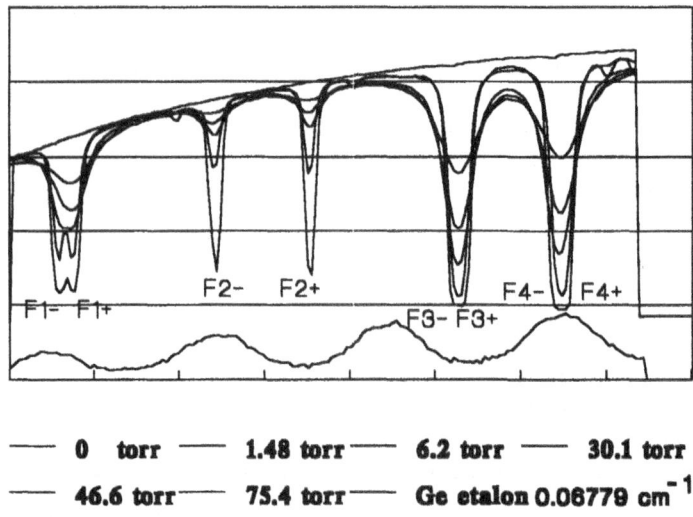

— 0 torr	— 1.48 torr	— 6.2 torr	— 30.1 torr
— 46.6 torr	— 75.4 torr	— Ge etalon 0.06779 cm^{-1}	

Figure 4. Nitrogen broadening in the ν_3 band of NO_2 at 1603 cm^{-1} .

The doublet splitting for investigated lines near 1603 cm^{-1}, shown in Fig.4, varies between $\Delta \cong 4 * 10^{-3}$ cm^{-1} for even at low pressure unresolved doublets (F3 and F4 in Fig.4 and Table 1) and $\Delta > 10^{-2}$ cm^{-1} for full resolved doublets (F2 and F5). The relation between widths of neighbouring lines (doublets) and their spacing is very important for the evaluation of broadening parameters.

If the doublet components are closely spaced, the behaviour of their broadening ought to be like one of a single line, exactly only at relative high pressure, if

$\gamma_{xy} * p >> \Delta = | \bar{\nu}(F+)-\bar{\nu}(F-) |$. So we have tried to determine the broadening parameter for doublets F3 and F4 (Fig.5) only for pressure values greater than 10 torr.

Table I. Broadening coefficients for NO_2 doublets about 1603 cm^{-1*}).

	$\bar{\nu}$/cm^{-1}	γ_{xy}/cm^{-1} atm^{-1}	N, K_a, K_c -->	N, K_a, K_c
F1-	1603.43783	0.0682	12, 3, 10-	13, 3, 11-
F1+	1603.44386	0.0682	12, 3, 10+	13, 3, 11+
F2-	1603.49665	0.0676	8, 5, 4-	9, 5, 5-
F2+	1603.53214	0.0674	8, 5, 4+	9, 5, 5+
F3-	1603.58922	0.068	14, 1, 14-	15, 1, 15-
F3+	1603.59356	0.068	14, 1, 14+	15, 1, 15+
F4-	1603.62682	0.065	13, 2, 11-	14, 2, 12-
F4+	1603.63084	0.065	13, 2, 11+	14, 2, 12+
F5-	1603.69316	0.071	10, 4, 7-	11, 4, 8-
F5+	1603.70691	0.069	10, 4, 7+	11, 4, 8+

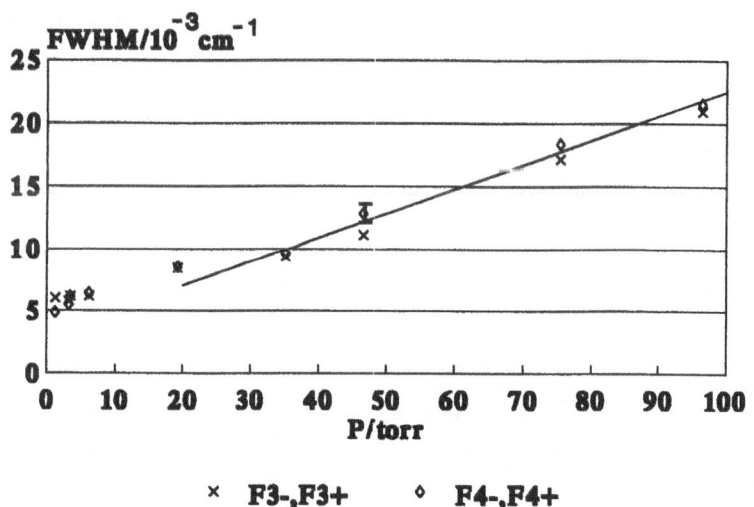

Figure 5. Pressure broadening of the unresolved doublets F3 and F4 (see Table 1).

The measured nitrogen broadening of the other lines is given by Fig.6 and Table 1.

---

* absorption line identification from A.Perrin [2]

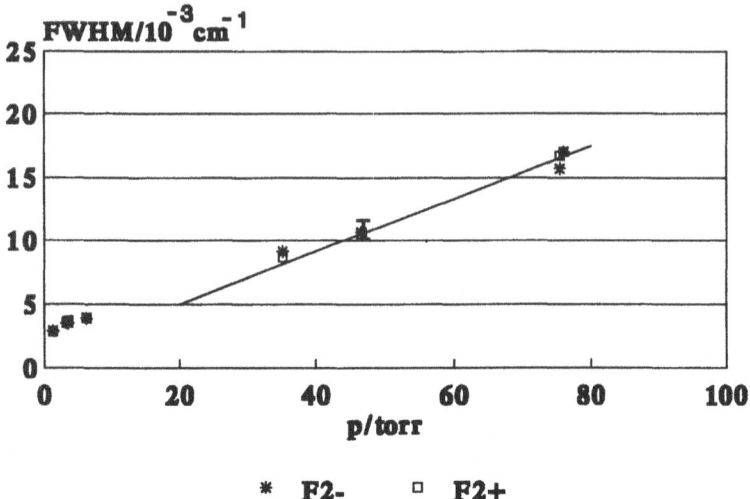

                    * F2-   □ F2+

Figure 6. Pressure broadening of the F+ and F- components of the doublet
        (8,5,4)--(9,5,5), see Table 1.

In Fig.7 are compared the broadening coefficients for different quantum numbers N: the dependence $\gamma_{xy}(N)$ seems to be weakly decreasing with increasing N. To make a more precise statement the study of individual linebroadening in the same band with the other N's is required.

Usually lines with high N-values are weak, but their investigation is of great interest for the theory of the NO_2 spectrum. Using a multipass cell of the Herriott type we can study lines and structures having small intensities (see Fig.8).

Additionally we report here about our measurements in which the Herriott cell has been used in order to estimate the minimal detectable concentration under atmospheric conditions. The behaviour of the strong lines and of neighbouring water line were of most interest. Simultaneously we have registrated a few weak absorption lines at the beginning of the DL pulse, where the tuning rate D was too high to make the correct lineshape analysis (absorption curve "c" in Fig.8). Due to the good diode laser mode stability it was possible to extend our experiment

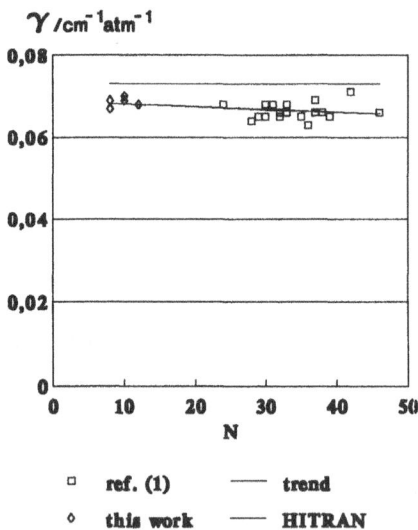

        □   ref. (1)   ——— trend
        ◇   this work   ——— HITRAN

Figure 7. Quantum number (N) dependence of broadening coefficients in the ν_3-band of NO_2.

to study the self-broadening of the weak absorption lines with a single pass 30 cm cell at pressure 10 Torr (the curve "b" in Fig.8). These experiments show, that Herriot-cells are useful in TDLAS study of individual broadening parameters of weak absorption lines.

The diode laser used in these experiments (made at IPM Freiburg) has shown an excellent time stability of the mode structure. In Fig.8 we compare measurements made within a time difference of five months. Both tuning rate and intensity within the TDL mode remained identical (under the same operation conditions).

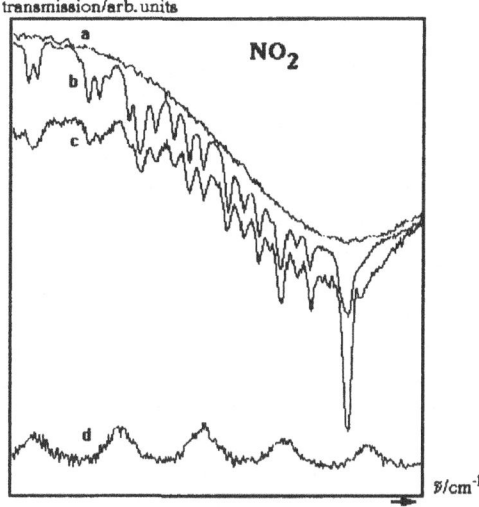

transmission/arb. units

NO$_2$

$\tilde{\nu}$/cm^{-1}

Figure 8. Registration of the weak NO$_2$ lines in the vinicity of a water line at 1603.320196 cm^{-1} (the line near the 5th etalon maximum), using a multi-pass Herriott cell.

Additionally we investigated the broadening of the $2\nu_2$ band of the nitrous oxide and of the nitricoxide fundamental. Here we present only two examples of these experiments.

4. N$_2$O

The linear 3-atomic molecule of nitrous oxide has a well resolved infrared spectrum. The broadening coefficients of the single lines are easily determinable. This is demonstrated in Fig.9, where the self-broadening of the single P(6) line in $2\nu_2$ band of the N$_2$O at 1163.132 cm^{-1} (Fig.9) was obtained to

$\gamma_{xx}=0.35\pm0.01$ cm^{-1}/atm

at 298 K (this work, Fig.9), what is slightly over the value

$\gamma_{xx}=0.32$ cm^{-1}/atm

at 300 K (Tubbs et al.,reference [3]).

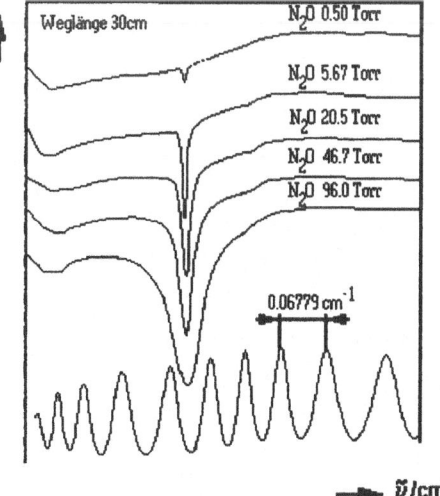

Transmission/arb.units

Weglänge 30cm

N$_2$O 0.50 Torr

N$_2$O 5.67 Torr

N$_2$O 20.5 Torr

N$_2$O 46.7 Torr

N$_2$O 96.0 Torr

0.06779 cm^{-1}

$\tilde{\nu}$/cm^{-1}

Figure 9. Determination of the self-broadening coefficient for P(6) transition in $2\nu_2$ band of N$_2$O at 1163.132 cm^{-1}. Path length 30 cm.

5. NO

In the spectrum of the NO molecule, containing an unpaired electron, spin-splitting in the order of magnitude 10^{-3} cm^{-1} occurs. In Fig.10 the components of the doublet at 1935.495 cm^{-1} are not resolved, even at low pressure because of the Doppler-broadening effect.

The nitrogen-broadening coefficient of the $^2\Pi_{3/2}R(18.5)$ transition in the 1-0 band of NO we measured to

$$\gamma_{xy}=0.057\pm0.003 \text{ cm}^{-1}/\text{atm}$$

at 298 K (this work, Fig.10). The result is in a good agreement with FTS data of Ballard et al.[4]

$$\gamma_{xy}=0.0526\pm0.0016 \text{ cm}^{-1}/\text{atm}$$

at 296 K.

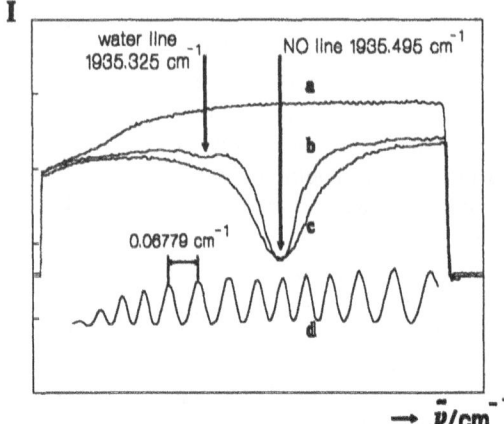

Figure 10. Nitrogen-broadening of the $^2\Pi_{3/2}R(18.5)$ transition in the NO fundamental. Total pressure 760 torr, path length 3.5 cm.

CONCLUSION

The TDLAS in pulsed operation mode with resolution $4*10^{-4}$ cm^{-1} allowed the investigation of individual broadening parameters of absorption lines.

In this work are presented the results on nitrogen-broadening for five doublets (10 lines) of the ν_3 band of NO_2 with low quantum numbers N from 8 to 14, on the self-broadening of the good isolated P(6) line of N_2O ($\gamma_{xx}=0.35$ cm^{-1}/atm at 298 K) and on the nitrogen-broadening of the $^2\Pi_{3/2}R(18.5)$ doublet of the NO fundamental ($\gamma_{xy}= 0.057$ cm^{-1}/atm at 298 K). The possibility for the registration of weak lines (NO_2) with a multipass Herriott cell was also demonstrated.

The combination of TDLAS with multipass Herriot cells is a powerful tool for trace simulation and direct pollutant measurements under atmospheric conditions. Thus we are able to check the results of our measurements and calculations directly.

REFERENCES
[1] V.Malathy Devi et al., J.M.S. **93**, 179 (1982)
[2] A.Perrin, priv. comm.
[3] L.D.Tubbs, D.Williams, J.O.S.A. **63**, 859 (1973)
[4] J. Ballard et al., J.M.S. **127** (1988)

THE ν_3 BAND OF SO_2 - LINE PARAMETERS FOR ATMOSPHERIC MONITORING

F. KÜHNEMANN, Y. HEINER, B. SUMPF, Ka. HERRMANN
N.V. LEMECHOV[*], E.V. STEPANOV[*]

Abt. Molekülphysik/Photobiophysik, Fachbereich Physik
der Humboldt-Universität zu Berlin
Invalidenstr. 110, Berlin, O-1040, Germany
[*] Institute of General Physics, U.S.S.R. Academy of Sciences,
Vavilov str. 37, Moscow, 117942, U.S.S.R.

SUMMARY

The ν_3 band of SO_2 near 7.3 μm, which is overlapped by the ν_2 band of H_2O, was studied with Tunable Diode Laser Spectroscopy using pulsed operated lead chalcogenide lasers. The measured line positions agree very good with previous Fourier Transform data, but more lines could be resolved with TDLS in the studied wavenumber windows. Line strengths measurements showed, that the vibrational-rotational interaction is not neglectable. The determined dipole moment derivative $\partial\mu_z/\partial Q_3 = (0.32 \pm 0.03)$ D agrees with literature data from integrated band strengths measurements.

Line broadening measurements showed a decreasing of both air- and self- broadening coefficients with increasing J". The air-broadening values (between $\gamma_{air} = (0.120 \pm 0.005)$ cm^{-1}/atm for J"=13 and $\gamma_{air} = (0.074 \pm 0.002)$ cm^{-1}/atm for J"=51) are in good agreement with theoretical literature data, derived with the Anderson-Tsao-Curnutte theory.

The improvement in the description of the atmospheric SO_2 spectrum with the present results is illustrated.

1. INTRODUCTION

The molecule SO_2 is an important atmospheric pollutant. The monitoring of trace gases such as SO_2 with infrared spectroscopic techniques demands a precise knowledge of the rovibrational spectra of these gases for the selection of suitable measurement procedures and conditions. That includes the determination of optimum wavenumber "microwindows" and the estimation of possible sensitivity limits due to system performance and/or overlapping from other atmospheric species.

Parts of the ν_3 band of SO_2, located near 7.3 μm, have been studied with pulsed operated Tunable Diode Laser Spectroscopy (TDLS). Line positions and intensities are compared with other experimental (Fourier Transform Spectroscopy) and theoretical data. Air and self broadening coefficients have been measured and compared with theoretical results.

2. EXPERIMENTAL DETAILS

The measurements were made with two different TDL spectrometers in pulsed operation with pulse lengths up to 2 ms for the wavenumber tuning. Both setups and the method of spectroscopy with pulse driven TDL are described in detail by Kosichkin et al. (1) and Kühnemann et al. (2). For signal registration a fast A-D-converter and a digital sampling oscilloscope, respectively, were used . Line locking for stabilisation provides a relative wavenumber accuracy and stability of $2 * 10^{-4} cm^{-1}$. Spectra were measured in direct absorption. Absolute wavenumber calibration was achieved using as a reference H_2O lines (Guelachvili (3)) in an additional gas cell or strong and rather isolated SO_2 lines as measured by Guelachvili (4). For relative wavenumber calibration the interference fringes of an Germanium Fabry-Perot-Etalon were used.

After the wavenumber calibration the line positions $\tilde{\nu}_i$ and the absorption coefficients were determined. An example is given in Fig.1: In the upper part (a) the recorded laser mode (averaged over 256 pulses) is shown together with the interference fringes of the Germanium etalon. In each pulse the laser was tuned through this mode within 100 μs. Fig.1b shows the derived absorption cross section χ (absorption coefficient α normalised to the SO_2 pressure).

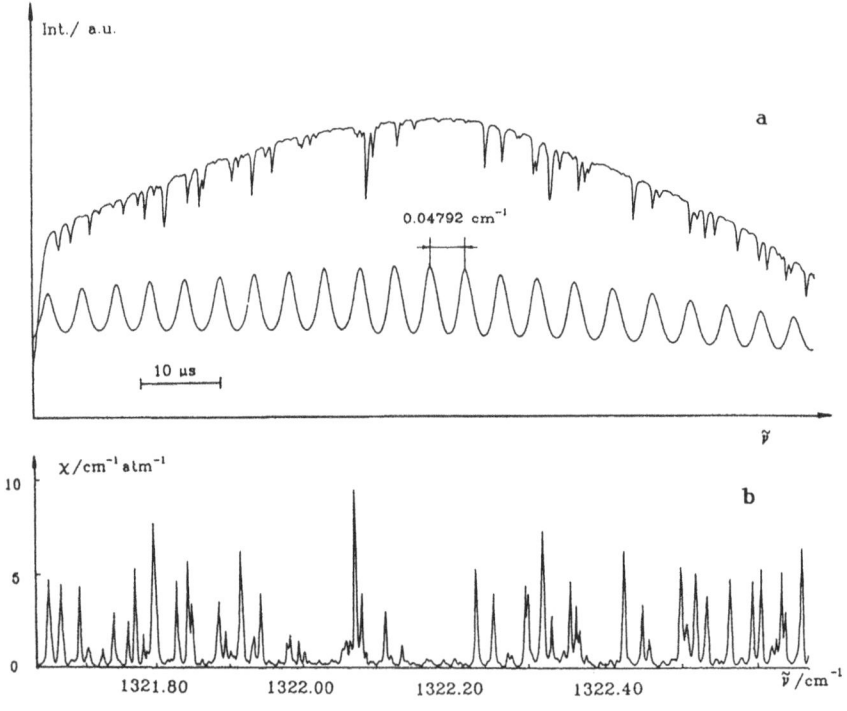

Fig.1: Part of the SO_2 spectrum near 1322 cm^{-1}
 a) 0.08 Torr SO_2, cell length 2m, additional Germanium etalon fringes
 b) spectrum calibrated and normalised to absorption

3. LINE POSITIONS AND IDENTIFICATION

Five wavenumber regions including 364 lines in the P branch of ν_3 (between 1321 and 1340 cm^{-1}) have been analysed. Comparison with the Fourier Transform data (Guelachvili(4)) showed, that more lines appear in the TDL spectrum than in the FT spectrum. That results from the lower resolution of the FT measurement (0.0054 cm^{-1}) in comparison with our Doppler limited TDLS (0.0021 cm^{-1} for the SO$_2$ measurements) and the fact, that in our TDL measurement the SO$_2$ pressure was set individually for the five wavenumber regions in dependence on the line strengths appearing.

Line identification was achieved using a calculated spectrum (Ulenikov/Naumenko, (5)) based on the experimental data from Guelachvili (4). The analysis included the subbands ν_3 (^{32}SO$_2$,^{34}SO$_2$) and $\nu_2+\nu_3-\nu_2$ (^{32}SO$_2$). In addition, two lines were found belonging to the $2\nu_2+\nu_3-2\nu_2$ subband of ^{32}SO$_2$.

Statistical analysis showed a very good agreement between the two sets of experimental data (TDLS and FTS) giving a rms of ±0.00031 cm^{-1}, which is close to the experimental uncertainty of the TDLS (±0.0002 cm^{-1}). For the TDLS and the calculated spectra, the equivalent value is 0.0015 cm^{-1}. In addition, the greatest deviations between experimental and calculated line positions occure for lines with K_a values > 20. Both these results indicate the necessary improvements in the theoretical description of the rotational parameters.

4. LINE STRENGTHS

Experimental line strengths were determined from
$$\alpha(\tilde{\nu}) = S_i * f(\tilde{\nu} - \tilde{\nu}_i) * p_{SO2}$$
with $f(\tilde{\nu} - \tilde{\nu}_i)$ - line shape function, p_{SO2} - SO$_2$ partial pressure. The contribution of pressure broadening (even at sample pressure 1 Torr) was taken into account using the value of the Voigt line shape function at the line centers with the broadening coefficients presented later.

Using different SO$_2$ partial pressures in the 5 wavenumber windows it was possible to cover a line strength range of more than three orders of magnitude with sufficient accuracy.

For a comparison theoretical values from Ulenikov/Naumenko (5) were used: data from a direct calculation (without vibrational-rotational interaction) leaving the dipole moment derivative $^z\mu_3 = \partial\mu_z/\partial Q_3$ as the only parameter. An example for the result is given in Fig.2 with S_{exp} vs. S_{calc} for the wavenumber window around 1338 cm^{-1}. The contributions from the subbands are shown. For the determination of $^z\mu_3$ the measured line strengths were weighted with their experimental uncertainty resulting from the noise of the laser signal (SNR=300:1) in direct absorption. This uncertainty is indicated in Fig. 2 by the dashed lines. For the lines inside such a window (which are measured under constant pressure conditions) the scattering of the S_c/S_{exp} ratio is much greater ($> 10\%$) than the experimental uncertainty (see also Fig. 3). In addition, the direct calculation gives too small values for S_c for lines with $K_a > 20$. These facts indicate, that a more precise description of the line strength demands the consideration of the vibrational-rotational interaction.

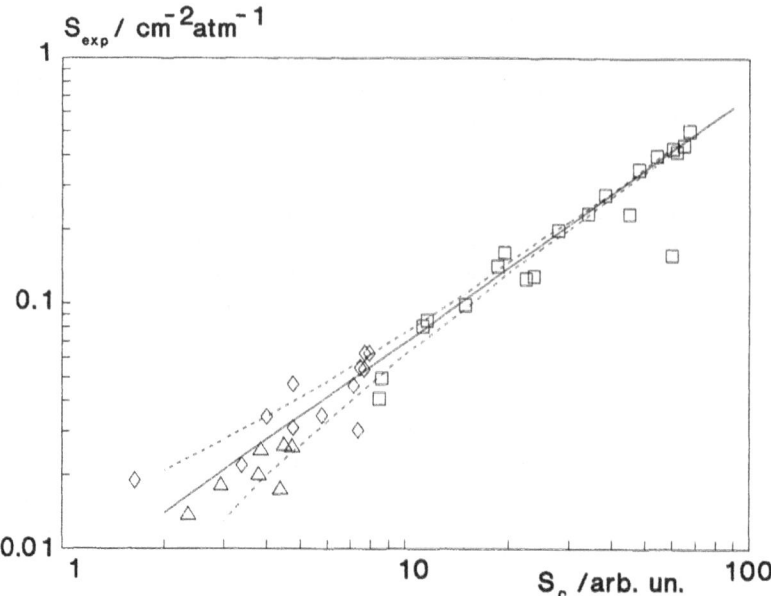

Fig.2: Measured and calculated line strengths around 1338 cm^{-1}
(□) 001← 000 ($^{32}SO_2$)   (◇) 011← 010 ($^{32}SO_2$)   (△) 001← 000 ($^{34}SO_2$)

Taking into account these facts the value for the dipole moment derivative $^z\mu_3$ was determined as $^z\mu_3=(0.32\pm0.03)$D, which is in good agreement with other experimental data derived from integrated band strength measurements (Secroun (6), Tejwani (7)).

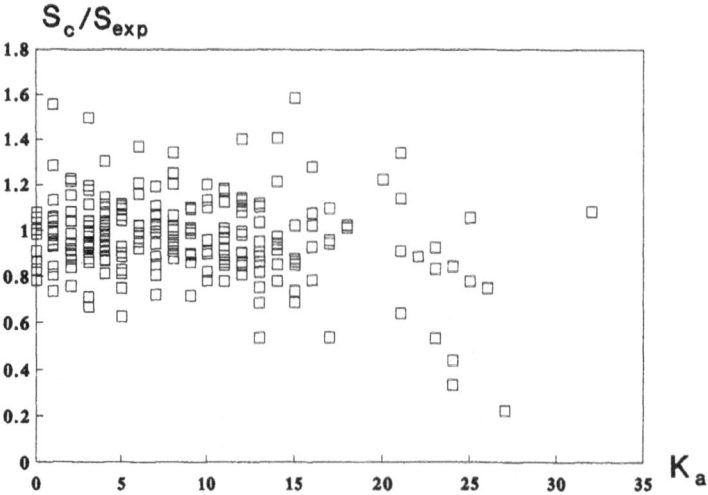

Fig.3: K_a dependence of the relation between calculated and experimental line strengths

5. LINE BROADENING

Air-and self-broadening was studied for P branch transitions with J" values between 13 and 51. Groups of closely spaced lines (with $J"_{max} - J"_{min} \leq 3$) were analysed using a Fourier analysis technique (see Ref. (1)). That allowed the separation of the Doppler and Lorentzian contribution to the linewidths and the derivation of an averaged broadening coefficient for all lines included in such a fit.

The results (see Ref. (2)) show an decrease of both air-and self-broadening coefficient with increasing J" in the studied range ($13 \leq J" \leq 51$). They are given in detail in in Table I. From that one can conclude, that a single constant air-broadening coefficient (as e.g. used in the HITRAN data base (8)) is not sufficient for the description of the whole band and should be substituted by the J" dependent values.

Table I: Broadening coefficients in cm^{-1}/atm as measured with TDLS in the present study (Ref.(2), by courtesy of Academic Press, Inc.)

J'	K_a'	K_c'	J"	$K_a"$	$K_c"$	$\tilde{\nu}$ /cm^{-1}	$2\gamma_{SO2-air}$	$2\gamma_{SO2-SO2}$
13	2	11	14	2	12	1352.569 ⎫		
13	4	9	14	4	10	1352.577 ⎬	0.24 ± 0.02^a	1.1 ± 0.1^a
12	6	7	13	6	8	1352.625 ⎭		
27	6	21	28	6	22	1342.545	0.187 ± 0.004	
27	6	21	28	6	22	1342.545 ⎫		
27	3	24	28	3	25	1342.569 ⎬	0.191 ± 0.005^a	
28	3	24	29	3	25	1342.578 ⎭		
34	9	29	35	4	30	1337.704 ⎫		
35	2	33	36	2	34	1337.753 ⎬	0.180 ± 0.018^a	0.71 ± 0.05^a
36	1	34	37	1	35	1337.831 ⎪		
37	0	37	38	0	38	1337.870 ⎭		
49	9	40	50	9	41	1325.653 ⎫	0.148 ± 0.005^a	0.71 ± 0.12^a
50	5	45	51	5	46	1325.665 ⎭		

[a] averaged values derived with the Fourier analysis technique, the first columns give the strongest lines in the individual line groups for orientation

The experimental data have been compared with calculations performed by Tejwani (Ref. (9,10)) using the Anderson-Tsao-Curnutte theory (see Fig. 4). The error bars in the J" direction arise from the averaging process in the Fourier analysis technique. Taking into account the experimental uncertainties there is a overall good agreement between the experimental and theoretical values for both air-and self-broadening. Nevertheless, experimental accuracy and the data analysis (with the Fourier analysis technique) aren't yet sensitive enough for a more detailed discussion of the applicability limits of the Anderson-Tsao-Curnutte theory in the case of SO_2. These problems will be subject to further work.

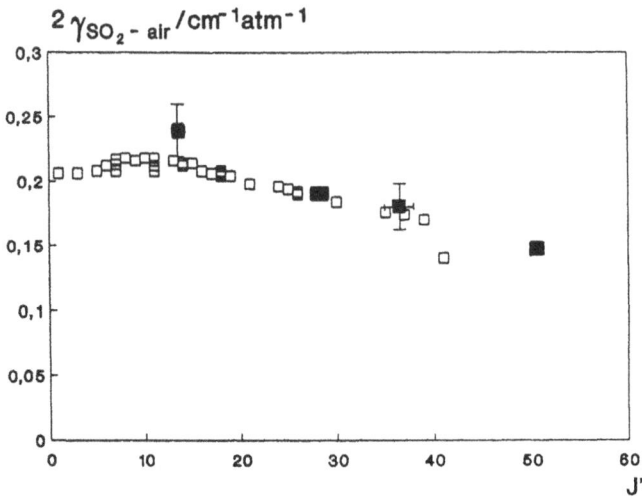

Fig.4: J" dependence of the air-broadening coefficient
(□) theoretical values (Tejwani, Ref.(9,10))
(■) experimental values (this study)
(Ref.(2), by courtesy of Academic Press, Inc.)

6. CONSEQUENCES

As a summary we want to outline the consequences of the present spectroscopic results on SO_2 for atmospheric monitoring and the modelling of atmospheric spectra. Our results show, that the following improvements have to be made in comparison with previous data:
- The present HITRAN value for the air-broadening of SO_2 ($2\gamma_{SO2-air}$=0.304m^{-1}/atm) is up to 2 times higher than the experimental and theoretical values, reported here, and neglects the J" dependence. To illustrate this difference, the SO_2 absorption spectrum around 1322 cm^{-1} has been synthesized with (a) $2\gamma_{SO2-air}$= 0.130 cm^{-1}/atm (our results) and (b) $2\gamma_{SO2-air}$= 0.304 cm^{-1}/atm, (Ref.(8)) respectively, for 760 Torr. For line positions and strengths the data from the present study were used (see Fig.5). The smaller broadening coefficient results in a greater contrast between continuum and resonance absorption.
- In addition to the 001← 000 transition of $^{32}SO_2$ other subbands have to be taken into account for the description of the SO_2 absorption around 7.3μm. This became obvious especially during the measurements of the low frequency P branch wing (wavenumber region around 1322 cm^{-1}). Hot and isotopic band centers are shifted to lower wavenumbers relative to the cold band. Due to that all three subbands exhibit lines with

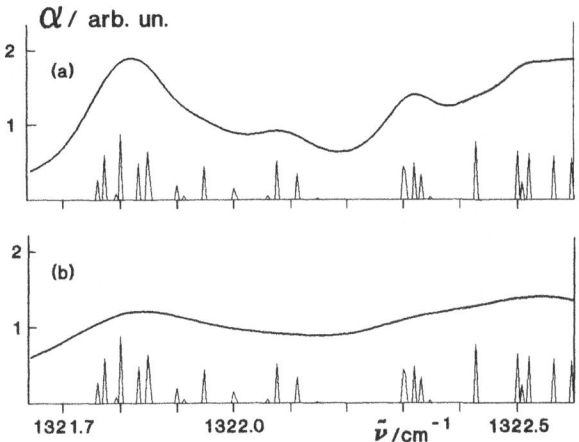

Fig.5: Synthesized SO_2 spectrum at 760 Torr. For a) and b) see text. Positions and strengths of the lines are given.

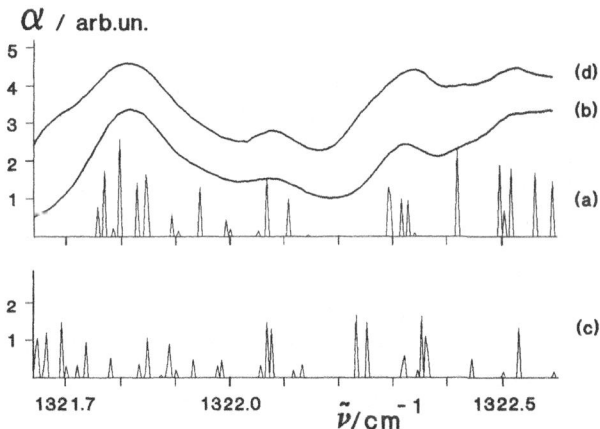

Fig.6: SO_2 spectrum around 1322 cm^{-1}
   a) line positions and strengths of the cold band (001← 000, $^{32}SO_2$)
   b) calculated spectrum at 760 Torr from these lines
   c) line positions and strengths of the hot and isotopic band (011← 010, $^{32}SO_2$,
      (001← 000, $^{34}SO_2$)
   d) synthesized spectrum at 760 Torr including the lines from a) and c)

comparable strengths despite the low isotopic concentration ($4.2\%\,^{34}SO_2$) and the population of the 010 state of $^{32}SO_2$ (8% at 300K). The result of the modelling around 1322 cm^{-1} at 760 Torr is shown in Fig.6: in a) and c) the experimental line positions and strengths of the cold band (001← 000) of $^{32}SO_2$ and of the hot band (011← 010 of $^{32}SO_2$) and the isotopic band (001←000 of $^{34}SO_2$), resp., are given. The synthesized spectrum at 760 Torr is shown in b) for the cold band transitions only and in d) including the lines of all three subbands.
- The line strengths data showed, that the vibrational-rotational interaction has to be taken into account for an accurate description of the experimental data. The analysis of this subject is still in progress.

Finally we want to state, that the improvement of spectral paramaters such as line positions, strengths and broadening coefficients should lead to a higher reliability of IR methods for atmospheric trace gas measurements and remote sensing.

ACKNOWLEDGEMENT

The authors are indebted to O.V.Naumenko and O.N.Ulenikov, Tomsk, U.S.S.R., who provided their calculated data on SO_2 line positions and strengths.

REFERENCES

(1) Yu.V.KOSICHKIN, A.I.NADEZHDINSKII, E.V.STEPANOV, J.Quant. Spectrosc. Radiat. Transfer 43, 499-509 (1990)

(2) F.KÜHNEMANN, Y. HEINER, B. SUMPF, Ka. HERRMANN, J.Mol.Spectrosc. (accepted)

(3) G.GUELACHVILI J.Opt.Soc.Am.73, 137 (1983)

(4) G.GUELACHVILI, O.V.NAUMENKO, O.N.ULENIKOV, J.Mol.Spectrosc. 125, 128-135 (1987)

(5) O.V.NAUMENKO, O.N.ULENIKOV, private communication

(6) C.SECROUN, A. BARBE, P.JOUVE, J.Quant.Spectrosc. Radiat.Transfer. 13,1325 (1973)

(7) G.D.T.TEJWANI, K.FOX, R.J.CORICE Jr., Chem.Phys.Lett. 18,365 (1973)

(8) L.S.ROTHMAN, R.R.GAMACHE, A.GOLDMAN, L.R.BROWN, R.A.TOTH, H.M.PICKETT, R.L.POYNTER, J.-M.FLAUD, C.CAMY-PEYRET, A.BARBE, N.HUSSON, C.P.RINSLAND, and M.A.H.SMITH Appl. Opt. 26, 4058-4097 (1987) and the 1991 edition of HITRAN

(9) G.D.T.TEJWANI, J.Chem.Phys.57,4676-4681,(1972)

(10) G.D.T.TEJWANI, unpublished results

PURE ABSORPTION SPECTROSCOPY OF MOLECULAR OXYGEN USING A CW AlGaAs LASER

M. de Angelis*, F. Marin**,F. S. Pavone, G. M. Tino*** and M. Inguscio****

European Laboratory for Non-linear Spectroscopy

*Istituto di Cibernetica, CNR, Arco Felice, Napoli
**Scuola Normale Superiore, Pisa
***Dipartimento di Scienze Fisiche dell'Università di Napoli
****Dipartimento di Fisica dell'Università di Firenze

Summary

A direct investigation of the forbidden $b^1\Sigma_g^+ \leftarrow X^3\Sigma_g^-$ band of O_2 by means of a AlGaAs diode laser at 760 nm is performed. Frequency control by optical feedback makes the quantitative analysis of lineshapes possible. Pressure broadening parameters are determined for different temperatures and rotovibrational components.

1. INTRODUCTION

Molecular oxygen detection is of great interest in many fields. Its analysis by magnetic or potentiometric techniques is routinely performed in medicine, biology and industry. In many cases a spectroscopic detection would be desirable for reasons of response time, sensitivity or absence of contact with the sampling gas.

As for as absorption spectra concern, allowed electronic transitions generate spectra in the UV, while rotovibrational transitions in the infrared are electric dipole forbidden by the molecule symmetry. However magnetic-dipole electronic transitions are responsible for a very weak absorption band ($b^1\Sigma_g^+ \leftarrow X^3\Sigma_g^-$) in the visible-near infrared, as originally evidenced in pioneristic works performed in the Earth atmosphere [1,2]. In spite of an absorption coefficient of only $\sim 10^{-6}$ cm^{-1}, these transitions are now made accesible to direct laboratory investigations thanks to high sensitivity laser spectroscopy. In the early eighties, the group at Stanford [3, 4, 5] started investigations using intracavity absorption in broad-band dye lasers. Sensitivities and detection limits of intracavity lasers spectroscopy are also discussed by Harris [6].

More recently, GaAlAs diode lasers have been successfully used as sources for absorption spectroscopy of the forbidden atmospheric band [7, 8]. Here the necessary high sensitivity was achieved either using a multipass configuration [7] or applying an external magnetic field on the sample for Zeeman modulation of the absorption and consequent phase sensitive detection [8].

It is worth noting that single mode diode lasers can have an amplitude noise level reduced to few parts in 10^6 , i.e. 10-20 dB over the shot noise limit, expecially when they are stabilized by optical feedback in extended cavity configuration [9]. This opens the possibility of a direct investigation of the forbidden band over a broad pressure range, with a careful quantitative analysis of the lineshape.

In this work we report measurements with oxygen pressures down to 10 Torr (1.5 meter path lenght) and investigate collisional broadening as a function of rotational quantum number and temperature.

2. EXPERIMENTAL APPARATUS

The experimental apparatus is schematically shown in Fig.1.

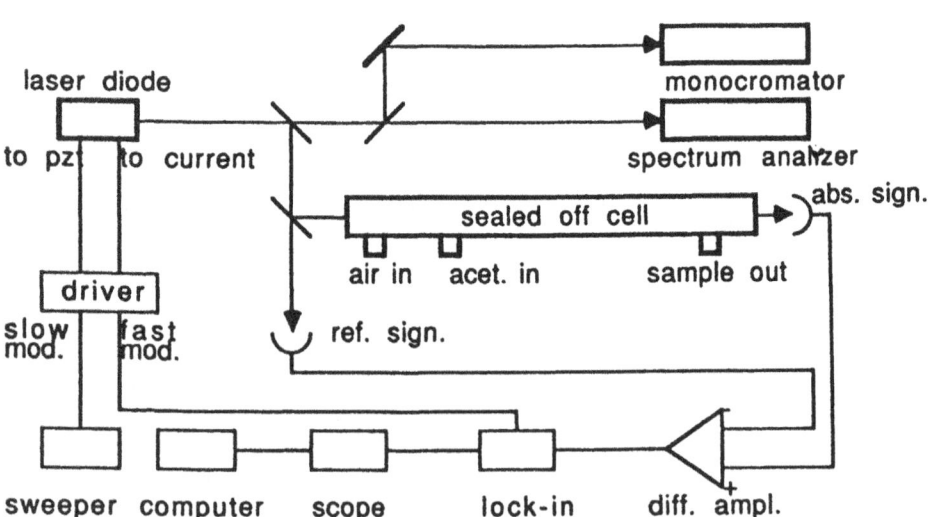

FIG. 1: Experimental apparatus

The general scheme is similar to the one used for overtone spectroscopy [9], with two main differences:

a) No room temperature diode laser are available at the desired wavelenght (760 nm), with good spectral qualities to allow lineshape narrowing and frequency control. Hence we utilized a diode laser at 780 nm (Sharp, model LT024MD0), operated at low temperatures, as already reported in [8]. In addition, we took advantage of the extended cavity configuration allowing line narrowing and frequency control by means of optical feedback. In particular, the laser was attached to a small copper block in thermal contact with a Peltier cell and all the mountings, collimating lens included, were positioned in a under vacuum box with an output AR window. We operated the laser at -30 Celsius degrees and completed the coarse tuning to the 760 nm region by means of the grating. The tuning range overlapped the whole vibro-rotational structure of the

$b^1\Sigma_g^+ \leftarrow X^3\Sigma_g^-$ band of O_2.

Note that the total wavelenght gap could have been covered simply by temperature tuning, should high spectral purity have not been a stringent requirement. Indeed from Fig. 2 it is possible to have an idea of the tuning of the laser output for temperatures down to 10 K [10].

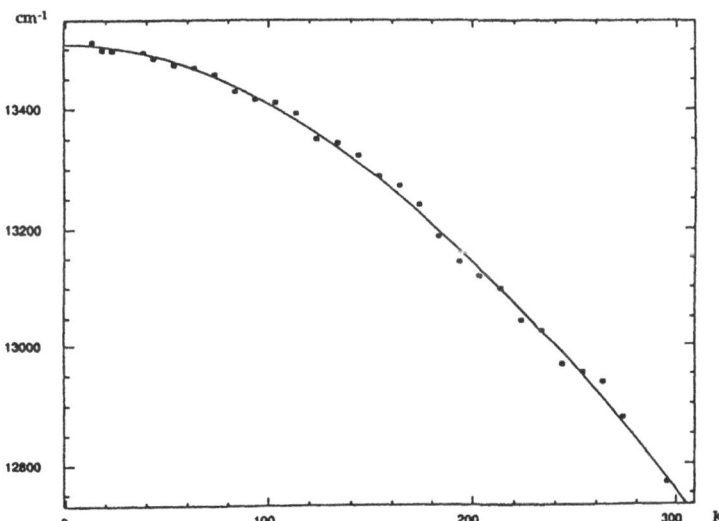

Fig. 2: Coarse frequency output of a AlGaAs diode laser as a function of the temperature. In this work grating tuning was added for improved spectral purity operation.

b) The stainless steel absorption cell was completely contained in a dewar in order to allow operation at low temperatures by means of immersion in liquid nitrogen (measurements were also performed leaving the liquid nitrogen to evaporate and keeping in the dewar the cool vapors, in order to have the sample at temperatures above 77K). The actual O_2 temperature was determined "a posteriori" from the doppler contribution to the molecular transition lineshape.

Fine frequency scans across the selected line, achieved by simultaneous injection current and grating tuning , were calibrated by means of a high-finesse Fabry-Perot confocal resonator with a free spectral range of 75MHz. Gas pressure was measured by means of an absolute capacitance manometer. The absorption signal was subtracted from the reference signal to avoid the background slope due to amplitude modulation.

3. RESULTS AND ANALYSIS

Doppler limited lineshapes could be recorded for several rotovibrational

components of the $b^1\Sigma_g^+ \leftarrow X^3\Sigma_g^-$ band. A typical room temperature recording is shown in fig. 3 for the ^PP (N =9) component at 13094.75 cm-1.

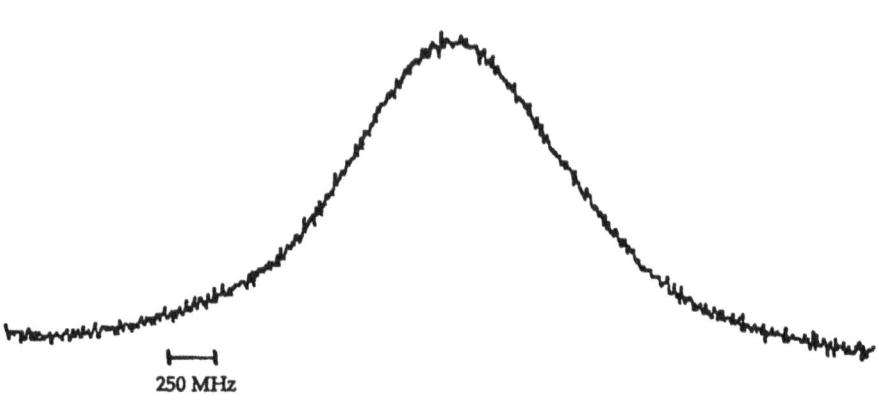

250 MHz

Fig 3: Direct absorption detection of the ^PP (N =9) component of the forbidden O_2 band (pressure 120 Torr, absorption path lenght 150 cm).

Lineshapes are determined by Gaussian Doppler and Lorentzian collisional broadenings. The linewidth (FWHM) is plotted as a function of the pressure and the two contributions are determined by fitting the resulting behaviour using Padé approximation [11]. This is shown in Fig. 4 for two different rotational components of the ^PP branch.

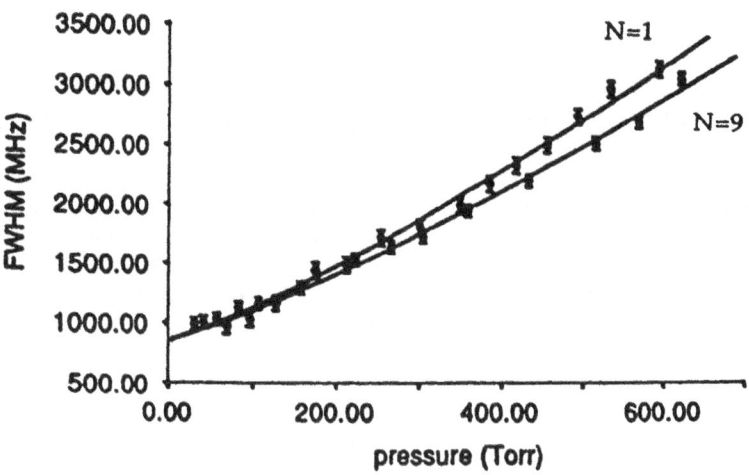

Fig. 4: FWHM Voigt profile versus pressure a) PP(1) component, b) PP(9)
   component

Pressure broadening (FWHM) parameters are then reported for N = 9, 3, 1
in TABLE 1.

	^{P}P branch		
component	9	3	1
press. broad value MHz/Torr	4.2 (1)	4.73 (7)	4.83 (8)

Table I: pressure broadening parameter values for different rotational
   quantum number (N).

These values slightly depend on the orbital angular momentum value N,
and in particular the pressure broadening parameter increases as N
decreases. This behaviour is consistent with previous observations on the

microwave spectrum [12]   (4.4, 4, 3.4 MHz/Torr for N = 1, 3, 9 respectively).
The values are also in agreement   with those obtained at higher
frequencies in the microwaves [13]   and in the far infrared [14, 15]. A
quantitative comparison is not straightforward since in our case the upper
level of the transition involves a different electronic state.

An additional problem for a quantitative comparison of pressure
broadening parameters obtained under different experimental conditions
arises from the strong and not very well established dependence on the
temperature [13]. As a consequence we decided to undertake a study of the
temperature dependence of the pressure broadening. We chose the lowest
N rotational component in order to take advantage of the signal increase
at low temperature. Indeed at 77 K the signal was a factor 10 larger than at
room temperature as it is shown in Fig. 5.

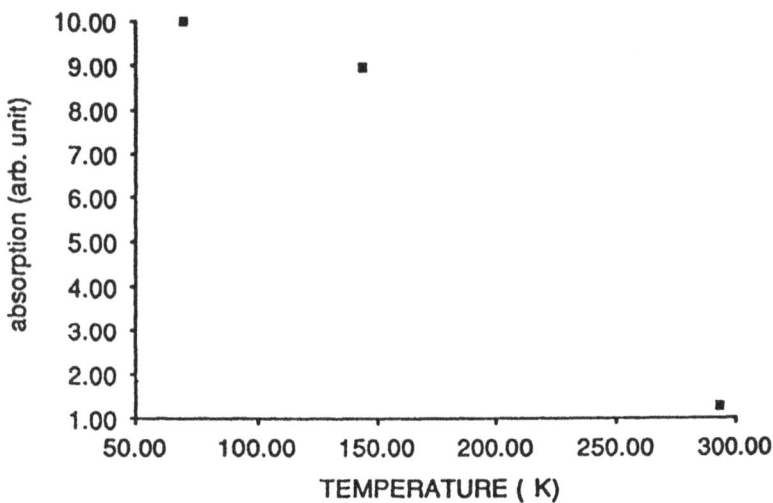

Fig. 5: signal absorption versus temperature on the PP(1) component

Different temperatures were set and   determined as discussed in the
previous section. For each temperature the Voight profile FWHM was
studied as a function of pressure as already discussed for the room
temperature measurements. From the extrapolated Doppler width, actual
temperature could be determined with an accuracy of about 10%. The
pressure broadening parameter as a function of the temperature is shown
in Fig. 6 .

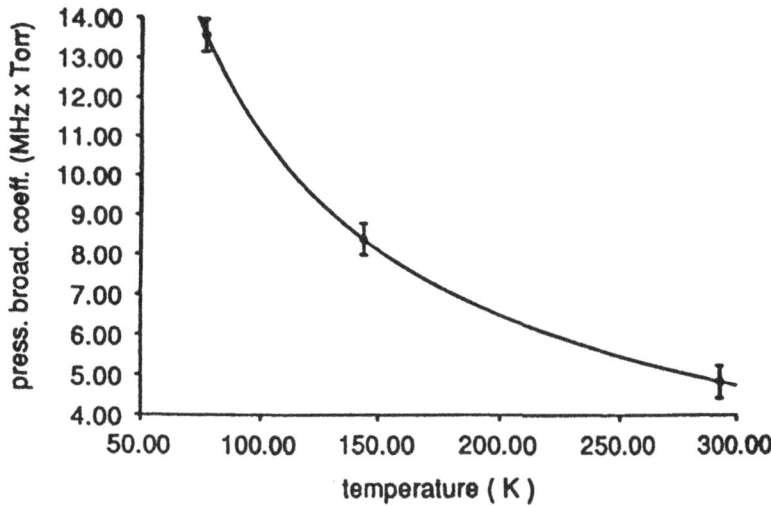

Fig. 6: pressure broedening parameter versus temperature

According to [16] the experimental points are fitted to the theoretically predicted curve $\gamma(T) = T^{-n}$ where n should range between 0.5 and 1 . From the fit we obtain a preliminary value n = 0.77 (4) suggesting a very short range intermolecular force low.

4. CONCLUSIONS

We have detected the electronic magnetic-dipole $b^1\Sigma_g^+ \leftarrow X^3\Sigma_g$ band of oxygen also studyng the effect of the temperature on the intensities and the collisional lineshapes. We think that in particular the temperature dependence of pressure broadening can be of great interest because of theoretical studies [13] and in view of molecular oxygen detection in the stratosphere.

REFERENCES

[1]   G. H. Dieke and H. D. Babcock. Proc. Nat. Acad. Sci. U.S.A. 13, 670 (1927).
[2]   H. D. Babcock and L. Herzberg, Astrophys. J. 108, 167 (1948).

[3]   W. T. Hill III, R. A. Abreu, T. W. Hansch and A. L. A. Schawlow, Opt. Comm. 32, 96-100 (1980).

[4]   W. T. Hill III, T. W. Hansch, and A. L. Schawlow, Appl. Opt. 24, 3718 (1985).

[5]   W. T. Hill III and A. L. Schawlow , J. Opt. Soc. Am. B, 5, 745-748 (1988).

[6]   S. J. Harris, Appl. Opt.,23, No. 9, 1311-1318 (1984).

[7]   R. Grisar, Proc. of the Int. Symposium: Monitoring of Gaseus Pollutants by Tunable Diode Lasers, Freiburg, F.R.G. 17-18 October (1988).

[8]   H. Kanamori, M. Momona and K. Sakurai, Can. J. Phys. 68, 313 (1990).

[9]   F. S. Pavone, F. Marin, M. Inguscio, K. Ernst, G. DI Lonardo, 'Sensitive Detection of Acetylene Absorption In The Visible Using a Stabilized AlGaAs Diode Laser', and References Therein, This Conference Digest.

[10] F. Marin, 'Spettroscopia ad alta risoluzione dell'atomo di   ossigeno con   laser a semiconduttore ad alta purezza spettrale ', Thesis, University   of Pisa (1990)

[11] P. Minguzzi and A. di Lieto, J. Mol.Spectros 109, 388-394, (1985).

[12] H. J. Liebe, G. G. Gimmestad and J. D. Hopponen, IEEE Trans Antennas and Prop. AP-25, 327 (1977).

[13] H. M. Pickett, E.A. Cohen, and D.E. Brinza, Astrophys. J. 248, 49-51 (1981).

[14] D.A. Jennings, K.M. Evenson and M.D. Vanek, Geophys. Res. Lett., 14, No. 7, 722-725 (1987).

[15] L. R. Zink and M. Mizushima, J. Mol. Spectros. 125, 154 (1987).

[16] C. H. Townes and A. L. Schawlow, 'Microwave spectroscopy', Mc Graw-Hill (1955), Dover Publ. Inc. (1975).

SPECTROSCOPIC MEASUREMENTS OF THE $\upsilon_2+2\upsilon_3$ band of CH$_4$ with a 1.3 μm InGaAsP diode laser.

R.A. ROOTH

N.V. KEMA
P.O. Box 9035
6800 ET, Arnhem
The Netherlands

SUMMARY

The development of sensitive spectroscopic techniques as for example TTFM (two-tone frequency modulation) spectroscopy opens the possibility for trace gas detection with near-infrared diode lasers. One of the species of interest is methane as it is a major greenhouse gas. This paper presents measurements of the $\upsilon_2+2\upsilon_3$ band of methane by both FTIR and diode laser spectroscopy. Using a InGaAsP diode laser, the R(2) to R(6) line of the band have been studied in detail at pressures of 13, 130 and 1000 mBar. Concluding from the absorption coefficients of the lines and the interferences by H$_2$O and CO$_2$, the R(2) line seems to be the best choice for monitoring in the 1.33 μm $\upsilon_2+2\upsilon_3$ band.

1.INTRODUCTION

The study of the greenhouse effect has had much attention in the last decade, because of its possible impact on global environment. These studies require equipment to monitor the concentration of greenhouse gases as H$_2$O, CO$_2$, N$_2$O, CH$_4$ etc. This paper deals with the detection of CH$_4$. Until now, routine methane monitoring employs gas chromatography, which is an accurate method, but not very fast. Development of a fast, sensitive methane monitoring device would be very helpful. Diode laser absorption spectroscopy offers this perspective.

Sensitive monitoring of trace gases is usually done by derivative spectroscopy employing lead-salt diode lasers as they can be manufactured with any wavelength between 3 and 30 μm, in which range many gaseous species show strong fundamental absorption bands. For methane, such bands are located around 7.7 μm (υ_4) and 3.3 μm (υ_3). Reid[1] showed that 1 ppb methane could be detected at atmospheric pressure over a 32 meter path with a liquid Helium cooled lead-salt diode laser at 7.4 μm. Koga employed an open-gas, portable monitor employing a 7.4 μm diode laser and achieved a sensitivity of 3 ppm in a measurement time of 10 ms. and a 1 meter path length[2]. More recently, they reported improvements on their system, now achieving a sensitivity of 40 ppb*m for atmospheric methane employing a 1 meter path[3]. In the 3.3 μm band, the accidental coincidence of the He-Ne laser emission and a methane absorption line has also been employed. Uehara achieved a minimum detectable methane concentration of 0.4 ppm for a 1 meter optical path and a 1 second averaging time with a dual wavelength laser[4]. This approach was, however, abandoned because of stability problems with the experimental set-up.

The use of lead-salt diode lasers has a number of drawbacks, especially when they have to be employed outside the laboratory. They have to be cryogenically cooled, requiring bulky cooling equipment, their output power is relatively low and the beam quality is poor. This has stimulated the research directed at the development of sensors based on NIR diode lasers, which are high quality devices, due to the interest of the communications industry, and relatively cheap. Their wavelength is within the range from 0.8-1.8 μm. A system based on such diodes lacks the need for cooling of the detector or the laser. The absorptions in this region are overtone absorptions however, so the absorption cross sections are one to two orders of magnitude less than those of the fundamental bands. For this reason, earlier attempts were directed towards the use of NIR diodes in devices with a limited sensitivity, but of a very simple and attractive design, for example as safety devices or leak detectors. Chan et al. used a LED[5,6,7], Mohebati and King a multimode laser diode at 1.33 μm[8]. The latter obtained a sensitivity of about 2000 ppm for a 1 meter optical pathlength. In an evaluation on the use of NIR diode lasers for gas detection, Cassidy[9] indicates that 5 ppm of CH_4 can be detected over a path length of 1 meter (employing second derivative detection and using AFGL data: 1.6 μm R(4) transition, 6057.08 cm^{-1}, $\alpha = 0.1$ $cm^{-1}atm^{-1}$) if the detection sensitivity is $5*10^{-5}$.

In recent years however, great improvement in the sensitivity of diode laser spectroscopy has been achieved with the technique of FM spectroscopy, especially two-tone (TTFM) FM spectroscopy[10,11,12]. This technique is especially well adapted to be used with NIR diode lasers as these devices are designed for high frequency modulation. Detection sensitivities down to 10^{-7} are reported[12,13]. This opens the possibility for high-sensitivity monitoring of trace gases using NIR diode lasers.

2. MEASUREMENT OF METHANE ABSORPTION LINE SPECTRUM AT 1.33 μm ($v_2 + 2v_3$ BAND)

Unfortunately, from literature, little detailed information on the CH_4 $v_2 + 2v_3$ absorption band is available. Some research has been done by Chan et al.[7], but their measurements have a resolution of only 0.3 cm^{-1}. The Hitran database does not contain information on the 1.3 μm absorption band of methane either. So, for the assessment of the problem whether 1.3 μm laser diodes can successfully be applied for the sensitive monitoring of atmospheric methane, some additional research is needed.

The methane absorption was studied by two methods.
1) FTIR spectroscopy in the near infrared with a Bomem DA3-002 spectrometer.
2) Direct absorption measurements using a multipass cell and 1.3 μm Philips CQF63 InGaAsP diode lasers.
The first method was considered especially convenient, as this method does not rely on the tuning range of the diodes, has good wavelength accuracy and resolution, and it is also possible to measure other methane absorption bands. The results by this method are given first.

Methane absorption measurements by FTIR.

For the experiments with the FTIR spectrometer, the methane was contained in a 10 cm-long cell. Analysis of the data taken with the cell filled with 1 atmosphere of pure methane and with a resolution of 0.1 cm^{-1} yields the data shown in Table 1.

The observed wavelengths show a significant discrepancy with the values reported by Chan[7]. The difference is towards longer wavelengths and often more than

Designation	Line position [cm⁻¹]	[nm]	Chan data [nm]
R(6)	7585.38	1318.33	1318.0
R(5)	7574.54	1320.21	1319.9
R(4)	7563.75	1322.10	1321.8
R(3)	7552.81	1324.01	1323.7
R(2)	7542.05	1325.90	1325.6
Q	7510.0	1331.56	1331.0
P(2)	7489.32	1335.23	1335.0
P(3)	7478.75	1337.12	1336.7
P(4)	7467.87	1339.07	1338.3
P(5)	7458.14	1340.80	1340.7

Table 1: Absorption line positions measured by FTIR.

0.2 nm (1.1 cm⁻¹). This is 40 times the accuracy suggested by Chan.

The wavelength accuracy of the FTIR spectrometer was checked as described in appendix A. The conclusion is that the FTIR is accurate to within its resolution, making the Chan[7] wavelengths very inaccurate and useless when interference by other gases is to be studied. Our results were independently confirmed by unpublished results of K. Uehara and K. Katakura[17] who recorded the $\upsilon_2 + 2\upsilon_3$ band of methane at 5 Torr using a DFB diode laser. The limit to their resolution was the Doppler width of methane (700 MHz FWHM). The stated accuracy was 0.002 nm. A table of methane lines with their measured values is included as table 2.

Assignment	Measured wavenumber [cm⁻¹]	Assignment	Measured wavenumber [cm⁻¹]
P 4 A1	7467.836	R 1 F1	7531.419
P 4 E	7467.836		
P 4 F1	7467.836	R 2 F2	7542.070
P 4 F2	7468.556	R 2 E	7542.070
P 3 F1	7478.620	R 3 F1	7552.739
P 3 F2	7478.710	R 3 F2	7552.807
P 3 A2	7478.995	R 3 A2	7552.882
P 2 E	7489.290	R 4 A1	7563.564
P 2 F2	7489.335	R 4 F1	7563.678
		R 4 E	7563.759
P 1 F1	7499.861	R 4 F2	7563.936
R 0 A1	7520.855	R 5 F1 0	7574.461
		R 5 F2	7574.524
		R 5 E	7574.633
		R 5 F1 1	7575.035

Table 2: List of 1.3 μm methane absorption lines, taken from Uehara and Katakura (reference 17).

Methane absorption measurements with a 1.3 μm diode laser.

Description of experimental setup.

The setup that was used for doing the absorption measurements is shown in figure 1. The laser, a Philips CQF63 is mounted in a temperature head (AC-9500), controlled by a TC-5000 thermoelectric controller, and driven by a LD-2310 current supply, all manufactured by Seastar Optics. The light is collimated, chopped and pas

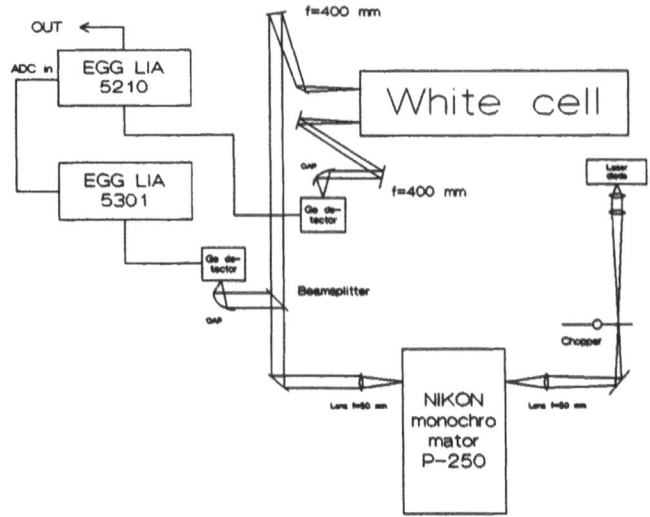

Figure 1: Experimental setup for direct absorption measurements using diode lasers.

sed through a Nikon P-250 monochromator. A quartz plate beamsplitter samples some of the transmitted radiation, which is directed to a Hamamatsu B2144 Ge photodiode by means of an off-axis paraboloid (OAP). The beam transmitted by the beamsplitter is focused into a White cell. Finally the exit beam is recollimated again and directed to a second Ge photodiode, Hamamatsu B1720, using an OAP. The signal of the diode before the White cell is recorded by an EGG model 5301 lockin amplifier (LIA) and the output signal is directed to the ADC input channel no. 1 of a EGG model 5210 LIA. This LIA also records the signal of the diode after the White cell and is operated in a ratioing mode so the final output is the ratio of the the input signal and the ADC channel 1 input.

The output signal of the LIA model 5301 can be described as

$$S_r = A_r*T_r(\upsilon)*R(\upsilon)*I_o*D_r(\upsilon) \tag{1}$$

whereas that of LIA model 5210 is given by

$$S_i = A_i*T_i(\upsilon)*(1-R(\upsilon))*I_o*\exp(-\alpha(\upsilon)L)*D_i(\upsilon) \tag{2}$$

where S is the recorded output in volts, A the amplification factor between the output and input of the LIA's, T the transmission of the optics, D the detector responsivities, all for the reference (subscript r) and signal channel (subscript i). $R(\upsilon)$ is the reflection by the beamsplitter, I_o the laser intensity impinging on the beamsplitter and $\exp(-\alpha(\upsilon)L)$ the transmission by the gas in the White cell when the pathlength is L. In the ratioing mode, the final signal output by LIA 5210 is given by

$$S = S_r/S_i \tag{3}$$

To obtain transmission spectra, this signal S is recorded with the cell both evacuated (in which case $\alpha(\upsilon)L=0$) and filled with a certain pressure of pure methane. Tuning of the diode laser is done by temperature change, which was controlled manually. The temperature could be set with a resolution of 10 mK. With 50 μm slit widths on the monochromator, the resolution was determined to be about 0.3 nm (1.7 cm^{-1} =^ 51 GHz at 1330 nm), so, the laser can be tuned approximately over this range without adjusting the setting of the monochromator.

If the ratio of the signals obtained with the cell filled with methane and those of the empty cell is taken, and no changes have been made to the setup between the two recordings, then the result is

$$T(\upsilon) = S_{gas} / S_{empty} = \exp(-\alpha(\upsilon)L) \tag{4}$$

as all the other factors cancel each other.

Absorption measurements at atmospheric pressure.

The recorded transmission curves at 1 atmosphere of methane for the R(2) to R(6) transition are shown in figure 2. In all cases, the path length was 160 cm.

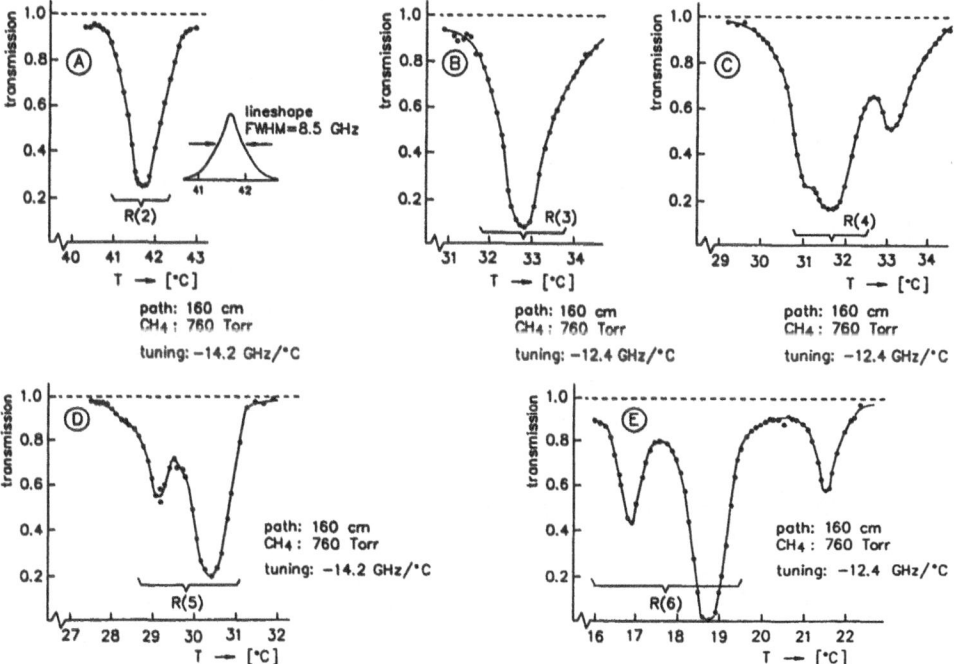

Figure 2: Recorded transmission spectra of the $\upsilon_2+2\upsilon_3$ band R-branch transitions of methane at 1 atmosphere using temperature tuning of the diode lasers.

Absorption measurements at reduced pressure.

In order to be able to study the possibility of interference reduction by reducing the pressure in the measuring cell, CH_4 absorption coefficients and linesha-pes were measured at reduced pressure as well. At four lines, R(2), R(3), R(5) and R(6), experiments were performed at 100 Torr. For the measurements on the R(2) and R(5), the White cell was replaced by a 50 cm long single pass cell. For all lines,

R(2) to R(6), measurements were performed at 10 Torr, using the White cell with path lengths of 8 meter or 16 meter. These results are shown in figure 3. The narrowest peaks observed have a width of about 0.8 GHz, which is somewhat larger than the Doppler width (0.7 GHz). This may be caused by small temperature instabilities. At these low pressures, the temperature must be stable to within 10 mK to keep the laser at the peak of an absorption line. This is about the limit of the temperature control system.

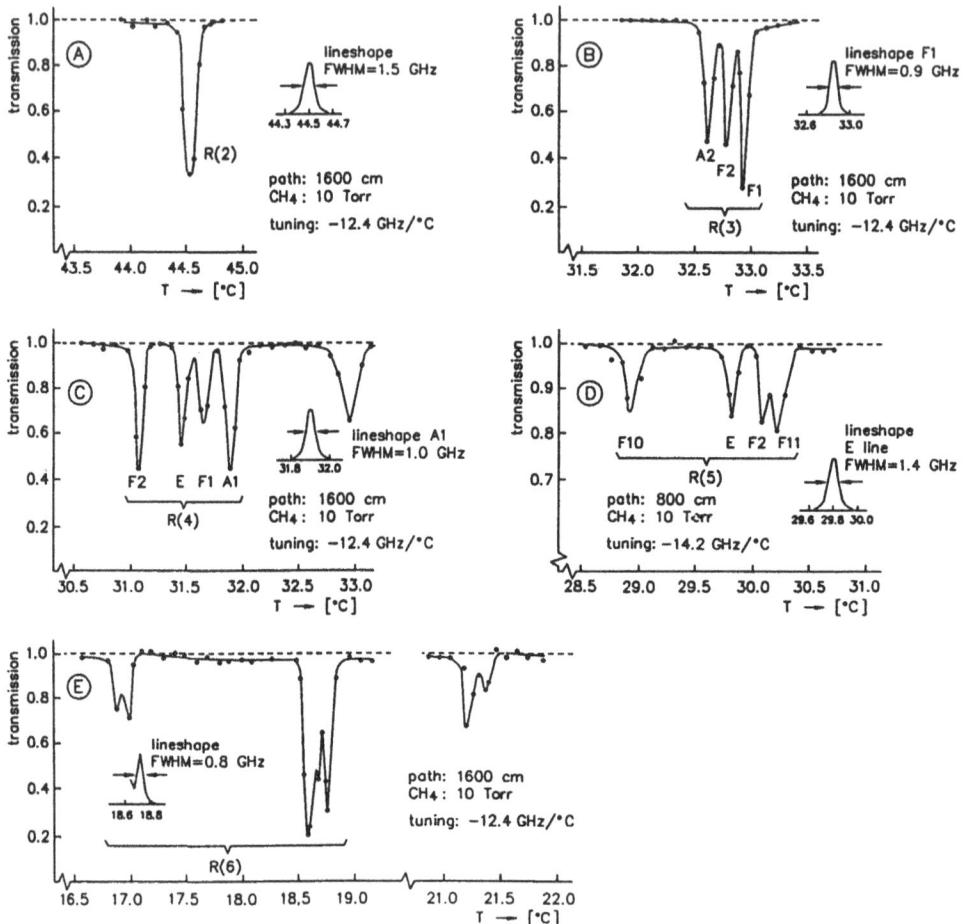

Figure 3: Recorded transmission spectrum of the R(2), R(3), R(4), R(5) and R(6) transition in the 1.3 µm band of methane at 10 Torr.

A summary of the absorption cross sections, obtained with the diode laser measurements is given in table 3. In the case where the peaks are not entirely resolved, the cross-section of the combined peak height is given. From this table can be seen that the R(2) is a relatively weak line at atmospheric pressure, but at reduced pressure it becomes comparable with the stronger lines because of the fact that it does not split.

Assignment	Wavenumber	σ [units of 10^{-22} cm²]		
		760 Torr	100 Torr	10 Torr
R(2) E	7542.07	} 4.0	} 15	} 19
F2	7542.07			
R(3) F1	7552.74	} 7.2	} 21	23
F2	7552.81			14
A2	7552.88		12	14
R(4) A1	7563.56	} 4.7		15
F1	7563.68			8
E	7563.76			11
F2	7563.94			15
R(5) F1 1	7574.46	} 4.5	} 13	9
F2	7574.52			7
E	7574.63		10	6
F1 0	7575.04		6	6
R(6)	7585.37	} 9.8	20	21
	7585.41		} 23	15
	7585.46			29
	7586.21	} 2.0		6
	7586.25			5

Table 3: Absorption cross sections of the v_2+2v_3 band of methane at various pressures. Wavenumbers and assignments according to Uehara[17].

3. MONITORING OF ATMOSPHERIC METHANE, INTERFERENCE PROBLEMS

As a rough estimate, both at 1 atmosphere and 100 Torr, the absorption coefficient on one of the studied R-branch lines will be about $\alpha = 0.01$ cm^{-1}. ($\alpha = \sigma^* n$). So, with a detection limit of for example 10^{-6} and a path length of 50 meter, the detection limit would be 20 ppbv. This is about 1% of the ambient value and would meet the requirements for a detector. But the required detection limit is tough and calls for sophisticated detection techniques as for example two-tone frequency modulation spectroscopy[12].

To study the interference problem, calculations were performed on absorption by H_2O and CO_2 in this region using the Hitran database. The results show that the best choice for a transition to monitor CH_4 in this band is the R(2) line. As for the Q-branch, this seems to be not so bad either, but a detailed study of the many lines of which this transition is composed is needed to draw a conclusion. It is necessary to bear in mind that sensitive detection techniques do not employ direct absorption but some kind of derivative spectroscopy. So, while having a relatively high absorption coefficient, its broad nature makes it less adapted to these sensitive techniques, at least at atmospheric pressure.

4. CONCLUSIONS AND SUGGESTED SETUP FOR 1.65 μm CH_4 MONITORING

A good system for the sensitive monitoring of methane at NIR wavelengths can be similar to that described by Johnson et al[12], employing two-tone frequency modulation spectroscopy. Recent developments in diode laser technology however, allow the use of a DFB laser at 1.65 μm[15]. These lasers are a big step forward to the construction of reliable and simple monitoring instruments.

As for the methane $2\upsilon_3$ spectral band, it can be seen from the literature and from the data in this report, that for all branches, the absorption coefficients for the transitions are about a factor of 5 higher than those in the $\upsilon_2+2\upsilon_3$ band. As for the interference by other gases, this is almost negligible at a number of lines, for example R(4)[9] or Q(6)[15], even at atmospheric pressure. This is all very advantageous.

The proposed setup is shown in figure 4.

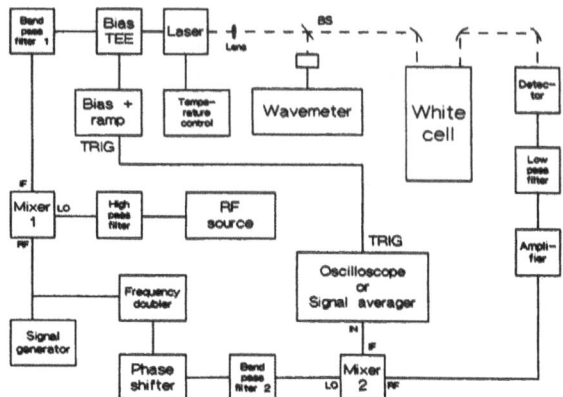

Figure 4: Two-tone FM spectroscopy setup using a 1.65 μm DFB laser for atmospheric methane monitoring.

APPENDIX A: CHECK OF BOMEM DA3-002 FTIR WAVELENGTH ACCURACY.

The wavelength accuracy of the FTIR spectrometer was checked by comparing the results of methane absorption measurements at around 6000 cm^{-1} ($2\upsilon_3$-band, Q branch) and of H_2O absorption around 7415 cm^{-1} with calculations by Fascode based on the 1986 Hitran database release[14]. The Hitran data are accurate within 0.05 cm^{-1} [7]. At 6000 cm^{-1}, the FTIR resolution was set to 0.03 cm^{-1} and for all the 9 peaks between 5998 and 6005 cm^{-1}, the difference between the calculations and the measurements was 0.03 cm^{-1} or less. At 7415 cm^{-1}, the resolution of the H_2O vapour absorption spectrum was 0.1 cm^{-1}. Again, comparison between measurement and calculations of all the peaks that could be unambiguously identified yielded differences smaller than the FTIR resolution. From these results it is concluded that the FTIR wavelength readings are most likely accurate to within the resolution in the region from 7400-7600 cm^{-1} as well.

REFERENCES

1) J. Reid, R.L. Sinclair, W.B. Grant, R.T. Menzies, High sensitivity detection of trace gases at atmospheric pressure using tunable diode lasers, Opt, Quantum Electr., vol. 17, p. 31 (1985).

2) R. Koga, M. Kosaka, H. Sano, Field methane tracking with a portable and real-time open-gas monitor based on a CW-driven Pb-salt diode laser, Opt. Laser Techn., p. 139 (1985).

3) R. Koga, Absorption spectrometry for atmospheric methane: techniques for higher sensitivity, Proc. Int. Symposium on monitoring of gaseous pollutants by tunable diode

lasers, Freiburg (1988).

4) K. Uehara, Alternate intensity modulation of a dual-wavelength He-Ne laser for differential absorption measurements, Appl. Phys. B, vol. 38, p. 37 (1985).

5) K. Chan, H. Ito, H. Inaba, 10 km-long fibre-optic remote sensing of CH_4 gas by near-infrared absorption, Appl. Phys. B, vol. 38, p. 11 (1985).

6) K. Chan, H. Ito, H. Inaba, Remote sensing system for near-infrared differential absorption of CH_4 gas using low-loss optical fiber link, Appl. Opt., vol. 23, pp. 3415-3420 (1984).

7) K. Chan, H. Ito, H. Inaba, Absorption measurement of v_2+2v_3 band of CH_4 at 1.33 μm using an InGaAsP light emitting diode, Appl. Opt., vol. 22, pp. 3802-3804 (1983).

8) A. Mohebati and T.A. King, Remote detection of gases by diode laser spectroscopy, J. Mod. Opt., vol. 35, p. 319 (1988).

9) D.T. Cassidy, Trace gas detection using 1.3 μm InGaAsP diode laser transmitter modules, Appl. Opt., vol. 27, pp. 610-614 (1988).

10) L.G. Wang, D.A. Tate, H. Riris, T.F. Gallagher, High sensitivity frequency-modulation spectroscopy with a GaAlAs diode laser, J. Opt. Soc. Am. B, vol. 6, pp. 871-876 (1989).

11) C.B. Carlisle, D.E. Cooper, Tunable diode laser frequency modulation spectroscopy through an optical fiber: High-sensitivity detection of water vapour, Appl. Phys. Lett., vol. 56, pp. 805-807 (1990).

12) T.J. Johnson, F.G. Wienhold, J.P. Burrows, G.W. Harris, Frequency modulation spectroscopy at 1.3 μm using InGaAsP lasers: a prototype field instrument for atmospheric chemistry research, Appl. Opt., vol. 30, pp. 407-413 (1991).

13) R.A. Rooth, Modulation techniques in optical spectroscopy, internal communication, (1991).

14) L.S. Rothman et al, The Hitran database: 1986 edition, Appl. Opt., vol. 26, pp. 4058-4097 (1987).

15) K. Uehara, H. Tai, Remote detection of methane with a 1.66 μm diode laser, to be published in Applied Optics.

16) H. Sasada, K. Yamada, Calibration lines of HCN in the 1.5 μm region, Appl. Opt., vol. 29, pp. 3535-3547 (1990).

17) K. Uehara, K. Katakura, Keio University, Yokohama, Japan. Unpublished results.

SENSITIVE DETECTION OF ACETYLENE ABSORPTION IN THE VISIBLE USING A STABILIZED AlGaAs DIODE LASER

F. S. Pavone, F. Marin*, M. Inguscio **
European Laboratory for Nonlinear Spectroscopy (LENS),
Largo E. Fermi 2, I50125 Firenze, Italy,

K. Ernst ***
Istituto Nazionale di Ottica (INO), largo E. Fermi 6, I50125 Firenze, Italy, and

G. Di Lonardo
Dipartimento di Chimica Fisica ed Inorganica, viale Risorgimento, 4 I40136
Bologna, Italy.

*   Also: Scuola Normale Superiore, Piazza dei Cavalieri 7, I56100 Pisa,
        Italy.
**   Also: Dipartimento di Fisica, Largo E. Fermi 2, I50125 Firenze, Italy.
*** On leave from: Institut of Experimental Physics, University of Warsaw,
        Poland.

Summary

Overtone transitions of C_2H_2 around 789 nm are investigated by means of a AlGaAs laser operating in external optical cavity configuration. Relative amplitude noise is of a few parts in 10^6 and allows an absorption detection limit of 0.2 ppm per km. Self- and air-broadening is measured for two components of the observed band (12646.966 and 12688.699 cm-1).

1. INTRODUCTION

Recent progresses of semiconductor diode lasers, and in particular their power, reliability and large spectral coverage, have been of great importance for a continuously increasing use both in pure and applied spectroscopy. In particular, the extension of operation from the infrared to the visible has made accessible the wide and important field of absorption measurements concerning the overtone molecular transitions. This is of relevant importance because only few molecules have electronic absorption bands in the visible.

It is worth noting that, moving from the infrared fundamental vibrational transitions to the overtones in the visible, absorption coefficients decrease several orders of magnitude. On the other hand, diode lasers show a great amplitude stability, approaching the shot noise limit [1], as compared to other laser sources. This is, of course, of great importance when weak absorptions have to be detected. Furthermore, stabilization of semiconductor lasers by optical feedback, developped in our group for line-narrowing and high resolution spectroscopy [2], has also been found to be a powerful tool for the general control of amplitude and frequency laser characteristics.

Purpose of the present work is to investigate the applicability of this tecnique to the overtone spectra of molecules of environmental interest, with emphasis to the sensitivity of the measurements, which is one of the most important criteria when one is dealing with the analysis of minor constituents of the atmosphere. We have chosen C_2H_2, present in small amounts both in urban atmosphere and in the troposphere. In addition, acetylene being a prototype of nonpolar linear poliatomic molecules, has been widely investigated and comparisons with different techniques are made possible. The visible overtone spectrum has been investigated at moderate pressures (~Torr) by Scherer et al. [3], using intracavity dye-laser spectroscopy combined with longitudinally resonant optoacoustic cell. More recently [4], commercial Ga-As diode lasers at 0.85 μm were demonstrated to be useful for the detection of C_2H_2 at 50 Torr gas pressure in a multipass cell. In the present work we show that the stabilization of the laser diode allows to conveniently operate in a simple single pass configuration and at three orders of magnitude lower pressures. In addition, air broadening is investigated, which also is crucial in the frame of an application to environmental optics. Moreover, the laser stabilization is achieved by means of the feedback from a grating which also allows an easy tuning on a large variety of constituents of the atmosphere.

2. EXPERIMENTAL

The experimental apparatus is schematically shown in Fig.1.

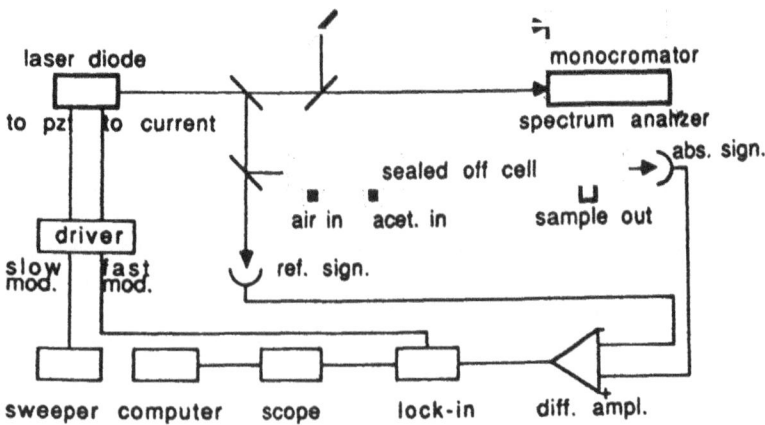

Figure 1: Experimental apparatus scheme

As a laser source, we used a Sharp LT024 AlGaAs/GaAs diode emitting at 780 nm at room temperature. The laser was attached to a small copper block in thermal contact with a Peltier cell. Since the laser is provided with a reduced reflection coating on the output facet, it could be used in an extended cavity configuration: the first order diffracted beam from a 1200 lines/mm ruled grating , mounted in the Littrow configuration, was fed-back into the laser diode (Fig. 2). This allowed to select and tune one of the modes of the cavity within a range of about 10 nm, at fixed and stable (0.001 K) temperature, and to reduce the laser linewidth to less than 1MHz.

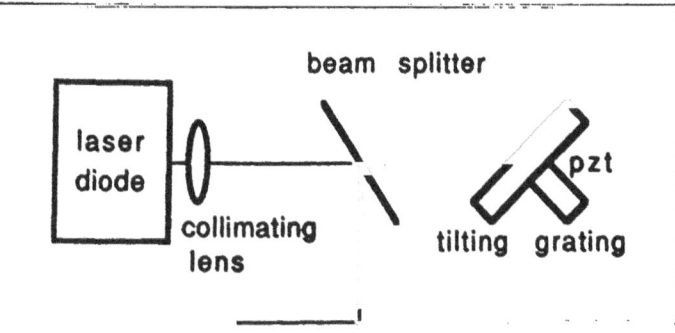

Figure 2: Extended cavity diode laser configuration

The tuning range overlapped the whole vibro-rotational structure of the $\nu_1+3\nu_3$ band of C_2H_2. Fine tuning to the transition of interest was then achieved by slightly changing the temperature and the injection current. Light from an intracavity beam-splitter (R~30%) was used for the experiment. As for the amplitude stability, we measured at 1 kHz (1Hz of bandwidht) a S/N ratio of 10^6.

Frequency scans were calibrated by means of a high-finesse Fabry-Perot confocal resonator with a free spectral range of 75MHz, while a double monocromator (resolution 0.3 cm^{-1}) was used to easily tune the laser wavelength to the spectral line of interest.

The sample cell was 1.5 meter long. The cell was pumped out to 10^{-4} Torrs and then filled with either pure acetylene or with a mixture of acetylene and air at the required pressure. Gas pressure was measured by means of an absolute capacitance manometer. The absorption signal was subtracted from the reference signal to avoid the background slope due to amplitude modulation. Molecular transitions could be observed either in pure absorption or in a phase-sensitive detection scheme. In the latter case, the lock-in derivative signals (10 msec time constant) were recorded by introducing a 2kHz modulation on the cavity length by means of the piezoelectric transducer on the grating.

3. RESULTS AND DISCUSSION

A typical derivative signal is reported in Fig. 3 for the **12646.966** cm^{-1} P(11) line observed at two diffrerent gas pressure. We measured a S/N value of a few thousands and an absorption of about 6% at 30 Torr, and a **S/N = 6 at 36 mTorr.**

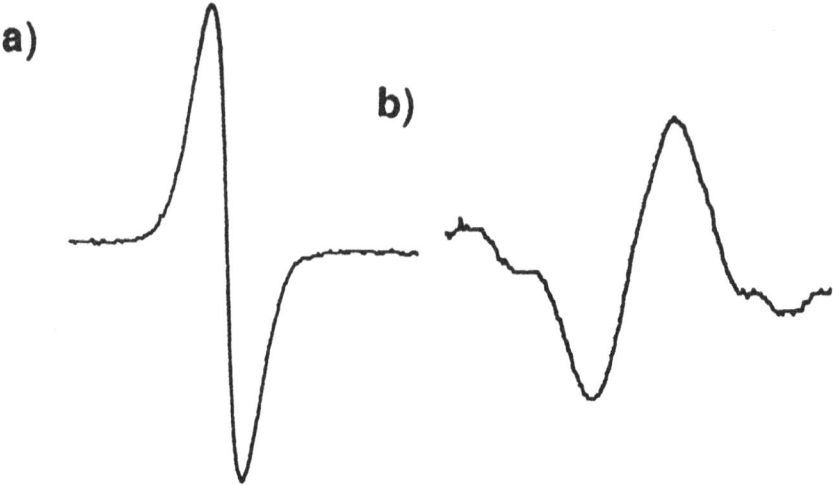

Figure 3: P(11) line derivative absorption signal at 10 Torr (a) and 36 mTorr (b)

While the phase sensitive detection scheme was used for evaluating the sensitivity of the technique at very low pressure, the pure absorption scheme (in Fig. 4 it is shown the absorption signal on the R(9) component at 10 Torr) was on the contrary preferred for the pressure broadening measurements operating in either pure C_2H_2 or in a C_2H_2 /air mixture.

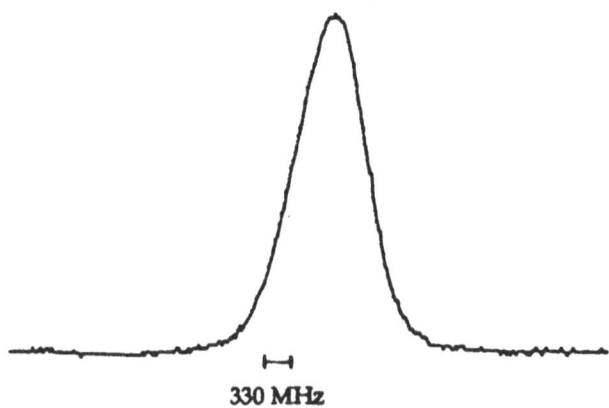

330 MHz

Figure 4: R(9) absorption signal at 10 Torr

The data analysis was performed using the Pade' approximation [5]. Systematic measurements of the self broadening on the P(11) line allowed us to fit the data with the Pade' curve at different pressures, as shown in Fig. 5, and to determine a pressure broadening parameter for the full widht γ = 11.0 (4) MHz / Torr . This is a reasonable value if compared with self-broadenig measurements on different overtone transitions [4].

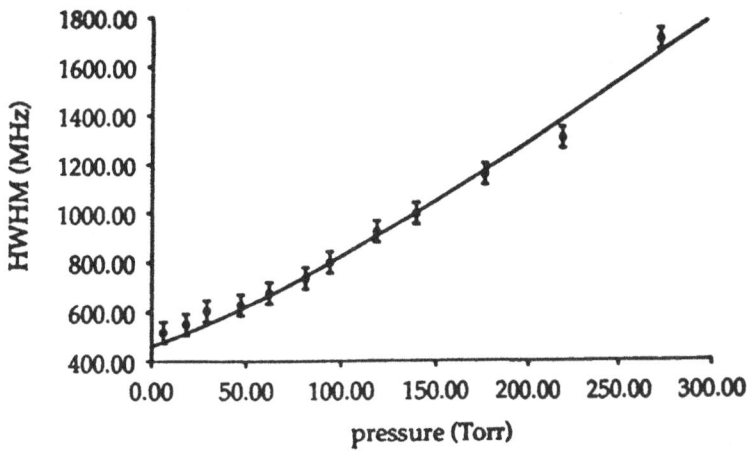

Figure 5: HWHM Voigt profile versus pure C_2H_2 pressure for the P(11) line

The results we could obtain for air broadening are, however, more important for environmental applications . **The air broadening** measurements have been performed for both the (12646.966 cm^{-1}) P(11) and (12688.699 cm^{-1}) R(5) lines with air broadening coeffecents of 8(1) MHz/Torr and 9 (1.3) MHz/Torr, respectively, as shown in Fig. 6 .

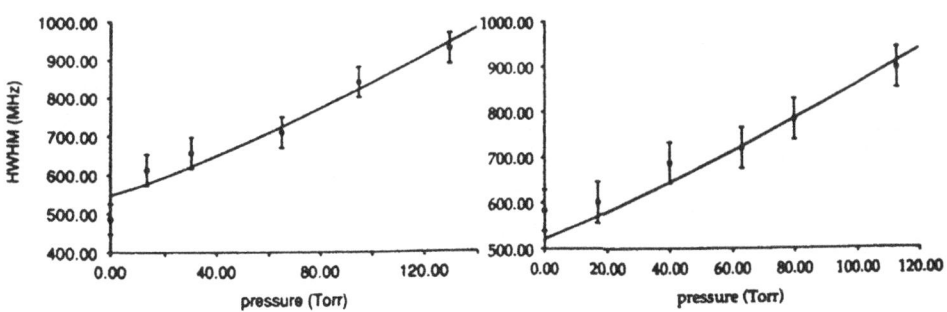

Figure 6: HWHM Voigt profile versus air pressure (at fixed C_2H_2 pressure for the p(11) (a) and R(5) (b) line

In the latter case the error is larger because the R(5) line (lower J value) is weaker than the other line (higher J value), so that the uncertainty due to the signal to noise is greater.

Our air broadening direct measurements at two different J values are in agreement with the behaviour of the air broadening coefficent extrapolated by C_2H_2-N_2 and C_2H_2-O_2 pressure broadening measurements for the fundamental vibrational transitions [6].

Measurements at different pressures can also provide information on the detection limit of the apparatus. It has been measured on the P(11) line at low pressure (30 mTorr) an absorption of a few parts in 10^5, corresponding to an apparatus sensitivity of a few parts x 10^{-7}/cm. This limit value of the detected absorption is consistent with the amplitude stability of the laser, of the order of 10^{-6} as previously discussed.

We have also experimentally studied the behaviour of the S/N ratio at different acetylene pressures or at different air pressures and fixed C_2H_2 pressure concluding that with the simple absorption apparatus we are able at atmospheric pressure to detect 0.2 ppm of acetylene x km.

4. CONCLUSIONS

We have demostrated that by means of extended cavity stabilized diode laser tuned on transitions of the $v_1 + 3v_3$ ovetone with origin at 12676 cm^{-1} it is possible to detect C_2H_2 in pure absorption.

With the apparatus described above we are able to perform pollution measurements on acetylene, one of the combustion gases present in small amounts in the atmosphere, with concentrations down to 0.2 ppm x km. The advantages of using a diode laser spectrometer with respect to other apparatus is due to the compact dimensions, easy functionning and low cost of the laser, and it allows a non contact, real time measurements of exhaust pollutant components.

The S/N measured value forsee the possibility of locking the laser to one of the C_2H_2 lines as was recently demonstrated for a fiber laser operating at 1.5 μm [7]. In our case the stabilization of the diode laser could be important to be used in conjunction with injection control of pulsed solide state laser in LIDAR [8] measurements making it possible to perform detection in larger space dimensions with improved sensitivity and the same stability characteristics.

Finally, as future research in the field of interesting molecular species detection by means of NIR diode lasers, we have listed in Table 1 a few overtone transitions:

MOLECULE	λ (nm air)	TRANSITION
CO_2	789	$00^\bullet 0 \to 02^\bullet 5$
HCN	791	$00^\bullet 0 \to 00^\bullet 4$
H_2O	796	$000 \to 013$
C_2H_2	848	$000 \to 211$
C_2HD	785	$00000 \to 40010$
NH_3	793	$4\nu_1$
CH_4	782	$2\nu_1 + \nu_2 + 2\nu_3$
$HC=C-C=CH$	787	$\nu_1 + 3\nu_4$
C_6H_6	869	$\Delta v = 4$
HI	836	$\Delta v = 6$
HCl	750	$\Delta v = 5$
HOOH	793	$5\nu_{OH} + \nu_{OO}$
C_2D_2	837	$000 \to 401$
CH_3D	861	$2\nu_1 + 2\nu_4$

TABLE 1: some molecular overtone transitions

ACKNOWLEDGEMENTS

Authors thank Drs. L. Hollberg (NIST, Boulder , USA) and G.M. Tino (University of Napoli, Italy) for many stimulating discussions concerning the operation of stabilized diode lasers.

REFERENCES

[1] C. Wieman and L. Hollberg, "Using diode laser for atomic physic", Rev. Sci. Instr. 62, 1-20(1991).

[2] G.M. Tino, L. Hollberg, A. Sasso, M. Inguscio and M.Barsanti, "Hyperfine structure of the metastable $^5S_2$ state of ^{17}O using an AlGaAs diode laser at 777nm", Phys. Rev.Lett. 64, 2999-3002 (1990).

[3] G.J. Scherer, K.K. Lehmann, and W. Klemperer, "The high-resolution visible overtone spectrum of acetylene", J.Chem.Phys. 78, 2817-2831, (1983).

[4] Y. Ohsugi and N. Ohashi, "0.85 µm diode laser spectroscopy of C_2H_2", J.Mol.Spectros. 131, 215-222 (1988).

[5] P. Minguzzi and A. di Lieto, J. Mol. Spectros. 109, 388-394, (1985).

[6] D. Lambot and G. Blanquet, "Diode laser measurements of collisional broadening in the ν_5 band of the C_2H_2 perturbed by O_2 and N_2", J. Mol Spectros. 136 , 86-92 (1989)

[7] S.L. Gilbert, "Frequency stabilization of a tunable erbium-doped fiber laser", Optics Lett. 16, 150-152 (1991)

[8] P. Rairoux,"Measure par diode de la pollution athmospherique et des parametres metereologiques", These, EPFL, Lausanne, (1991).

FAST SCANNING LASER DOAS, AN ULTRASENSITIVE TECHNIQUE FOR MONITORING TROPOSPHERIC TRACE GASES

W.ARMERDING, A.HERBERT, M.SPIEKERMANN, J.WALTER and F.J.COMES

Johann Wolfgang Goethe–Universität
Institut für Physikalische und Theoretische Chemie
Niederurseler Hang, W–6000 Frankfurt 50, Germany

SUMMARY

An ultrasensitive technique for monitoring atmospheric trace gases is discribed which combines the advantages of differential absorption spectroscopy (DOAS) with a fast scanning procedure to avoid atmospheric fluctuation. The light source is a tunable laser. The output power of the laser system is stabilized. Due to the fast scanning of the laser and the power stabilization the fluctuation of the atmosphere and of the laser itself are minimized. Thus a detection limit better than 10^{-5} can be achieved for atmospheric long path absorption measurements. The performance of this technique was demonstrated by successful monitoring of tropospheric OH during a field campaign on the Schauinsland, Black Forest, Germany. "Fast scanning laser DOAS" can be applied to all tunable lasers in the uv, the vis and the ir independent of the special properties of the active material.

1. INTRODUCTION

The last twenty years have shown that high resolution absorption spectroscopy with tunable diode lasers (TDLAS) using frequency modulation (FMS) or photoacustic techniques (PAS) is a powerful instrument for a wide variety of scientific applications, especially for the measurement of atmospheric trace gases (1, 2). Diode lasers can be scanned very rapidly in an easy way, essential for atmospheric measurements to avoid atmospheric fluctuation, and they cover the 2 to 15 micron region where most gases of atmospheric interest have well resolved rotational–vibrational absorption structures. But the use of TDLAS is — at present time — restricted to wavelengths above 550 nm. Thus, a number of very important tasks in atmospheric research cannot be realized by tunable diode lasers. The most prominent example is the measurement of the hydroxyl radical which plays the central role in tropospheric photochemistry. Due to severe water interfereces in the ir the monitoring of OH by an absorption technique requires measurements in the uv (at 308 nm). In order to realize ultrasensitive absorption measurements of trace gases in this wavelength region we have developed a modified DOAS (differential optical absorption) technique with a fast scanning laser as light source. In this paper we will describe the technique and demonstrate its performance.

2. BASIC IDEA OF "FAST SCANNING LASER DOAS"

The laser DOAS technique utilizes the absorption of radiation from a laser beam propagating through the atmosphere as the measurable parameter to determine the concentration of the trace gas under investigation. The absorbance is measured directly and converted into a concentration according to Lambert–Beer's law

$$I_a = I_u \exp\left(- \int N \, \sigma \, dL\right) \tag{1}$$

where I_a is the intensity of the light after traversing the absorption path. N is the number density of the trace gas under investigation, σ its absorption cross section and L the absorption path length. $I_u(\lambda, t)$ takes into account all atmospheric fluctuations and interferences, and in addition the starting conditions I_0 given by the light source.

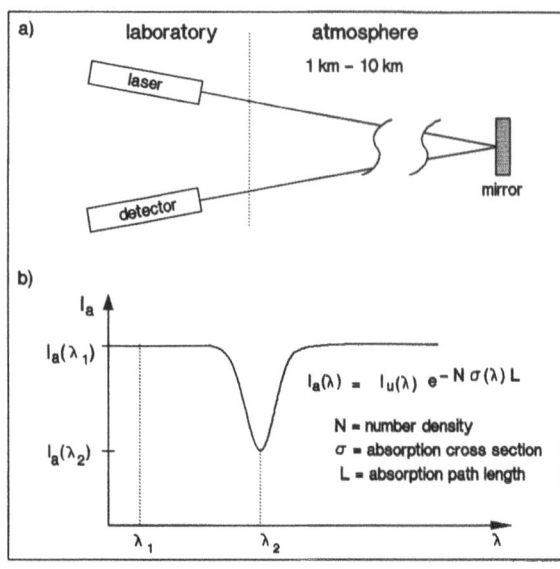

Fig.1:   DOAS method
         a) setup   b) analysis

Expression (1) is valid under the condition that no saturation effects will occur due to the intensity of the radiation and that the linewidth of the light source is small compared to the width of the spectral structures. The spectral variation of the absorbance must be specific for the trace gas under study. The unknown values of I_u and N are determined by measurements on different wavelength positions with different absorbances (see Fig.1). The required number of measurements for a complete data set is N_i+2 where N_i is the number of interferences. It must be ensured that $I_u(\lambda,t)$ is constant in time for all measurements belonging to the same data set.

The most important adventage of DOAS method compared with other techniques is that it is direct, absolute and a fingerprint technique. Knowing once the absorption cross section of the trace gas for the given experimental conditions (e.g. pressure and Doppler broadening) and the linewidth of the light source the system is self calibrating; hence, no further adjustment to a standard is necessary when the system measures at two wavelength positions with a different absorbance. The unequivocal identification of the species under study is based on its spectral characterisation. Furthermore, DOAS technique has in principle an excellent sensivity. But under field conditions, the detection limit of the conventional DOAS method is essentially determined by random fluctuations of the atmosphere due to turbulence effects and light scattering through dust and aerosoles which leads to an unknown attenuation of the laser radiation along the absorption path. A further source of problems are the fluctuations and the modulation of the laser radiation itself. In addition, tropospheric trace gases, whose spectra show narrow absorption structures in the same spectral region as the molecule under study, can cause severe interference effects.

In order to overcome these difficulties we have developed an instrument based on "fast scanning laser DOAS". The fundamental idea of this technique is to tune the wavelength of a laser rapidly over a spectral region characteristic for the observed species in such a short time that the atmosphere can be assumed as stationary and changes in the light intensity due to atmospheric influences during the aquisition time become negligibly small. These scans are repeated many times and averaged in order to minimize the random noise. In contrary to the FMS technique, the whole spectral range under investigation is scanned with high speed in every single run (see Fig.1). The wavelength region has to contain enough spectral informations such as line shapes and absolute positions which makes the molecule easy to identify. Furthermore, these informations can help to separate the absorption signal from the background noise and will allow to improve the overall signal to noise ratio by using integrating signal processing methods. Due to the serial generation of subsequent wavelengths positions, time dependent fluctuations and wavelength dependent modulations of the laser output power — and as a consequence noise and systematics — can be reduced substatially by a negative feedback modulation of the laser output power. As a consequence, this technique is a one beam method with all related advantages. No calibration problems with reference measurements utilizing different channels for data acquisition with different devices will occur.

3. REALISATION OF "FAST SCANNING LASER DOAS"

Proper spectral characteristics for DOAS measurements are single rotational —vibrational lines or narrowbanded structures of overlapping lines. All atmospheric trace gases of interest have suitable absorption features. The fundamental rotational —vibrational lines in the mid infrared, the "fingerprint" region of molecules, are well resolved in most cases or partially overlapping for heavier and more complex molucules. In the uv and the vis, only diatomic or small polyatomic molecules have resolvable characteristic absorption features due to electronic transitions with vibrational—rotational structures .

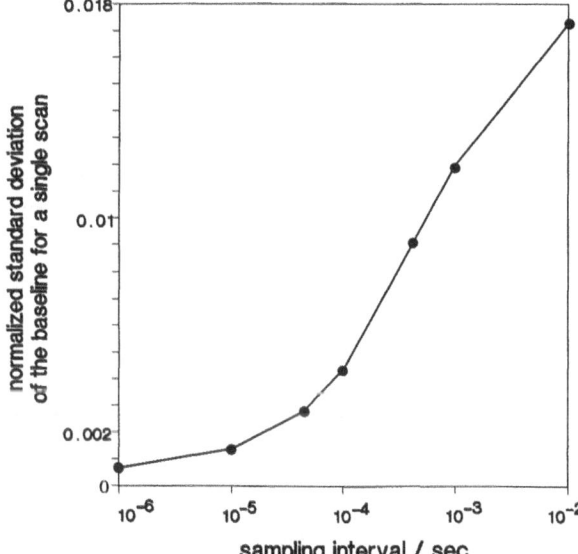

The analysis of the spectral features of the most important atmospheric trace gases shows that spectral intervals of 10 cm^{-1} are sufficient to select some rotational—vibrational lines which makes interferences from other trace gases very unlikely. In the case of overlapping structures mostly enough information is still available to separate the different contributions.

Fig.2: time dependence of atmospheric fluctuations

Preceding studies have shown that in the frequency region above 10 kHz corresponding to a scanning time of less than 100 μs for one spectrum the influence of the atmosphere on the noise level is negligibly small compared with the other sources of error. The result of these measurements is shown in Fig. 2. Therefore, the realization of fast scanning laser DOAS requires a scanning time not exceeding 100 μs for a scanning range of 10 cm^{-1}.

The wavelength of tunable lasers can be changed in principle by the following three procedures : the resonator modes can be shifted by a variation of the resonator length or — depending on the active material — by a shift of the gain profile; the resonator modes can be selected using an intracavity etalon. A continuous tuning over a longer spectral range of e.g. 10 cm^{-1} is only possible if a synchronous combination of the three different procedures is applied. But — due to problems in synchronisation — the realization is only possible for a slow scanning procedure.

Fig.3 shows schematically the different methods of wavelength tuning. The main procedure for a continuous wavelength tuning — the changing of the resonator length by a tweater or by brewster plates — leads to a continuous laser tuning with high spectral resolution due to mode competition. But the laser oscillation is unstable in a sense that modes on other wavelength positions can start oscillations in case of perturbations. In addition, in short times only small spectral ranges can be scanned, for example 30 GHz in 1 ms. The other possibility for mode shifting — the shift of the gain profile — depends strongly on the properties of the active material of the laser in use. For example, diode lasers can be scanned continuously and fast enough. But without a change of the resonator length the spectral range is insufficiently small for "fast scanning laser DOAS". Only the methode of mode selection by an etalon, if necessary in combination with a shift of the gain profil, offers a high scanning speed and a wide tuning range. But it should be noted, that this procedure is discontinuous because the oscillating modes are selected, not shifted. The spectral resolution depends on the resonator length and the properties of the etalon, especially the spectral width and the free spectral range, not on the properties of the active material.

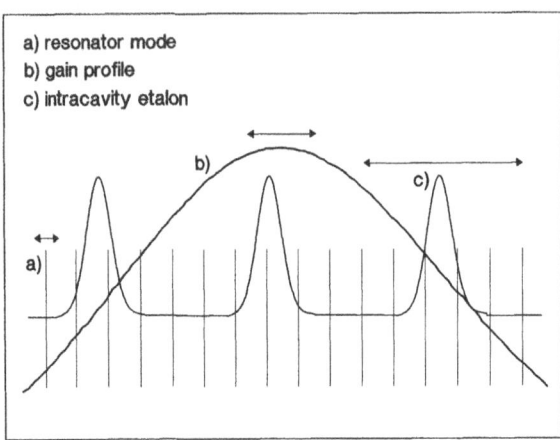

a) resonator mode
b) gain profile
c) intracavity etalon

Fig.3:   (fast) wavelength tuning

The easiest way to realize an etalon tuning device is by angle tuning. In this case a thin etalon (e.g. 0.5 mm) with low reflectivity (e.g. 30%) is mounted on a galvanometer drive. The change of the wavelength due to the variation of the angle is given by

$$\Delta\lambda \sim \lambda_0 \, \varphi^2 \qquad\qquad (2)$$

where $\Delta\lambda$ describes the change in the wavelength, λ_0 the starting condition and φ the tuning angle. The square function shows that the scanning speed can made be very high. Scanning velocities of 80 cm^{-1} ms^{-1} and higher were obtained. Compared to the other scanning methods etalon tuning results in an enlarged linewidth of the laser because a number of laser modes can oscillate simultaneously. But in most cases the spectral resolution is sufficient for the analytical requirements of atmospheric trace gas measurements. If necessary, the linewidth of the laser system can be improved using a resonator with an increased length and a scanning Fabry Perot with a high spectral resolution.

A fast scanning laser system was developed in our laboratory using an intra cavity scanning Fabry Perot etalon. As a crucial test of the performance of " fast scanning laser DOAS" tropospheric OH measurements were carried out (3 – 5).

4. TROPOSPHERIC OH MEASUREMENTS

The in situ determination of the OH radical plays a central role in tropospheric research. First it is generally accepted that OH measurements form a key and unique test of the modelling of fast tropospheric photochemistry. Second – related to the

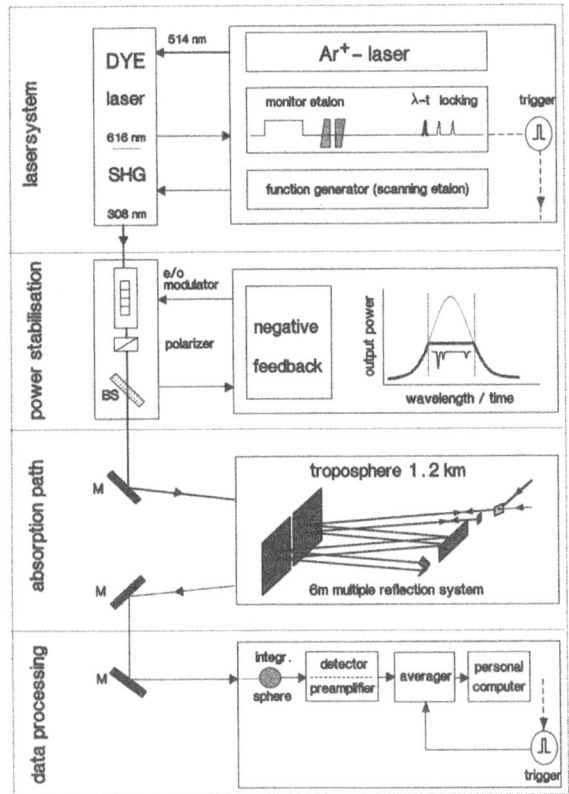

Fig.4: general experimental design

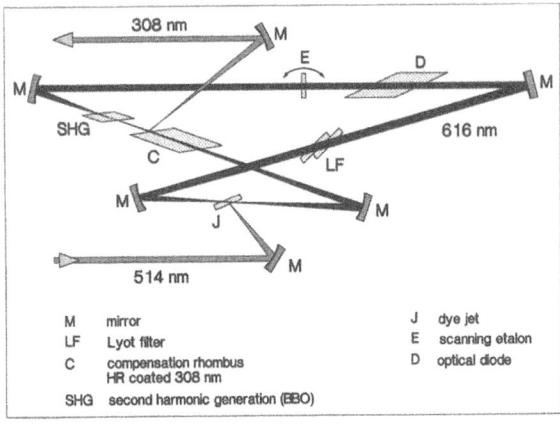

M	mirror	J	dye jet
LF	Lyot filter	E	scanning etalon
C	compensation rhombus HR coated 308 nm	D	optical diode
SHG	second harmonic generation (BBO)		

Fig.5: dye laser setup

problems of the cleansing processes of the troposphere – the prediction of future trents of the OH concentration have to be verified by tropospheric measuments. In addition, OH is an excellent tracer for studies of basic problems related to tropospheric photo-chemistry. But, compared to measurements of other tropospheric trace gases measurements of the OH radical are extremly difficult due to the very low troposheric OH concentration in the order of 10^6 OH cm^{-3}. Thus, tropospheric OH measurements are a crucial test for "fast scanning laser DOAS".

The analysis of the spectral properties of the OH shows that a spectral region in the uv at 308 nm is very suitable for monitoring OH by DOAS technique. This intervall contains some strong, sharp and well resolved rotational lines characteristic for OH. Most measurements use the $Q_1(2)$-line of the $(0-0)$-band in the $(A\ ^2\Sigma \leftarrow X\ ^2\Pi)$-transition at 307,9951 nm, which is one of the strongest lines, its weaker satellite, the $Q_{21}(2)$-line at 308,0006 nm, and the $R_2(2)$ line at 308,0231 nm. The so-called "fingerprint" region of the infrared is not suitable for tropospheric OH measurements because of the severe interference due to atmospheric water vapour. The experimental setup for OH LPA field measurements essentially consists of the light source, a modulator system for power stabilisation, the absorption path with the detector and the data acquisition and processing system (see Fig.4). The light source of the system is a modified cw dye ring laser (Coherent Cr–699, Rhodamin 6G) pumped by an 8W argon laser (Coherent 308) in the all line mode. Fig.5 shows the design of the dye laser in more detail. Due to the modest energy and water requirements of this small frame configuration the system is highly mobil. The intracavity frequency doubling is performed by a BBO crystal positioned inside the beam waist. The maximum cw output power at 308 nm is about 20 mW. The distortion of the outcoupled beam is compensated by a cylindrical mirror. The dye laser is rapidly tuned by a thin etalon mounted on a galvanometer drive. A spectral range of 7 cm^{-1} in the uv is scanned in less than 100 μs. The scans are taken with a repetition rate of 1.3 kHz. The quasi cw operation of the dye allows a very high sampling rate which is limited only by data acquisition and processing. For the reported configuration, the dynamical bandwidth of the laser system in the uv is about 4 GHz in the fast scanning mode. In addition, the system is spectrally locked using a Fabry Perot etalon as spectral reference. The output power of the system is stabilized and normalizied by a closed–loop control system using an 4–crystal electrooptical modulator system. With this real–time demodulation of the outgoing signal the flatness of the baseline can be reduced to the order of 10^{-4} using a double stage system. The overall signal to noise ratio of the stabilized system is of the order of 8×10^{-4} for a single scan. The absorption path is folded by a Whitecell–type multiple reflection system. An open path set–up was developed using a special design for tropospheric measurements (6). This instrument has a length of 6m and the diameter is 0.3m. The absorption path–length is 1.2 km. Due to the special design the influences of tropospheric fluctuations to the signal are below the detection limit. All optical elements in the system, particulary windows, lenses, polarizers and detector elements can act as etalons. In order to avoid problems with etalon fringes all lenses were replaced by mirrors, all plano–plates were replaced by plates with a slight wedge between faces, and polarizors with a special design are in use. Problems related to detector non–unifomity and fluctuations of the beam axis are reduced by using a detector system consisting of an integrating sphere with high grade uv reflection coating and a photomultiplier. Because the required low noise level cannot be reached in a single scan, a large number of repeated measurements is necessary to reduce the detection limit given by the standard deviation. Hence, for real time measurements a fast signal averager with a sampling rate of 20 MHz and 12 bit is used for data acquisition.

Due to the negative feedback modulation the noise and the systematics affecting the detection limit are substantially reduced. For a integration period of 2 min, a noise level of 2–3 10^{-6} has to be expected due to the signal to noise ratio of a single scan and averaging of about 150.000 scans. The (wavelength correlated) systematics are caused by the laser and the optics. They can be reduced to a level better than 10^{-5}. As main limitation of the detection limit remains the atmospheric background spectrum caused by interfering trace gases. Thus, the detection limit of the instrument for an actual experimental situation depends on the environmental conditions, especially the composition of the air. Measurements in Frankfurt have shown that here the SO_2 molecule is the most important interfering species for OH measurements. An air spectrum taken in March 91 in Frankfurt/M is shown in Fig.6. The very strong SO_2 absorbance demonstrates that OH measurements in Frankfurt will be problematic. But this measurement demonstrates in addition that the fast scanning uv spectrometer is a very good instrument for monitoring SO_2.

In order to realize OH measurements with reduced atmospheric background a field campaign was carried out in summer 1991 on the Schauinsland, Black Forest, Germany. The experimental conditions were good because this area is distinguished by a slightly polluted air and well characterized air masses. The ozone concentration is high in the 100 ppb region and the area has the most sunny days in Germany. In addition, the NO_x concentration is moderate in the 5ppb region, and the concentration of the SO_2 molecule – the strongest interfering species – is very low and of the order of 1ppb. During this campaign we have achieved very interesting results related to the problem of the OH linewidth – important in general for the calibration of the DOAS

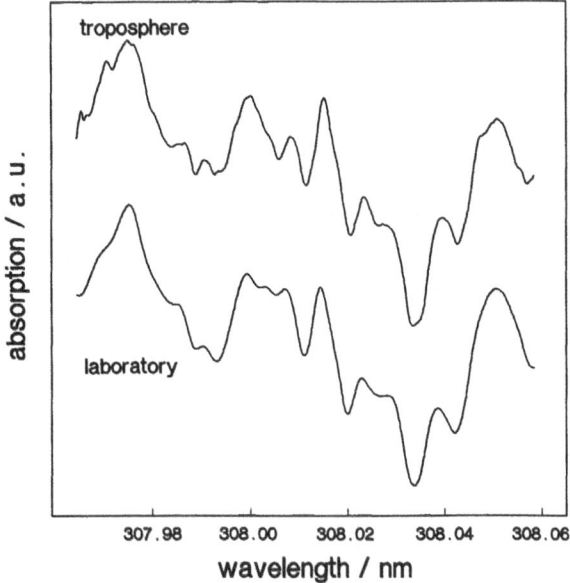

Fig.6:   tropospheric SO_2 spectrum
           (20 ppbv; $\sigma_{max} \approx 1 \times 10^{-20}$ cm^2)

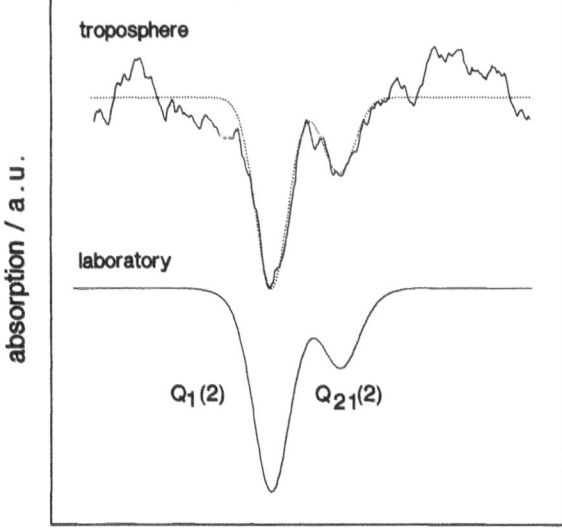

Fig.7:   tropospheric OH spectrum
           (0.2 pptv; $\sigma_{max} \approx 2 \times 10^{-16}$ cm^2)

method – and to the self generation of OH caused by the laser beam itself. Both problems are essential for the applicability of the DOAS method. A tropospheric OH spectrum is shown in Fig.7. Taking into account the linewidth of the laser, the picture shows that the linewidth of OH under tropospheric conditions is 8.5±1 GHz. In addition, a large number of data sets of diurnal variations of OH were obtained. The results of the OH measurements during this campaign will be published in more detail elsewhere (7, 8).

5. CONCLUSION

A method for the ultra–sensitive monitoring of tropo–spheric trace gases has been developed which is based on DOAS technique and which uses a fast tunable laser as light source. The excellent performance of this instrument was demonstrated in tropo–spheric OH measurements. A further advantage of this technique is its applicability in the uv, vis and ir. It is planned to realize a multi–component device based on fast scanning laser DOAS for the use in atmospheric research.

ACKNOWLEDGMENT

The work has been perfor–med as part of a program of the Bundesminister für Forschung und Technologie (BMFT). We thank B. Kinzer for his help in the development of the electronic equipment, and we are very grateful to the UBA for the help in realizing our field campaign 1991.

REFERENCES

(1) GRISAR, R., PREIER, H., SCHMIDTKE, G., and RESTELLI, G. (Eds)
Proc Int Symp on Monitoring of Gases Pollutants by Tunable Diode Lasers,
Freiburg 1986, D.Reidel Publishing Company (1987)

(2) GRISAR, R., SCHMIDTKE, G., TACKE, M., and RESTELLI, G. (Eds)
Proc Int Symp on Monitoring of Gases Pollutants by Tunable Diode Lasers,
Freiburg 1988, Kluwer Academic Publishers (1989)

(3) ARMERDING, W., HERBERT, A., SCHINDLER, T., SPIEKERMANN, M.,
and COMES, F.J. (1990). In situ measurements of tropospheric OH radicals — a
challenge for the experimentalist. Ber.Bunsenges.Phys.Chem., 94, 776 — 781

(4) ARMERDING, W., HERBERT, A., SCHINDLER, T., SPIEKERMANN, M.,
and COMES, F.J. (1991). A long—path absorption method for monitoring OH
and other tropospheric trace gases with a rapidly tunable uv—laser.
Proc EUROTRAC Symp 1990, SPB Academic Publishing bv, The Hague, The
Netherlands (1991), 469 — 473

(5) ARMERDING, W., HERBERT, A., Walter, J., SPIEKERMANN, M.,
and COMES, F.J. (1991). Fast scanning laser DOAS — a very promising
technique for monitoring OH and other tropospheric trace gases.
Fresenius J Anal Chem, 340, 654 — 660

(6) ARMERDING, W., Walter, and COMES, F.J. (1991). A White cell type
multiple reflection system for tropospheric research
Fresenius J Anal Chem, 340, 661 — 664

(7) ARMERDING, W., SPIEKERMANN, M., GRIGONIS, R., WALTER, J.,
HERBERT, A., and COMES, F.J. Fast scanning laser DOAS for local
monitoring of trace gases, in particular tropospheric OH radicals.
submitted to Ber.Bunsenges.Phys.Chem

(8) COMES, F.J., ARMERDING, W., GRIGONIS, R., HERBERT, A.,
SPIEKERMANN, M., and WALTER, J. Tropospheric OH: local measurements
and their interpretations.
submitted to Ber.Bunsenges.Phys.Chem

ENCLOSIVE FLOW COOLING:
CONCEPT OF A NEW METHOD
FOR SIMPLIFYING COMPLEX MOLECULAR SPECTRA

S. Bauerecker[1], F. Taucher[2], C. Weitkamp[2], W. Michaelis[2], H. K. Cammenga[1]

[1]Institut für Physikalische und Theoretische Chemie der TU Braunschweig,
Hans-Sommer-Straße 10, W-3300 Braunschweig, Germany

[2]Institut für Physik, GKSS-Forschungszentrum Geesthacht GmbH,
Max-Planck-Straße 1, W-2054 Geesthacht, Germany

SUMMARY

Molecules with more than four atoms generally exhibit relatively complex vibration-rotation spectra at standard conditions. Therefore they are difficult to detect in trace concentrations by absorption spectroscopy. A simplification of the spectrum can be achieved by supercooling the gas.

For this purpose a novel cooling scheme is presented. A cooling gas is sucked from the outside through the porous wall of a sinter metal hollow cylinder towards its axis. A laminar axially symmetrical cooling gas flow develops to which the sample gas is added through a heatable tube at the cylinder front end. The cooling gas encloses the sample gas beam, confines it to form a narrow column, cools it down and advects it along the cylinder axis, preventing it from wall adsorption. Optical absorption, e.g., of a laser beam can be observed along the cylinder axis where the sample gas concentration is highest.

Because of the relatively low flow velocity on the order of 0.01 to 1 m/s the optical absorption efficiency of the sample gas per unit of gas quantity is several orders of magnitude larger than with supersonic jet cooling. In the case of low temperatures and not too low pressure values a given quantity of sample gas may be compressed to 1/100 to 1/1000 of the cell volume of a White cell, thus an enclosive flow cooling cell may produce an optical absorption comparable to a multireflexion cell and so compensate the multireflexion advantage.

Smoke experiments in a transparent plexiglas model cell confirm the results of a computer simulation and demonstrate a promising behavior of both the enclosive flow and the sample gas flow.

1. INTRODUCTION

The vibration-rotation spectra of molecules that contain more than four atoms are quite complex under conditions of atmospheric pressure and ambient temperature. The reasons for this behavior are abundant rotational structures, pressure broadening, Doppler broadening, overlapping fundamental bands, and the presence of hot bands. In trace gas analysis by high-resolution optical spectroscopy such as tunable diode laser absorption spectrometry (TDLAS), the superposition of the spectra of the more abundant atmospheric gas species further complicates the situation. Supercooling of the sample gas greatly simplifies the structure of the spectra and must be considered as one of the prerequisites for successful spectrometric determination of minor components of gas mixtures. The present paper describes a novel device for efficient supercooling of gases for spectrometric purposes.

2. THEORETICAL CONSIDERATIONS

The pressure and Doppler widths, f_C and f_D, of a spectral line

$$f_C \sim n\sqrt{T} \quad \text{and}$$

$$f_D \sim \sqrt{T/m}$$

are both proportional to the square root of the temperature T; n is the particle density, m the particle mass. If we take acetaldehyde (CH_3CHO) as a model substance, its values of f_C and f_D at 300 K and 20 hPa are both about 0.004 cm^{-1}. This shows that both temperature and pressure must be reduced in order to sensibly reduce the total width of the spectral lines. A change from ambient to liquid nitrogen temperature at a few hPa pressure reduces the width by a factor of two.

Freezing in hot bands by cooling to 77 K is even more effective, as can be seen from Table 1. The energy of the lowest fundamental vibration of acetaldehyde corresponds to 150 cm^{-1} and is thus almost equal to the Boltzmann energy at room temperature; the population of the first vibrational state is thus about half as large as that of the ground state.

Due to the temperature dependence of the half-maximum energy of the Boltzmann distribution (kT ln 2), see Table 1, which is a measure of the spectral extension of the vibrational-rotational bands, cooling to 80 K (liquid nitrogen) reduces the band width by a factor of 4, cooling to 5 K (liquid helium) by a factor of 60.

The reduction of line density and line width of the sample gas are not the only effects of cooling. Equally important are these effects in the other components of the sample gas, thus increasing the selectivity of the optical analysis and rendering the detection of a trace gas feasible under conditions under which no spectroscopic detection is possible at room temperature.

Table 1:   Ground state population of acetaldehyde and half-maximum energies of the Boltzmann distribution.

Temperature / K	Population of the ground state	Half-maximum energies of the Boltzmann distribution / cm^{-1}
5	$\approx 100\ \%$	2.4
80	94 %	39
300	62 %	144

3. COOLING METHODS

In [1] a variety of cooling methods for simplifying complex molecular spectra are compared, with special emphasis on trace gas analysis by optical techniques such as TDLAS.

Supersonic flow cooling, i.e., adiabatic expansion of a gas through a nozzle [2], see Figure 1, is an elegant and widespread cooling method which requires relatively little expenditure. It is very often used for simplification of molecular spectra but in most cases with higher sample gas concentrations. Since high flow velocities of about 2000 m/s are necessary, the residence time of the sample gas molecules in the laser beam volume and therefore the optical absorption efficiency are very small. Compared with collisional cooling the detection efficiency is smaller by a factor 10^4.

In collisional cooling [3], see Figure 1, the sample gas is pressed through a heatable tube into a measuring chamber filled with a cold buffer gas, e.g., helium. The sample gas molecules are cooled down by collisions with the buffer gas. Part of them can be measured within the laser beam volume until they reach the wall by diffusion and are adsorbed. Thermal equilibrium is reached after about 100 collisions.

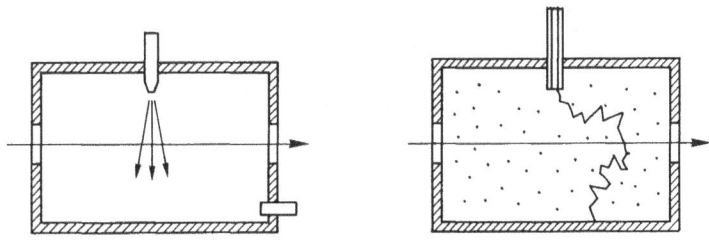

Figure 1: Schematic representation of supersonic flow cooling (left) and collisional cooling (right).

Diffusive trapping [4] is a variant of collisional cooling in which the sample gas is mixed with a buffer gas and injected into an evacuated cylindric chamber through a heatable inlet tube. The gas mixture is cooled by contact with the metallic cylinder wall which is in thermal contact with a liquid nitrogen bath. Again the sample gas molecules must be detected by the measuring laser beam before they reach the chamber wall. The elongated shape of the cylindric chamber ensures that most of the sample gas molecules adsorb to the cylinder wall and only very few of them remain to settle on the cell windows where they impair the measurement.

In matrix isolation spectroscopy the trace substance is trapped in a solid inert gas matrix. The absence of rotations of the sample gas molecules simplifies their spectrum and simultaneously increases the line intensities. If one uses atmospheric CO_2 for the matrix, an increase in preconcentration of the trace gases by a factor of about 3000 can be achieved. Water, which disturbs the spectra, must previously be removed from the gas [5]. The technique is suitable for trace gas analysis with Fourier transformation infrared (FTIR) spectrometry but is not the method of choice for high resolution diode laser spectrometry because the spectral line widths of sample gases within a matrix are quite large, about 0.1 to 2 cm^{-1}.

4. PRINCIPLE OF ENCLOSIVE FLOW COOLING

Among the cooling schemes mentioned above, collisional cooling and the diffusive trapping method are best suited for trace gas analysis. But either method has the disadvantage that only a small number of molecules is optically effective within the measuring laser beam. Each contact of a sample gas molecule with the chamber wall excludes it from spectrometric detection in the gas phase.

The principle of enclosive flow cooling is to add the sample gas to a cooling gas flow in such a way that the cooling gas, flowing from the internal cylinder wall to the axis, encloses the beam of sample gas molecules which are thus prevented from getting adsorbed by the cylinder wall, see Figure 2. The precooled cooling gas enters the cell through an inlet tube *(1)*, is distributed in the space between the inner and the outer hollow cylinder and is then sucked into the chamber through the porous lateral area of the inner hollow cylinder

(4). An axially symmetric flow profile develops in the chamber *(5)*. First the cooling gas flows almost perpendicularly towards the axis and then bends into the direction of the cylinder axis. The sample gas is added to this cooling gas flow through a heatable tube *(3)*. The cooling gas encloses the sample gas, supercools it below its boiling point within the first few millimeters, and advects it along the axis, preventing it from getting adsorbed to the wall. The sample gas is sucked out of the chamber with the cooling gas through a suction torus *(7,9)*. Boreholes drilled into the torus are distributed in a ring pattern on its inner side. Behind the torus a mirror *(8)* is positioned.

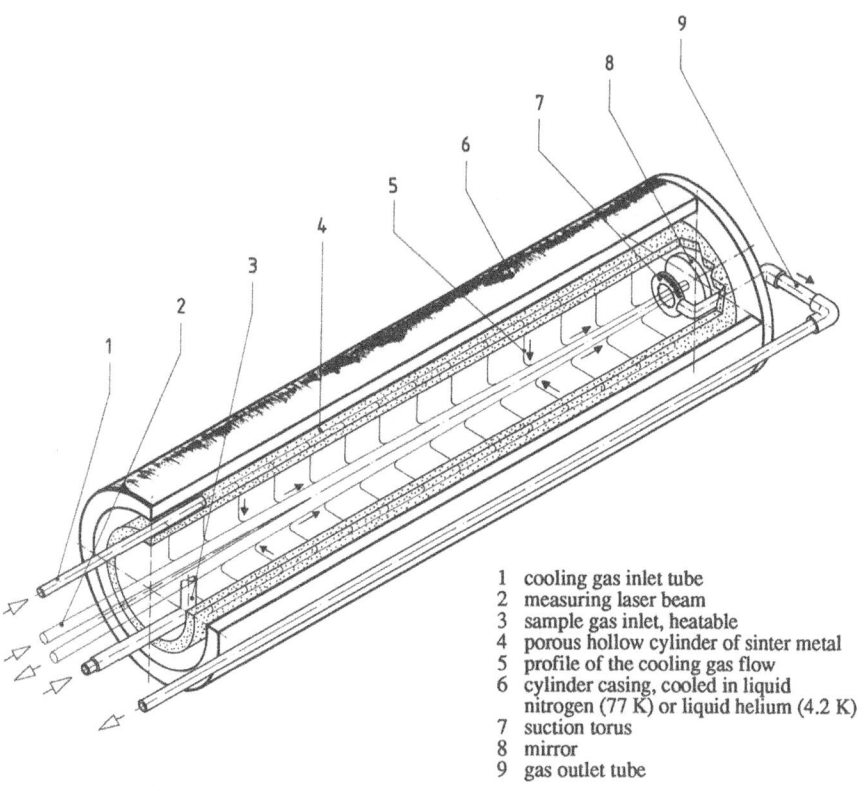

1 cooling gas inlet tube
2 measuring laser beam
3 sample gas inlet, heatable
4 porous hollow cylinder of sinter metal
5 profile of the cooling gas flow
6 cylinder casing, cooled in liquid
  nitrogen (77 K) or liquid helium (4.2 K)
7 suction torus
8 mirror
9 gas outlet tube

Figure 2: Sectional drawing of the enclosive flow cooling cell.

This arrangement allows to position the measuring laser beam *(2)* on the cylinder axis, i.e., on the flow path of the sample gas molecules. It passes through the suction torus and is reflected by the mirror, thus passing the measuring path twice. The outer cylinder *(6)* is immersed in liquid nitrogen (77 K) or liquid helium (4.2 K), or is in thermal contact with the coldhead of a He refrigerator. N_2, H_2 or He are appropriate cooling gases. The sample gas will be mostly some carrier gas (e.g. air, N_2, He) and for the smaller part the target or trace gas.

From kinetic gas theory, equations can be derived [6] describing the temperature and pressure dependence of the mean free path λ and the mean diffusion time τ of a gas phase molecule moving into a given direction x:

$$\lambda = \frac{kT}{\sqrt{2}\pi d^2 p}$$

$$\tau = \sqrt{\frac{3}{2}}\pi \frac{d^2 \overline{x^2} \sqrt{m}\, p}{(kT)^{3/2}}$$

(d effective molecular diameter, about 2×10^{-10} m for N_2, O_2 and acetaldehyde, m molecular mass 7.3×10^{-26} kg for acetaldehyde and 4.7×10^{-26} kg for N_2, $\overline{x^2}$ mean square translation in a given direction x). Hence for $T = 100$ K and $p = 10$ hPa λ is about 0.01 mm. Therefore thermal equilibrium (≈ 100 collisions) is reached after less than 1 mm of diffusion path and within less than 1 ms of diffusion time.

As an upper limit for the final temperature of the sample gas one can assume the mixing temperature of cooling gas and sample gas within the sample gas thread. With the same heat capacity of cooling gas (at $T = 77$ K) and sample gas (at $T = 300$ K) this final temperature is 99 K for 10 % sample gas in the mixture and 122 K for 20 %. If one takes into consideration the additional cooling effect by heat conduction from the inner flow thread to the surrounding area, the final temperatures will be even lower than the ones given above.

Table 2:   Temperature dependence of the time τ needed for an acetaldehyde molecule in a buffer gas to diffuse from the axis of a cylinder of radius 25 mm to the cylinder wall, at a pressure p = 50 hPa.

Temperature T / K	Diffusion time τ / s
4	320
10	80
50	5.5
80	3.5
100	2.5
200	0.91
300	0.50
400	0.31

The mean time a sample gas needs to diffuse from the cylinder axis to the wall (at a cylinder diameter of 50 mm) drastically depends on the temperature, see Table 2. These data, calculated for acetaldehyde, also hold approximately for O_2 and N_2.

The idea of enclosive flow cooling is that the diffusive motion of the sample gas molecules from the cylinder axis to the cylinder wall where they get adsorbed and lost is compensated by the radial flow from the wall to the axis. Although the radial flow speed is not constant along the cylinder radius and also depends on the diameter of the central axial beam, there is little doubt that at flow velocities of 0.01 to 1 m/s along the cylinder axis the radial flow speed easily compensates the diffusion velocities that can be calculated from the τ values of Table 2.

5. PRELIMINARY EXPERIMENTS WITH A TRANSPARENT MODEL CELL

To visualize the flow conditions, preliminary experiments were carried out with a transparent plexiglas flow cell using cigarette smoke at standard conditions. The inner hollow cylinder has four rows of 12 boreholes each. This model cell is 30 cm long and 5 cm in inner diameter. It has thus nearly the same geometry as the original cell. In spite of the restriction of discrete boreholes instead of multiple fine pores a characteristic laminar flow pattern with good radial symmetry was observed over a wide range of flow velocities.

In Figure 3 a longitudinal section of the enclosive flow is shown, made visible by the expanded beam of a HeNe laser. The mean axial flow velocity is about 0.02 m/s and the gas flow rate amounts to 2 l/min. After the passage of the boreholes the smoke first flows radially towards the axis and then relatively abruptly bends into an axial direction. When the flow velocity is increased this bending becomes sharper, and the bending point moves closer to the axis. The flow velocity on the axis increases from the sample gas inlet to the suction aperture. On the photograph one can clearly see several layers of smoke, with the smoke from the sample gas inlet region flowing in the middle as desired. The enclosive flow of cooling gas provides most of the total flow in the cell. The aim, however, is to maximize the contents of sample gas in the center part of the flow pattern along the cylinder axis.

It is equally important to visualize the flow of sample gas alone, see Figure 4. To do this smoke is added through the sample gas inlet close to the cylinder axis. The enclosive flow of cooling gas remains invisible. A very regular laminar smoke thread is formed with a diameter of 2 to 5 mm which deviates no more than about 3 mm from the cylinder axis. Particles of this smoke thread can be kept on the axis for up to 35 seconds.

The flow patterns are surprisingly stable against perturbing movements of the cell and against pressure fluctuations.

6. COMPUTER SIMULATION OF THE FLOW BEHAVIOR

To obtain further information for optimizing the flow behavior, a relatively simple computer simulation of the enclosive flow was performed. If an incompressible, laminar, irrotational and frictionless flow is assumed, the Navier-Stokes equations reduce to the Laplace equation. A numerical solution can be obtained by a simple relaxation method known as Liebmann's averaging method [7,8].

The axial symmetry of the cell reduces the problem to two spatial dimensions. The Laplace equation in cylinder coordinates that must be solved has the form

$$\frac{\partial^2 \phi}{\partial r^2} + \frac{1}{r}\frac{\partial \phi}{\partial r} + \frac{\partial^2 \phi}{\partial z^2} = 0$$

(ϕ velocity potential, r radial coordinate, z length coordinate). The boundary value problem is a mixed one (Neumann/Dirichlet problem): at the porous wall, which is the cooling gas inlet, the derivative of the potential (i.e., the flow) is given, at the gas outlet the potential itself (that is, except for a constant offset value, the pressure) is fixed. The cell is divided into a grid of mesh points. A starting or a boundary value is assigned to each of the mesh points. In an iteration step the potential of each of the cell volume elements is replaced by the average value of its four neighbours, weighted according to the cylinder geometry. The convergence of the iteration can be achieved either with small or with high starting values and can also be checked by approaching the final solution from the high and from the low side. For the 7 × 21 mesh point grid used here, typically 25,000 iteration steps are needed to reduce the deviation of the two solutions from each other to below 0.1 percent. The compu-

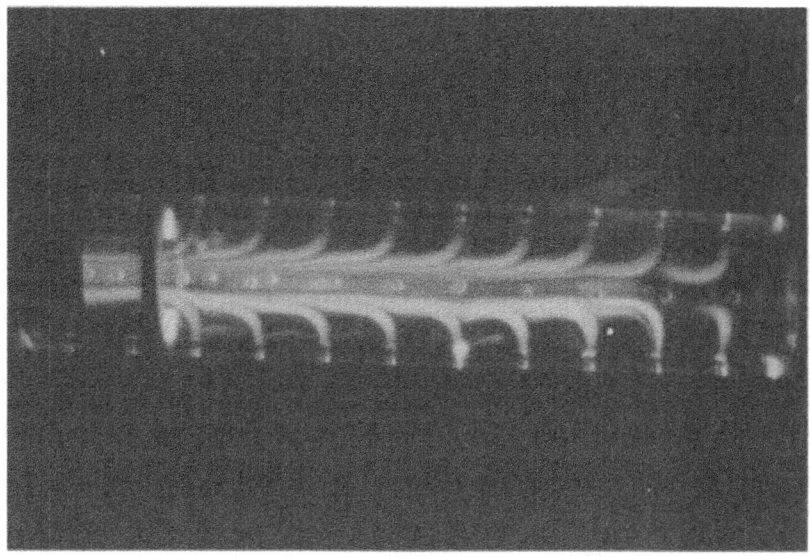

Figure 3: Longitudinal section of the enclosive flow consisting of cigarette smoke illuminated by a HeNe laser, flow direction is from right to left.

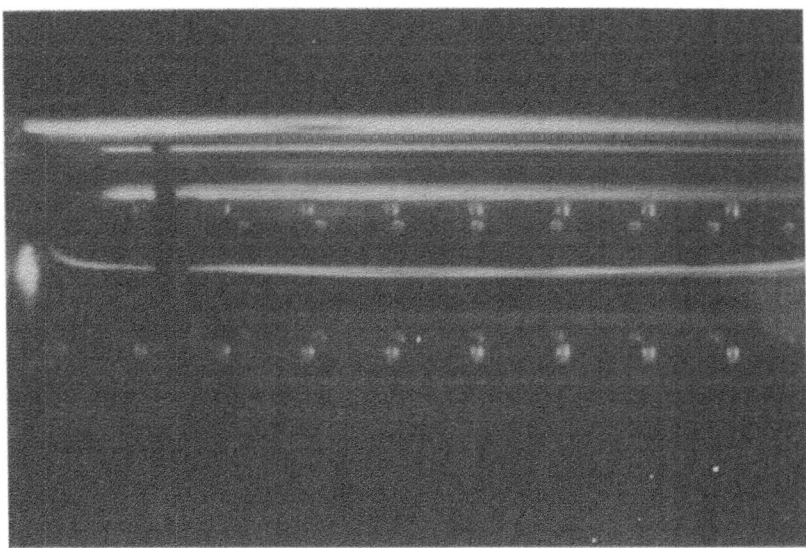

Figure 4: Longitudinal section of the sample gas flow consisting of cigarette smoke illuminated by a HeNe laser, flow direction is from right to left.

tations were performed by a Turbo Pascal program on a simple 386 personal computer in approximately one hour.

To further improve the results a friction correction was introduced. Because of the friction due to the gas viscosity, the radial velocity profile of a pure laminar flow through a cylinder has a parabolic shape (Hagen/Poiseuille):

$$v = \frac{p_1 - p_2}{4\mu l} (R^2 - r^2)$$

(v flow velocity, p pressure, l length of the tube, R radius of the tube, r radial coordinate, μ viscosity). To consider this the velocity field obtained by solving the Laplace equation numerically is corrected by a parabolic weighing function so that the continuity equation is fulfilled.

Figure 5 shows a longitudinal section of the calculated velocity field inside the cell. The directions of the lines indicate the velocity vectors, their lengths correspond to the amounts of the velocity. One can see that the flow velocity increases towards the cell end. Figure 6 shows the velocity potential of the four longitudinal sections of the calculated velocity potential, one for each axially parallel row of elementary cells. The slope is proportional to the flow velocity.

Flow Field

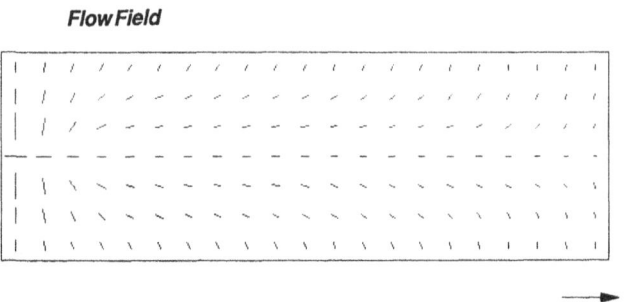

Figure 5: Longitudinal section of the computed velocity field, flow direction is from right to left.

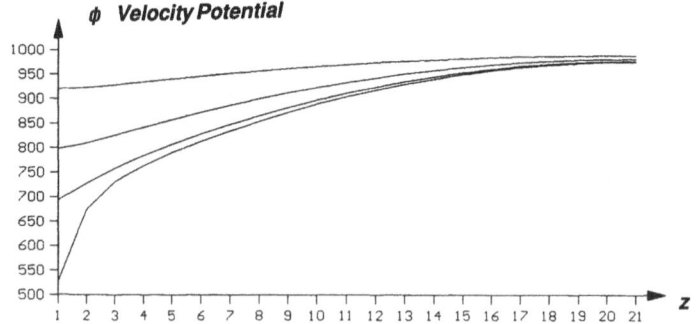

Figure 6: Four longitudinal sections of the computed velocity potential. Sequence of curves is from cylinder wall (top) to cylinder axis (bottom).

7. DISCUSSION AND FUTURE WORK

The enclosive flow patterns produced with smoke in the model cell and the patterns obtained from the computer simulation agree quite well in shape and velocity behavior. The results are promising as to sample gas cooling by flow enclosure. To extrapolate the observed flow patterns to cryogenic temperatures and low pressure values, Reynolds' similarity law can be applied which states that the same flow patterns are characterized by the same Reynolds number

$$Re = \frac{v\,d\,\rho}{\mu}$$

(v flow velocity in a distance far from the body, d characteristic length, ρ density of the fluid). The ideal gas law yields $\rho \sim p/T$, and from statistical thermodynamics one can obtain $\mu \sim T$ and $\mu(p)$ = const. (i.e., pressure independency). Hence it follows that

$$Re \sim \frac{v\,d\,p}{T^{3/2}} \; .$$

These flow conditions can be maintained if p and T are reduced simultaneously such that $p/T^{3/2}$ remains constant. Matching pairs of values are (p,T) = (125 mbar, 75 K), (20 mbar, 22 K), (10 mbar, 14 K).

In a first step the cell will be cooled to 77 K with liquid nitrogen. Experiments will show to what extent a further reduction of temperature is desirable. Calculations as well as simulations with the transparent cell will be carried out to investigate the following problems:

- optimization of the ratio diameter/length of the cell;
- optimization of the enclosive flow pattern by using a sinter metal hollow cylinder with a wall thickness that varies along its length;
- decision whether a commercially available sinter metal hollow cylinder or a hollow cylinder with discrete boreholes is preferable;
- examination of a hollow cylinder with slant boreholes to produce a rotating or a downstream-directed enclosive flow;
- study of the mixing behavior within the flow by using two reacting gases, e.g., NH_3 and HCl;
- optimization of the suction torus geometry.

At low temperatures and suitable pressure values the concentration of a given sample gas quantity can be measured within a narrow beam of 3 to 10 mm diameter, or 1/100 to 1/1000 of the volume of a conventional White cell. This can make multireflexion spectroscopy unnecessary, an advantage particularly important in the case of trace gas analysis with preconcentration, for which no unlimited amounts of sample gas are available.

In summary, a cooling method has been proposed that overcomes several limitations of conventional cooling schemes used in molecular spectroscopy. Possible applications of the technique include

- diode laser spectrometry, especially for trace gas analysis;
- FTIR spectrometry, especially for trace gas analysis;
- examination of cluster formation;
- isotope separation, e.g., of UF_6.

Cooling the enclosive flow cell will frequently require little additional experimental effort, especially in cases such as TDLAS for which cooling of the lasers and detectors has become part of the routine operation.

REFERENCES

[1] S. Bauerecker, W. Michaelis, H. K. Cammenga: Durchführbarkeitsstudie zur Technik der spurenanalytischen Bestimmung von Acetaldehyd mit dem Diodenlaserspektrometer - Ein Beitrag zur Waldschadensforschung, GKSS 91/E/69 (1991), 72 p.

[2] P. B. Davies, A. J. Morton-Jones: Evaluation of Jet-Cooled Laser Spectroscopy for Simplifying Infrared Spectra, Appl. Phys. B 42 (1987) 35–40

[3] D. R. Willey, D. N. Bittner, F. C. DeLucia: Very Low Temperature Spectroscopy: The Helium Pressure Broadening Coefficients below 4.3 K for the Higher Lying States of CH_3F, J. Mol. Spectrosc. 133 (1989) 182–192

[4] J. A. Barnes, T. E. Gough, M. Stoer: Diffusive Trapping: An Alternative to Supersonic Jet Cooling for Spectroscopic Experiments? Rev. Sci. Instrum. 60 (1989) 406–409

[5] D. W. T. Griffith, G. Schuster: Atmospheric Trace Gas Analyis Using Matrix Isolation Fourier Transform Infrared Spectroscopy, J. Atmosph. Chem. 5 (1987) 59–81

[6] F. Reif: Statistische Physik und Theorie der Wärme, Walter de Gruyter, Berlin, New York (1985) 567-573

[7] L. Collatz: Numerische Behandlung von Differentialgleichungen, Springer, Berlin, Göttingen, Heidelberg (1955) 320-434

[8] R. Günzel: SCAPOT, BAHN und FELDS - Ein Programmsystem zur Lösung der Laplace-Gleichung, zur Berechnung der Bahn von Ladungsträgern und zur Berechnung der Feldstärke an Elektrodenoberflächen, ZfK-372 (1978) 4-7

Section IV

SPECIAL APPLICATIONS

TDLS ANALYSIS OF WATER VAPOUR TRACES
IN SEMICONDUCTOR PROCESS GAS

R. KÄSTLE, R. GRISAR, M. TACKE

Fraunhofer-Institut fuer Physikalische Messtechnik
Heidenhofstrasse 8, 7800 Freiburg, Germany

and

D. DORNISCH, C. SCHOLZ

CS GmbH, Semiconductor- und Solar Technology
Gollierstrasse 70, 8000 Munich 2, Germany

SUMMARY

We report on the analysis of ultrapure semiconductor process gases by
infrared diode laser spectroscopy. For worst case demonstration, the
system H_2O in NH_3 was chosen, which can not be handled at trace levels of H_2O by conventional techniques. A detection limit of 300 ppb
was obtained with considerably lower values possible for optimized systems. The NH_3 cross sensitivity was found to be below the H_2O detection limit. The technique is potentially applicable to a large number of
trace and carrier gas species under difficult conditions.

1. INTRODUCTION

Purity of process gases in semiconductor manufacturing is a subject of
rapidly increasing importance. The required sensitivities can not be matched
by established analytical techniques. The high sensitivity of gas analysis by
infrared tunable diode laser spectroscopy (TDLS) with sub-ppb detection
limits at temporal resolution of some minutes has been proven earlier in
environmental investigations (1) and ppm sensitivities with millisecond time
resolution were obtained in car exhaust gas monitoring (2). The TDLS technique also appears most promising for solving problems of trace impurities in
ultrapure process gases. We report on detection of H_2O traces in NH_3 as a
demonstration of TDLS gas analysis chosen as a worst case example in
ultrapure semiconductor process gases (3).
Practically all standard measurement techniques fail on this pair of
gases which are chemically related. Mass spectroscopy, for example, is ham-
pered by the comparable masses of these gas species. Generally, standard
techniques are well developed for bulk gases. They are based on adsorption,
dew-point detection, micro balance, capacitive sensors and gas chromatography. All of them are, however, not applicable to H_2O in NH_3, since both
molecules condense readily and have high dipole moments.
Atmospheric Pressure Ionization Mass Spectrometry (APIMS) has proven ppt
sensitivities for H_2O in N_2 and Ar and has been applied successfully to ultrapure bulk gases. It can not be taken for H_2O in NH_3, since the necessary
accumulation of the trace component relies on secondary ionisation by the
main component. This works properly only, if the ionisation potentials are
considerably smaller for the trace component (4).

2. EXPERIMENTAL

The experimental setup consists of a TDLS trace gas system developed earlier for environmental research (1). The optics are illustrated in Fig. 1. In addition to the sample beam with second derivative signal processing, a reference beam enables a frequency scale calibration by transmission through high concentration gas and subsequent first derivative signal detection. Cooling of the required semiconductor diode lasers and detectors is performed by two commercial Stirling-type cryopumps. The system was provisionally adapted to the H_2O trace gas measurement by flooding the optical path outside the sample cell with dry liquid boil-off N_2 gas in order to reduce absorption by atmospheric H_2O, and by attaching an ultra-clean gas supply. The 100 m optical path cell (5) of the White type is made of quartz and provided with viton O-ring seals which are clearly not optimally suited for H_2O detection. The system was operated with a sample gas pressure of typically 50 mbar.

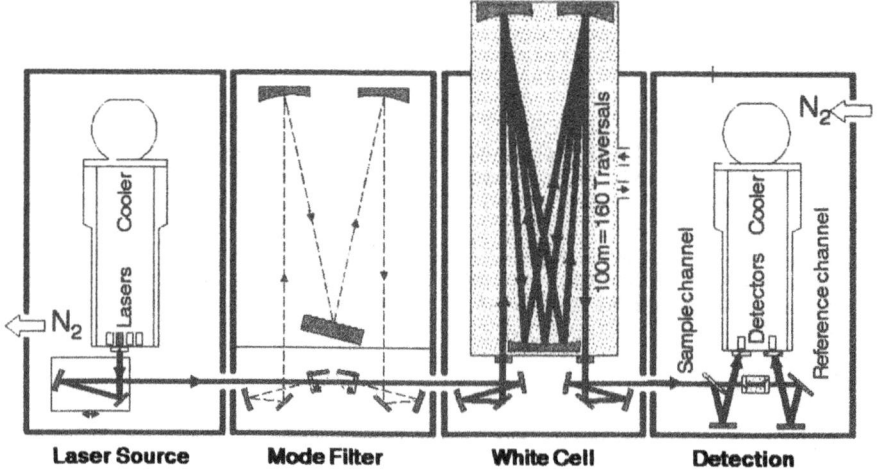

Fig. 1: Schematic of the trace gas analyzer optics

Two H_2O absorption lines with wavenumbers taken from an absorption line parameter compilation (6) to be 1923.162 cm^{-1} and 1922.342 cm^{-1} were identified to be suited for the task in intitial experiments, although they are not the optimum choice. The line strength, however, is sufficient for the envisaged ppm sensitiivity in conjunction with low NH_3 interference. According to the high resolution FTIR transmission spectra shown in Fig. 2, overlap with NH_3 lines is a crucial point in view of the abundance of strong NH_3 absorption lines in the spectral range of strongest water vapour absorption.

The experiments did not include a zero point test with dry NH_3, since such a gas can not be obtained at the purity level required, the highest commercial specification being 5.0. In order to generate lower H_2O concentrations, a mixture of 5% 5.0 NH_3 in a base of 6.0 N_2 was prepared which should contain less than 1.5 ppm H_2O.

A further advantage of the diode laser technique is the possibility to calibrate with mixtures of high concentration and a correspondingly shorter optical path. For this purpose, a 10 cm length calibration cell was inserted into the optical path of the sample beam. In the present case, the required calibration concentrations are scaled up by a factor of 1000 and a calibration at 10 ppm equivalent concentration may be performed with 1% H_2O. Care has been taken, not to modify the optical conditions under calibration in the potential presence of optical feedback by leaving the empty calibration cell in the optical beam also during the measurements.

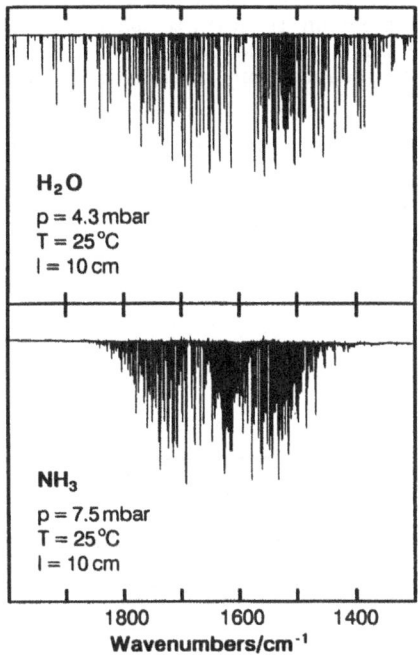

Fig. 2: High-resolution FTIR transmisssion spectra of H_2O and NH_3

3. RESULTS

As a first step, H_2O in N_2 was measured in order to demonstrate the sensitivity for H_2O. In a second step, the same procedure was applied to 5% NH_3 in N_2 and in the final experiment, a set of known mixtures of $NH_3/H_2O/N_2$ was analyzed.

H_2O in N_2

The sample gas for this first step was $N_2(5.0)$ with a H_2O content specified by the supplier to be 7.5 +/- 0.5 ppm. Since desorption of H_2O from the quartz walls of the sample gas cell was inevitable, different mass flow values were used. Assuming the wall desorption rate not to depend on the gas velocity and to remain constant during a measurement cycle, the real H_2O content (c_t) of the test gas can be extrapolated from the experimental data (c_{exp}) for infinite mass flow as

$$c_{exp} = f_t/(f_t + f_d) \cdot c_t + f_d/(f_t + f_d) \cdot c_d, \qquad (1)$$

where f_t and f_d is the mass flow in standard litres per minute (Sl/min) for test gas and desorbed gas, respectively, and c_d is the desorption concentration of 10^6 ppm, which enters for correct dimensions of the equation. In the limit of f_t large compared to f_d, this can be linearized to

$$c_{exp} = c_t + constant \, / \, f_t. \tag{2}$$

A set of four data points with the mass flow between 0.2 and 0.5 Sl/min is plotted in Fig. 3. Extrapolation to infinite mass flow results in a H_2O concentration of 11.9 +/- 2.2 ppm. The zero point fluctuations and hence the minimum detectable H_2O concentration in each individual measurement amount to some 50 ppb.

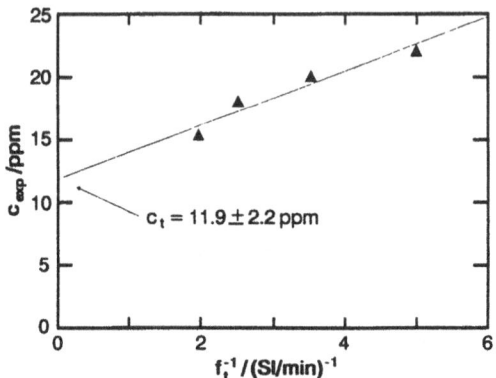

Fig. 3: Observed H_2O concentrations versus reciprocal mass flow of N_2 (5.0)

The experimental error was taken to be 3 times the standard deviation from the regression line. This error can be made smaller by using larger mass flow. The difference between our results and the gas specification is not unexpected, since prior APIMS measurements doubted the validity of dew-point measurements in the ppm range (7).

H_2O in NH_3

The same approach was used for determination of the H_2O content in a mixture of 5% NH_3 (5.0) in N_2 (6.0) which was performed with two different water vapour absorption lines. No independent concentration values could be gained for reference, since there is no proven analytic method for this mixture. Assuming that the complete specified trace content of both gases is due to H_2O, the upper limit of H_2O concentration is 1.5 ppm. The data were obtained by first increasing the mass flow in steps from 0.5 Sl/min to 2.5 Sl/min and then repeating the procedure with decreasing mass flow. The results for both H_2O absorption lines are shown in Fig. 4. By extrapolation, H_2O concentration values of 3.0 +/- 0.3 ppm and 3.1 +/- 0.3 ppm are obtained.

Two conclusions can be drawn from the experimental findings. The first one is based on the observation that the centre wavenumber and the shape of the H_2O absorption line in NH_3 are exactly the same as those of H_2O in N_2. Line positions and profiles of molecular transitions are extremely sensitive to the formation of complexes.g This observation of pure H_2O lines in NH_3 seems to be the first proof of the coexistence of molecular H_2O and NH_3, which settles a long-standing open chemical question.

The second conclusion regards cross sensitivity. NH_3 itself has a multitude of infrared absorption lines. The absolute strengths of these lines in the spectral range of water vapour line absorption around 1923 cm^{-1} of interest here are small. Due to the four orders-of-magnitude difference in concentration, however, the absorption of these NH₃ lines becomes comparable to that of the strong H₂O lines. Any accidental partial or full coincidence of respective lines will result in cross sensitivity. In principle, this should be verified by a measurement with absolutely dry NH₃, which is, however, not possible because such a gas is not available.

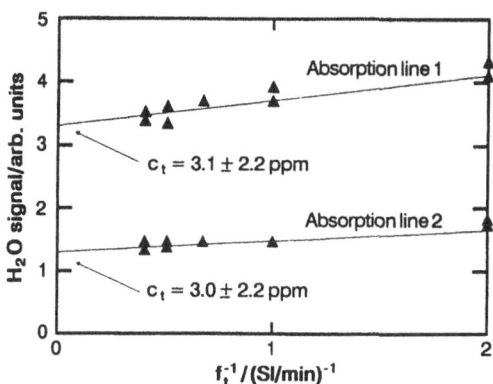

Fig. 4: Observed H₂O signal versus reciprocal flow of 5% NH₃(5.0)/N₂(6.0)

We can nevertheless give an upper limit for the cross sensitivity by the following consideration. The probability to hit the same NH₃ line overlap with two different H₂O lines is extremely small. The equality of the result for the H₂O concentration obtained with two different absorption lines thus indicates an eventual cross sensitivity to be below the experimental error of 0.3 ppm. In addition, the observed line shapes were symmetrical. For the case of overlap, this would only be possible for identical centre wavenumbers and two NH₃ lines with same line strength ratio to the respective H₂O lines. This is extremely unlikely.

The observed concentration of H₂O is again higher than specified, as was the case for H₂O in N₂ (Table 1). The discrepancy, however, is less distinct for NH₃ than for N₂.

Table 1: H₂O vapour concentrations in test gases

Test Gas	H₂O concentration as specified	H₂O concentration as measured
N₂(5.0)/7 ppm H₂O	7.5 +/- 0.5 ppm	11.9 +/- 2.2 ppm
5% NH3(5.0)/N2(6.0)	< 1,5 ppm	3.1 +/- 0.3 ppm

NH3 in wet N2

In order to ascertain these experimental findings, H₂O in NH₃ at different mixing ratios was analyzed. For that purpose, 5% NH₃ (5.0)/N₂ (6.0) and

12 ppm H_2O/N_2 were mixed with different mass flow ratios, the total mass flow kept constant at 1.0 Sl/min. The minimum content of 5% NH_3/N_2 was 50%.

As can be seen from Fig. 5a, the H_2O line absorption, and hence their concentrations, are in linear relation to the mixing ratio. This is not the case for three observed lines that we attributed to NH_3. From a line parameter compilation (6) one expects only one NH_3 line between the observed H_2O lines. Experimentally, a couple of lines were found, and three week ones among them were further examined. The result for one of them is given in Fig. 5b. With the dilution of the NH_3 content from 5% to 2.5%, the absorption does not decrease by 50%, but only by 25%. At present we can not explain this finding. It may be attributed to forbidden NH_3 lines that are induced by H_2O by collision or other interaction, or by the formation of complexes. This interesting phenomenon provokes the need for further experiments.

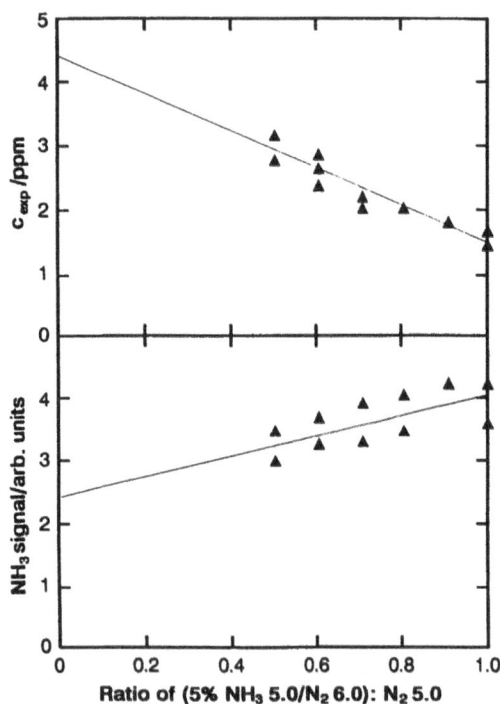

Fig. 5: Experimental results obtained for different mixing ratio of 5% NH₃ (5.0)/N₂(6.0): N₂(5.0);
a: measured H₂O concentration;
b: measured uncalibrated NH₃ signals

4. CONCLUSIONS

Applying infrared diode laser spectroscopy to the analysis of a mixture of H_2O traces in NH_3, molecular H_2O was clearly identified. This is evidenced by the existence of unchanged pure H_2O absorption lines in NH_3 as observed in H_2O/N_2 mixtures.

The theoretical detection limit for H_2O by diode laser gas analysis with 100 m optical path length can in principle be derived from the absorption line strength to be of the order of 10 ppt when taking one the strongest absorption lines. Experimental values for H_2O in bulk gases of 10 ppb were obtained earlier. This value, however, was limited by the residual H_2O concentration in the sample gas cell due to wall desorption (8). In the analysis of H_2O in NH_3 reported here, weaker H_2O absorption lines in a spectral region with low NH_3 absorption were selected in order to minimize potential cross sensitivity. This choice allowed for a detection limit of 300 ppb, as determined by the experimental error. There is potential for a considerable increase in sensitivity for H_2O in NH_3 down to the ppb level by choosing stronger absorption lines. The verification requires a more systematic study.

The successful detection of H_2O traces in NH_3 substantiates the fact, that diode laser spectroscopy is very well suited to trace gas analysis of ultrapure process gases. This statement holds not only for bulk gases, but also for reactive gases such as SiH_4, Cl_2 and HCl. TDLS is universally applicable with high potential sensitivities to most light molecules (Table 2) as well in inert as in corrosive carrier gases. Its high sensitivity, selectivity and speed make this technique well suited for in-line process control.

Table 2: Typical present sensitivity limits in ppb as realized by laboratory systems with 1 m and 100 m optical path

Gas species	atmospheric pressure		low pressure	
	1 m	100 m	1 m	100 m
CO	5000	50	20	0.2
CH_4	10	0.1		
SO_2	10000	10	30	0.4
NH_3	20000	20	3	0.03
HNO_3	20000	20	20	0.2

The high sensitivity and selectivity evidenced in these experiments is due to the low sample gas pressure of the order of 50 mbar. At higher pressure, the line width increases which reduces the performance. The limits of application then typically rise to some 10 ppb instead of 0.1 ppb at low pressure. With bulk gases, the cross sensitivity will still be negligible, but may become prohibitive for infrared absorbing carrier gases like NH_3. In a couple of cases, in-situ measurements within the gas supply lines and without gas sampling will be possible. This is another appealing feature of infrared diode laser gas analysis.

REFERENCES

(1) SCHMIDTKE, G., KOHN, W., KLOCKE, U., RIEDEL, W.J. and WOLF, H. (1989). Diode Laser Spectrometer for Monitoring up to Five Atmospheric Trace Gases in Unattended Operation, Applied Optics 28(17), 3665–3670.

(2) WOLF, H., GRISAR, R., KLOCKE, U., RIEDEL, W.J., and WISSLER, R. (1989). Dynamic Car Exhaust Gas Analysis Using Tunable IR Diode Lasers, in: Monitoring of Gaseous Pollutants by Tunable Diode Lasers, Volume 2, R. Grisar G. Schmidtke, M. Tacke and G. Restelli, eds., Kluwer Academic Publishers, Dordrecht, 61–67.

(3) FLAHERTY, E., FEROLD, C., WOJCIAK, W., MURRAY, D. and AMATO, A. (1987). Reducing the Effects of Moisture in Semiconductor Gas Systems, Solid State Technology, 30(7), 69–75.

(4) MITSUI, Y., IRIE, T. and MIZOKAMI, K. (1990). Mass Spectrometer for ppt-Trace Analysis, Ultra Clean Technology, 1(1), 3–12.

(5) RIEDEL, W.J. (1991). Optics for Tunable Diode Laser Spectrometers, SPIE Proc. 1433, 179–189.

(6) ROTHMAN, L.S., GAMACHE, R.R., GOLDMAN, A., BROWN, L.R., TOTH; R.A. PICKETT, H.M. POYNTER, R., FLAUD, J.-M., CAMY-PEYRET, C. BARBE, A. HUSSON, N., RINSLAND, C.P. and SMITH, M.A.H. (1987). The HITRAN Database: 1986 Edition, Applied Optics 26(19), 4058–4097.

(7) YABUMOTO, N., MINEGISHI, K., SAITO, K. and HARADA, H. (1990). An Analysis for Gases with APIMS, Ultra Clean Technology 1(1), 13–21.

(8) MANTZ, A.W. (1987). Application of High Resoluition Infrared Techniques to Semiconductor Processes, in: Monitoring of Gaseous Pollutants by Tunable Diode Lasers, R. Grisar, H. Preier, G. Schmidtke and G. Restelli, eds., D. Reidel Publishing Company, Dordrecht, 136–144.

DETERMINATION OF UNBURNED FUEL IN THE WALL BOUNDARY LAYER OF METHANE/AIR FLAMES

H. Eberius, Th. Just and M. Overhamm

Deutsche Forschungsanstalt für Luft- und Raumfahrt
Institut für Physikalische Chemie der Verbrennung
Pfaffenwaldring 38, Stuttgart
Germany

SUMMARY

The interaction of flames with cold walls can be a major source of unburned hydrocarbons in the homogeneous-charge spark-ignited engine. The low temperature in the combustion boundary layer adjacent to the cold walls reduces the rate of several important chemical reactions. In the direct vicinity of the cold walls flame propagation is quenched and a layer of the fuel/air mixture is left unburned at the wall. Some of the fuel in the boundary layer may be transported into the main flow of the flame gases by diffusion processes, where further oxidation may occur. The remaining unburned fuel may be discharged into the exhaust gases as residual hydrocarbon emissions. For the investigation of these processes laminar premixed methane/air flames have been stabilized on a porous plate between water-cooled walls. The residual fuel concentrations near the wall have been determined using high resolution laser infrared spectroscopy. The measurements show high methane concentrations in the boundary layer compared to the regions at a larger distance from the wall. In contrast to recent numerical predictions rather large concentrations of unburned fuel are found in the boundary layer far downstream of the flame zone.

1. INTRODUCTION

Flame quenching may contribute to pollutant emissions from combustion engines. This is a problem of increasing concern (1). At cold walls of combustion chambers there may be large heat fluxes from the flame gases to the wall (2,3,4). Besides radiation and molecular heat diffusion, the diffusion of chemical reactive species can also contribute to the energy transport to the cold wall. Additionally, the heat flux may be increased by flow properties such as forced convection or turbulence. The heat loss leads to a temperature decrease with the result that the combustion processes become slower in the area near the wall. For large heat losses the flame propagation may cease, a process which is called 'wall quenching'. Residual fuel, unburned intermediate products such as carbon monoxide and unburned hydrocarbons are left, representing a source of unburned hydrocarbons (UHC) additional to the emissions of the bulk gases.

These processes are important for combustion in piston engines. In spark-ignited (SI) engines there can be large emissions of residual hydrocarbons. Flame quenching in the crevices of the cylinder is argued to be a major source of unburned hydrocarbons. The retarded heat release at the cold walls and the heat losses of the flame gases to the walls are also relevant for engine combustion efficiency.

Various theoretical (5,6) and experimental (7) investigations have been performed concerning the emissions of unburned hydrocarbons from combustion engines. Results from closed vessel experiments demonstrated that crevices in the combustion chambers of SI engines have to be considered important sources of unburned hydrocarbons (8). Investigations related to the development of lean-burning engines showed that the rate of combustion in lean flames decreases considerably near the wall. Thus, within the time interval available for combustion within the engine cycle, the fuel cannot be oxidized completely, with the result that large amounts of unburned hydrocarbons are emitted.

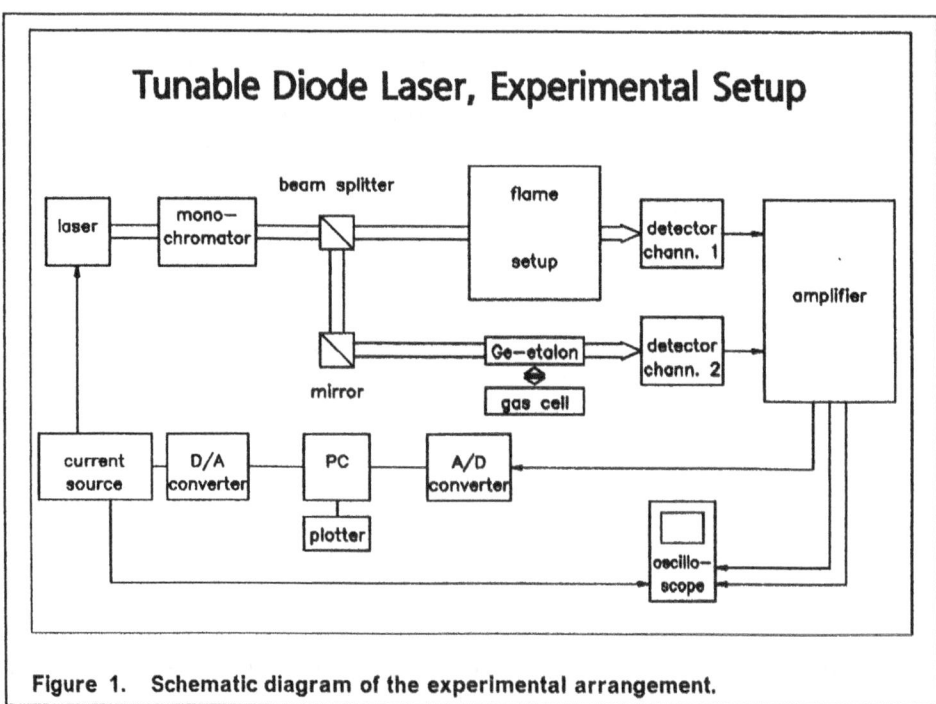

Figure 1.   Schematic diagram of the experimental arrangement.

The quenching of a flame front is of fundamental interest with respect to the processes of heat transfer, and to the chemical kinetics of the recombination of radicals at the wall (5,9,10). A better understanding of flame quenching processes in small crevices is of vital importance, in order to develop models to assess the influence of crevices on hydrocarbon emissions and to estimate these contributions (8,14,16). Several of the investigations have been conducted on stationary flames which are burning in a crevice (11,12) or in the vicinity of a cold wall (13).

In the first stage of the present investigations, stationary flames have been analyzed which are burning near a cold wall or in a crevice (14). The flames were fueled with methane. The concentration of the fuel is determined using laser absorption spectroscopy and an infrared (IR) diode laser. Investigating fuel con-

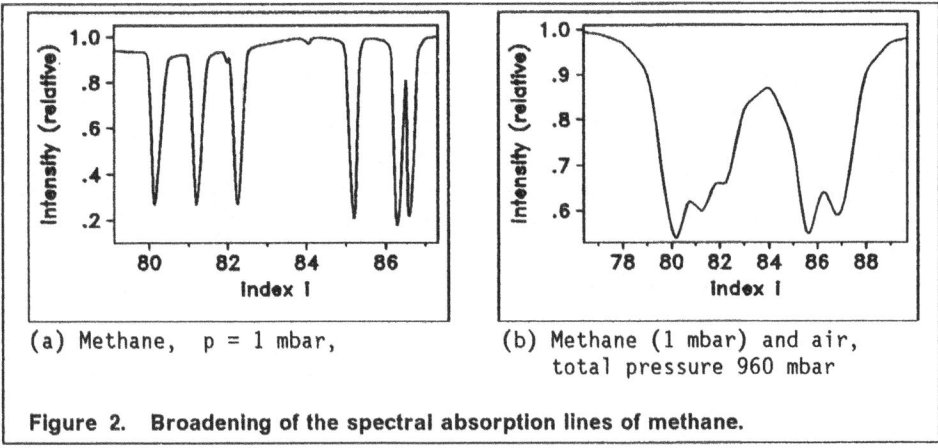

(a) Methane, p = 1 mbar,

(b) Methane (1 mbar) and air,
total pressure 960 mbar

Figure 2. Broadening of the spectral absorption lines of methane.

version in the boundary layer can provide a more accurate understanding concerning the interaction of the flame with cold walls in a crevice, and the reaction times needed for complete combustion of the fuel in the boundary layer.

2. EXPERIMENTAL

For these experiments premixed methane/air flames have been stabilized at the bottom of a cooled, u-shaped crevice of 5 mm width, 15 mm height and 20 mm length and a wall temperature of 56 °C. The walls are made of aluminia. The bottom of the crevice consisted of a porous, sintered bronze plate. The velocity of the methane/air mixture entering the bottom was in the order of 10 cm/s.

Figure 1 shows schematically the experimental setup used to calibrate the infrared absorption parameters and to determine the fuel concentrations in stationary flames. The main building blocks are the laser system, the burner, the detectors, the signal processing system, and the personal computer for controlling the experiment and for handling and evaluating the data.

A tunable lead salt diode laser provided the IR radiation for the absorption spectroscopic measurements. The different modes of the laser radiation could be separated by a 50 cm focal length monochromator. The laser has a very small spectral line width of about $3 \cdot 10^{-4}$ cm^{-1}. For the methane measurements the spectral region at the wavenumber 2958 cm^{-1}, the P(6) line in the ν_3 vibration band, was used. In this region the IR absorption of the methane is rather strong, and the absorption lines of water and carbon dioxide are comparatively weak.

Behind the monochromator, the laser beam is split into the main beam and the reference beam. A germanium etalon for frequency calibration, or a reference gas cell could be positioned in the path of the second beam. The first beam has been shaped to allow measurements with a lateral spatial resolution of 0.3 mm in the boundary layer. The combustion system could be traversed within the measurement area of the main beam. The calibration measurements were made with the same optical setup.

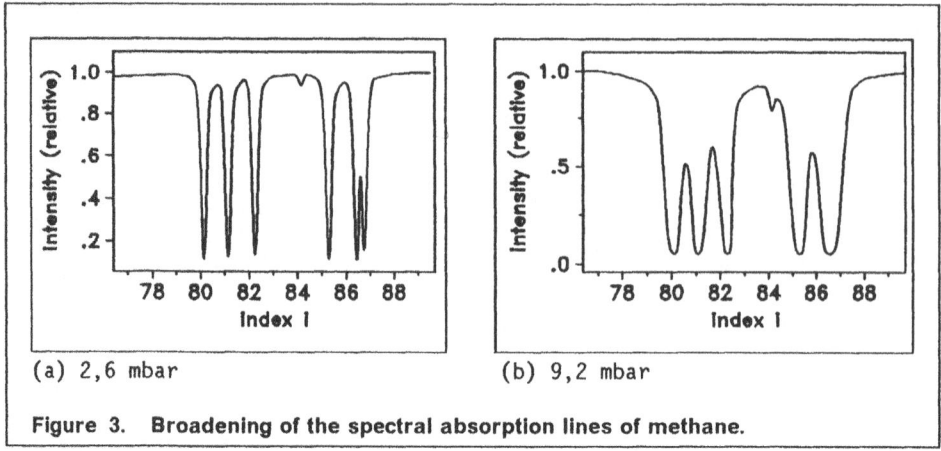

(a) 2,6 mbar   (b) 9,2 mbar

Figure 3.   Broadening of the spectral absorption lines of methane.

Copper doped germanium detectors were used for the main and reference beams. For stationary flames the signals are phase-sensitive amplified, using an intensity modulation of the laser beam of 400 Hz.

The absorption signals were calibrated using methane/nitrogen test mixtures with different concentrations of methane. In several experiments the dependence of the spectral absorption signal on concentration has been investigated at different conditions of temperature and pressure. A gas cell was built to investigate absorption at temperatures up to 250 °C. Starting from low pressures (about 0.1 mbar), where the spectral width of the absorption lines is predominantly deter-mined by the thermal motion of the molecules (Doppler width), the influence of pressure broadening on the shape of the absorption lines has been investigated up to pressures of about 1000 mbar. The influence of temperature has also been investigated.

Several preliminary experiments have been made in order to optimize the magni-tude of the absorption signal and the signal-to-noise ratio and to check cross sensitivities with various flame species.

3. RESULTS

Figures 2 and 3 show several spectra of the P(6) line at different conditions. The spectra, not yet corrected with regard to the baseline, are given as a function of an index-parameter i. The index is a characteristic parameter related to the current which drives the wavelength of the laser emission. For the spectrum in figure 2a the index i = 80 corresponds to the frequency $\nu = 2958.0$ cm^{-1}, and i = 87 corre-sponds to $\nu = 2958.7$ cm^{-1}. The frequency of the laser emission is to a good approximation a linear function of the index i. This can be monitored accurately by comparison with the fringes of the etalon in the reference laser beam. The absolute positions of the spectral lines are taken from the AFGL compilation (15). Figure 2 shows the pressure broadening of the spectral lines in air. The appear-ance of the absorption spectrum at 960 mbar is quite different compared to the spectrum at 1 mbar. The main differences are that the absorption lines are no longer well separated, due to line broadening and an increase in the base width.

These effects influence the sensitivity of the method in two ways, namely by decreasing the height of the absorption lines, and by making it more difficult to establish the base line corresponding to zero methane absorption intensity.

Figure 3 shows absorption spectra at different absolute pressures of pure methane. Higher methane concentrations also change the appearance of the spectra considerably due to self-broadening and to the higher absorption. These effects have to be considered very carefully, especially for gas mixtures which exhibit interfering spectra.

For an accurate determination of the concentration, the absorption lines have to be integrated with respect to the wave number. The calibration curve in figure 4 shows the good linear relationship between the integrated absorption and the methane concentration.

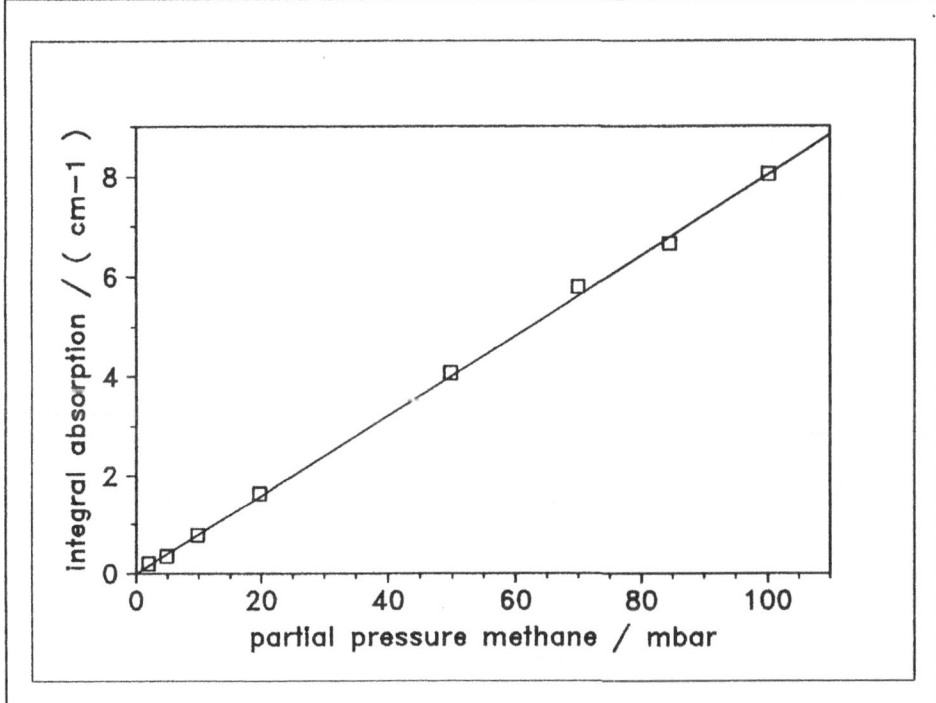

Figure 4.   **Dependence of the integral absorption on the partial pressure.** P(6) line at wave number $v = 2958$ cm^{-1}, total pressure p = 1000 mbar, temperature T = 295 K. Methane in nitrogen.

The intensity of a spectral line is also a function of temperature. The evaluation of this dependence allows the selection of a line where the influence of temperature on line strength is small. The choice of the spectral line for methane in the present case has been guided by the necessity to have good sensitivity and, at the same time, a small temperature dependence.

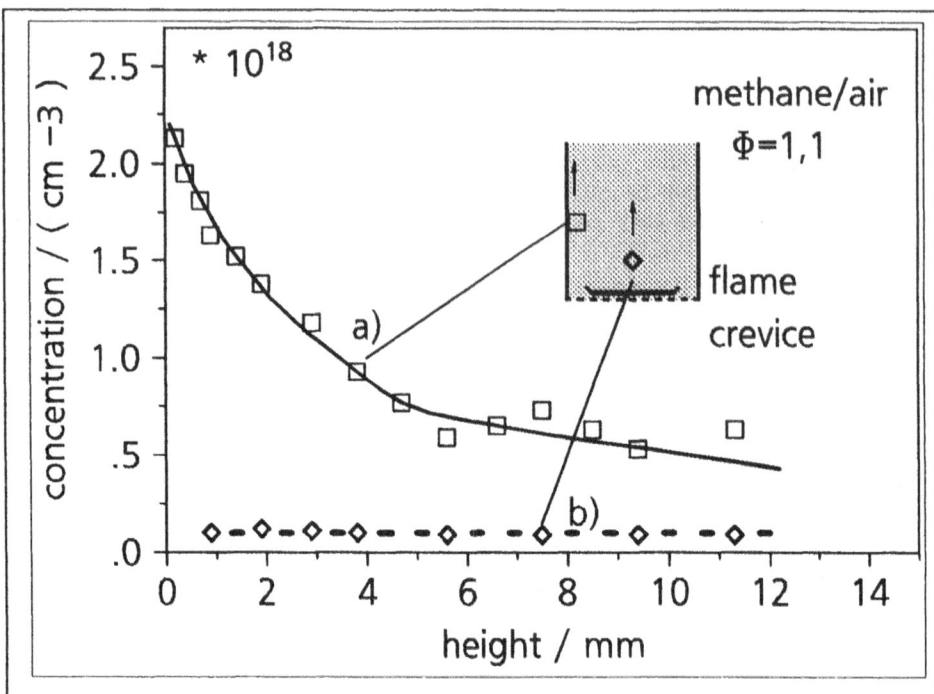

Figure 5. Residual methane concentration in the wall boundary layer. Methane/air flame in a crevice of 5 mm width. Concentrations (a) at the wall, (b) in the center of the crevice.

Calibration measurements have been performed where the integrated absorption is measured as a function of temperature up to 200 °C. Within this temperature regime, the absorption is only a weak function of temperature and changes by less than 10 percent. Thus, to a first approximation, the calibration factor can be set constant in this regime. It is also assumed that the rotational and vibrational energy levels of the methane molecule are in thermal equilibrium.

First measurements of methane absorption in the wall boundary layer of crevice flames showed higher methane concentrations in the boundary layer for fuel-lean flames than for rich flames, in contrast to the smaller fuel concentration in the unburned gas of lean mixtures compared to rich mixtures. Obviously fuel-lean flames are more sensitive to flame quenching, and the combustion boundary layer is more extended in lean flames compared to rich flames. This behaviour of lean flame boundary layers is also in accordance with the known measurements of large quenching distances of flames in the lean regime.

Figure 5 shows quantitatively the methane concentration (molecules/cm³) in the boundary layer (curve a) as a function of the height in the crevice. The concentrations have been calculated from the integrated absorption measurements. The influence of temperature on absorption is small, as already discussed, and was not taken into account. This assumption seems to be reasonable for the boundary layer near the wall, where the temperature is low. As a further assumption, the absorption length of 20 mm has been set equal to the length of the crevice. The absorptions still contain a small background value equivalent to a concentration of about $0.1 \cdot 10^{18}$ molec./cm^3. The stoichiometric ratio is $\Phi = 1.1$. The measurements have been made very near the wall within a layer less than 0.5 mm thick. In this layer no visible flame process has been observed.

The measurements in the center of the crevice (dashed curve in fig. 5) show only the small background signal. Above the flame the methane completely disappeared, as expected. Curve (a) shows a rather continuous decrease in methane concentration at the wall. At 2 mm height in the crevice, the decrease in methane concentration becomes smaller, and above 4 mm height even smaller. Thus, rather high concentrations of methane are still found in the wall boundary layer above 6 mm height. These concentrations are considerably higher than predicted by the theoretical examination in ref. 17. In that approach several assumptions have been made to obtain a model which could be solved numerically. The assumption of a large 'activation temperature' for the one-step combustion chemistry obviously resulted in overestimating fuel consumption in the boundary layer.

Such residual fuel in the boundary layer may be a potential source for hydrocarbon emissions. In technical combustion chambers the unburned fuel can be emitted from the crevices and contribute to the emission of unburned hydrocarbons.

4. CONCLUSIONS

Methane/air flames have been stabilized between cold walls with a separation slightly larger than the quench distance, and a diode laser based system has been developed, which allows measurements in the boundary layer with a local lateral resolution less than 0.5 mm. Monitoring of fuel concentration in the boundary layer showed high residual fuel levels in the boundary layer of the flames, and the more extended quenching layers in lean flames compared to rich flames. High fuel concentrations have been found in the boundary layer also far downstream of the flame zone. As the measured concentrations are larger than those of the theoretical examination, it would be appropriate to improve the model by adopting a more detailed chemical reaction mechanism, and then to compare the results with the measurements. The experiments also showed the feasibility of time-resolved measurements to investigate the non-stationary boundary layers of transient flames.

ACKNOWLEDGMENT

This work has been supported by the Ministry of Research and Technology (BMFT), which is gratefully acknowledged.

REFERENCES

(1) BOULOUCHOS, K. and EBERLE, M.K. (1991). Engine thermodynamics today. MTZ 52, p.575.

(2) LU, J.H., EZEKOYE, O., GREIF, R. and SAWYER, R.F. (1990). Unsteady heat transfer during side wall quenching of a laminar flame. 23rd Symposium (International) on Combustion, p.441, The Combustion Institute.

(3)   VOSEN, S.R., GREIF, R. and WESTBROOK, C.K. (1984). Unsteady heat transfer during laminar flame quenching. 20th Symposium (International) on Combustion, p.75, The Combustion Institute.

(4)   WOSCHNI, G. and SPINDLER, W. (1988). Heat Transfer With Insulated Combustion Chamber Walls and Its Influence on the Performance of Diesel Engines. Trans. ASME 110, p.482.

(5)   WESTBROOK, C.K., ADAMCZYK, A.A. and LAVOIE, G.A. (1981). A Numerical Study of Laminar Flame Wall Quenching. Combustion and Flame 40, p.81.

(6)   SLOANE, T.M. and SCHOENE, A.Y. (1983). Computational Studies of End-Wall Flame Quenching at Low Pressure: The Effects of Heterogeneous Radical Recombination and Crevices. Combustion and Flame 49, p.109.

(7)   ADAMCZYK, A.A., KAISER, E.W., CAVOLOWSKY, J.A. and LAVOIE, G.A. (1981). An experimental study of hydrocarbon emissions from closed vessel explosions. 18th Symposium (International) on Combustion, p.1695, The Combustion Institute.

(8)   BERGNER, P., EBERIUS, H., JUST, Th., POKORNY, H., HÄFNER, G. and LUTZ, W. (1983). Untersuchungen zur Kohlenwasserstoff-Emission eingeschlossener Flammen im Hinblick auf die motorische Verbrennung. VDI-Berichte 498, p.233.

(9)   HOCKS, W., PETERS, N. and ADOMEIT, G. (1981). Flame Quenching in Front of a Cold Wall under Two-Step Kinetics. Combustion and Flame 41, p.157.

(10) MÜLLER, H. and von WATZDORF, S. (1968). Der Gehalt an Kohlenwasserstoffen in der Nähe der Brennraumwand nach Abschluß der Verbrennung. MTZ 29, p.489.

(11) FAIRCHILD, P.W., FLEETER, R.D. and FENDELL, F.E. (1984). Raman spectroscopy measurements of flame quenching in a duct-type crevice. 20th Symposium (International) on Combustion, p.85, The Combustion Institute.

(12) ALY, S.L. and HERMANCE, C.E. (1981). A two-dimensional theory of laminar flame quenching. Combustion and Flame 40, p.173.

(13) SAFFMAN, M. (1984). Parametric Studies of a side wall quench layer. Combustion and Flame 55, p.141.

(14) EBERIUS, H. (1985). Untersuchungen zur Restkohlenwasserstoff-Emission von Flammen in zylindrischen abgeschlossenen Brennräumen. 1. TECFLAM Seminar, p.93. Stuttgart.

(15) ROTHMAN, L.S. et al. (1983). AFGL atmospheric absorption line parameters compilation: 1982 edition. Applied Optics 22, p.2247.

(16) Yoshida, M. (1980). Einfluß der Spaltgeometrie am Feuersteg des Kolbens auf die Kohlenwasserstoffemission bei einem Ottomotor. MTZ 41, p.93.

(17) CARRIER, G.F., FENDELL, F.E. and FELDMANN, P.S. (1984). Laminar flame propagation/quench for a parallel-wall duct. 20th Symposium (International) on Combustion, p.67, The Combustion Institute.

ANALYSIS OF TRACE COMPONENTS IN AUTOMOMOTIVE EXHAUST GAS

W.J. RIEDEL, R. GRISAR, U. KLOCKE, M. KNOTHE, H. WOLF

Fraunhofer-Institut fuer Physikalische Messtechnik
Heidenhofstrasse 8, 7800 Freiburg, Germany

and

P. SCHOTTKA, E. BESSEY, N. PELZ

Daimler-Benz AG, Abt. FVA/VE
Postfach 60 20 20, 7000 Stuttgart 60, Germany

SUMMARY

We report on a four-component TDLS system for highly sensitive real time car exhaust gas analysis with four diode lasers operated simultaneously in frequency multiplex. The sample gas cell with a long optical path of 100 m and a short one of 15 cm is kept at 190 C and 50 hPa. This enables analysis of a number of trace components like formaldehyde, ethene or methanol with detection limits in the 10 ppb range in correlation to major constituents. First test measurements on a dynamometer facility are presented.

1. INTRODUCTION

The demands on automobile exhaust gas analysis systems in view of sensitivity and number of gas constituents to be measured are constantly growing with the requirements of environmental protection activities and legislation. There is for example considerable interest to measure a couple of organic and inorganic molecule species at low concentrations simultaneously with common major components at a time resolution of some seconds. Conventional techniques imply time constants in excess of 100 s for the analysis of low-concentration species and therefore are not suited to observe correlations in instationary engine operation.

Tunable diode laser spectroscopy (TDLS) for gas analysis can be applied to all molecules with a small number of atoms which exhibit a discrete narrow absorption line structure. It features high detection sensitivity and small time constants. This has been demonstrated earlier for the cases of multicomponent real-time analysis of regulated (1) and unregulated (2) car exhaust gas constituents. Fast detection of CO and NO in car exhaust under dynamic engine operation with a response time of 3 ms has been achieved within a different project in our laboratory (3).

In order to obtain extremely low detection limits < 1 ppb with TDLS detection of minor constituents in environmental studies under clean air conditions, multiple-path reflection cells with optical path lengths of up to some hundred meters are used by numerous groups all over the world [4,5].

The aim of the development reported here was to extend the highly sensitive technique of tunable diode laser spectroscopy in combination with a multiple-reflection White cell to a four-component automobile exhaust gas analyzer with appr. 1 s response time. Details of the system including optics, signal processing and sample gas conditioning will be presented. First results obtained on a dynamometer test facility demonstrate the performance of this system.

2. SPECTROMETER SETUP

A block diagram of the diode laser spectrometer which was specially developed for muli-component trace gas analysis in car exhaust is shown in Fig. 1. Four diode lasers and three detectors are kept at low temperature by a common Stirling cooler unit. The variable temperatures of the individual diode laser stations are controlled independently with a stability of 0.003 K, whereas the detectors are kept at fixed temperature of appr. 80 K.

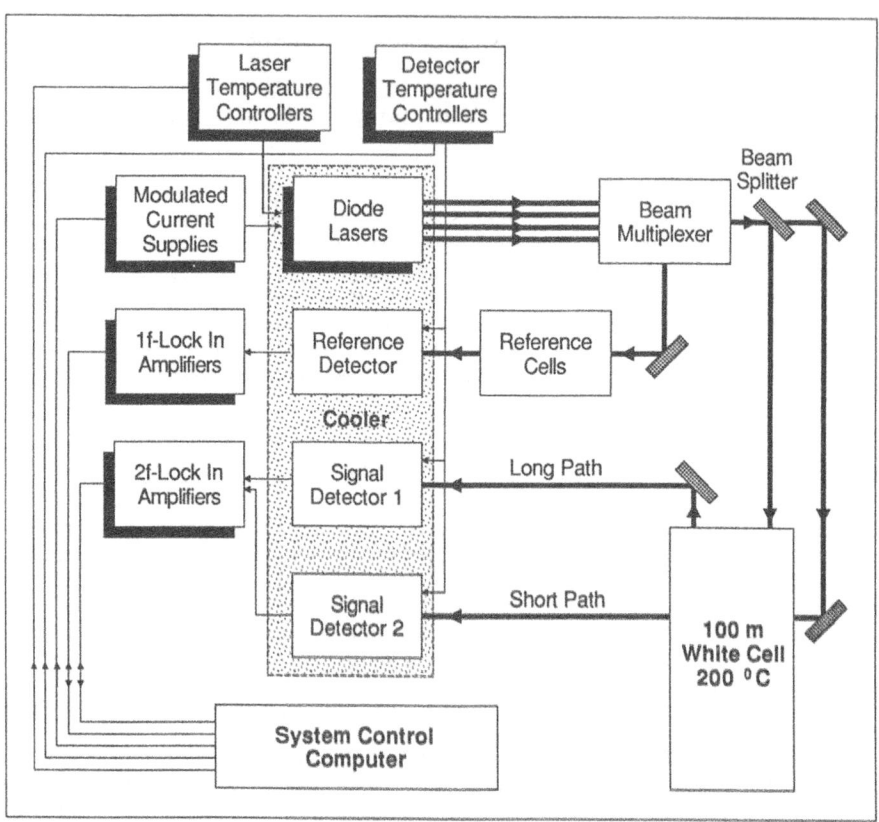

Fig. 1: Block diagram of the diode laser spectrometer electronics and optics

Diode laser and detector stations as well as the vacuum case with infrared windows are rigidly fixed to the optical base plate, as schematically illustrated in Fig. 2. The Stirling cooler with first and second stage connected by copper ribbons to corresponding copper bars in the cooler case is allowed to move. Laser and detector cold stations are again connected to the copper bars by copper ribbons (6). This double vibration decoupling design results in unsurpassed mechanical stability with respect to cooler vibrations, but also external mechanical shocks. In addition, this scheme is characterized by the absence of moving windows, which have been found to lead to opti-cal-noise induced modulation effects on the laser operation.

The current supply to the diode lasers includes four individual highly stable electronic units which allow for separate control of dc current and one of four different modulation waveforms at frequencies between 200 Hz and 10 kHz, enabling operation of the four lasers in frequency multiplex.

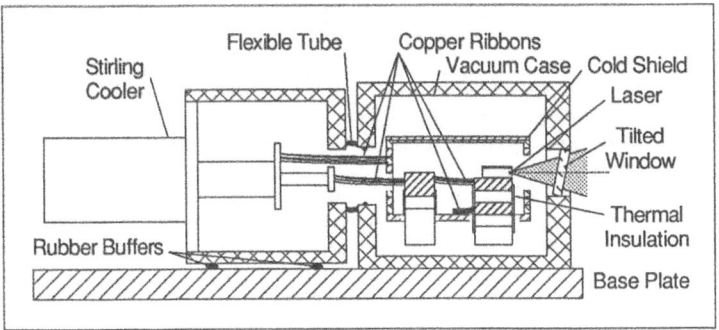

Fig. 2: Schematic of the Stirling cooler vibration isolation

Second harmonic signal processing is applied with the diode lasers frequency-locked to the absorption line centre, detected by first derivative technique in the optical reference channel. The resulting emission frequency stability of the diode lasers was found to be < 0.0001 cm^{-1}. Care was taken to have a set of non-rational modulation frequency ratios in order to minimize channel crosstalk. The detectors are photoconductive HgCdTe elements. Details on the multiplexer optics schematic, using beam splitters and mirrors, are given in Fig. 3.

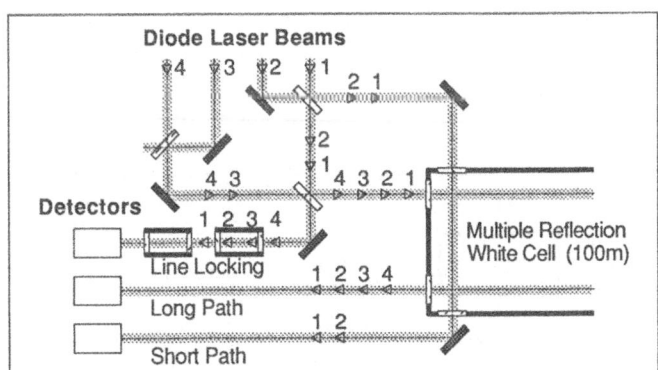

Fig. 3: Schematic of the optical setup and the beam multiplexing

The sample beam undergoes an optical path of up to 100 m within a multiple-reflection sample gas cell of the White type described elsewhere (7) before getting focussed on signal detector 1. This optical channel serves for high-sensitivity analysis of trace constituents. A second sample beam with the radiation of two of the lasers performs a short optical path of 15 cm within the White cell and is then incident on signal detector 2. This channel is suited for analysis of major gas constituents without the risk of saturation of absorption.

The sample gas concentration values are derived by second harmonic lock-in techniques, whereby the different laser absorption signals are demultiplexed. Problems of channel interference arise by the nonlinear response of the detectors at the relatively high signal levels. In general, photovoltaic elements could be a better choice because, in principle, the range of linear response can be extended to higher signal levels.

All electronics, including laser temperature and current supply, derivative signal processing electronics and gas concentration evaluation by use of special automated calibration cycles are controlled by a sytem control computer. Special software menus enable efficient operation of the complete exhaust gas analyzer system by personal which is not specialised in diode laser spectroscopy. In addition, external signals as e.g. speed or throttle valve position of the car under investigation can be recorded synchroneously to the measured concentrations. Using an external exhaust gas flow signal, the mass flow of the measured exhaust gas components can be computed online.

3. GAS SAMPLE CONDITIONING

As commonly done for automobile exhaust gas analysis, the gas sample and all gas-leading devices are kept at 190 °C in order to avoid condensation of water or other gas constituents. To this goal, the White cell is conceived to be heatable to temperatures of 200 °C without resulting mirror misalignment. This is achieved by a quartz glass body in combination with thermally compensated kinematic mirror mounts (7). Details of the gas sample conditioning system are illustrated in Fig. 4.

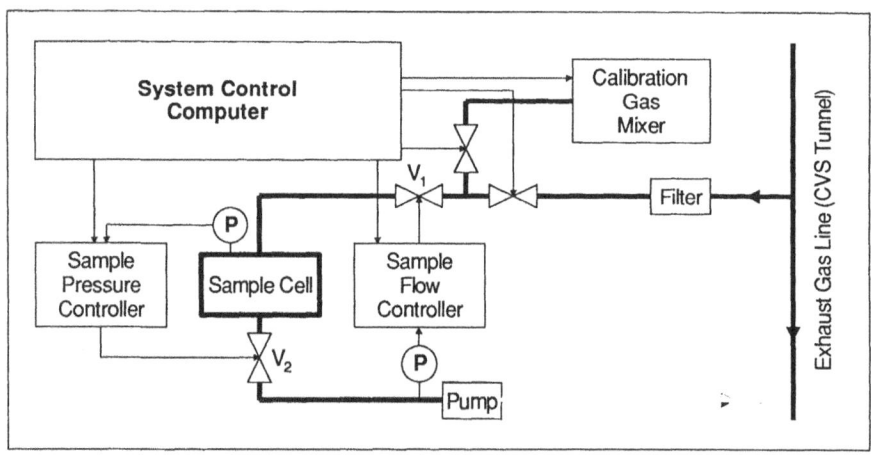

Fig. 4: Simplified schematic of the gas sample conditioning system

The sample gas is taken from the exhaust gas stream through a heated ceramic filter in order to remove aerosols. Sampling can be performed from any location in the undiluted exhaust gas stream. Alternatively, diluted gas from a constant volume sampling (CVS) device can be extracted. In a CVS sytem, the total gas flow, consisting of the total exhaust gas and dilution air, is kept constant.

Both, sample gas flow and sample gas pressure in the optical absorption cell are controlled separately. The gas flow, which is proportional to the pressure in front of the sampling pump, is essentially controlled by valve V_1, since the sample cell pressure is low compared to that in front of V_1. Sample flow rates range from 0.15 to 0.5 l/s NTP. Valve V_2 at the outlet of the sample cell serves to stabilize the sample pressure to a value of typically 50 hPa.

For calibration, a number of different gas concentrations are mixed from cylinders with high concentration gas by means of a combination of flow controllers. In the same way, pure nitrogen acting as zero gas, or ambient air for standby operation, can be passed through the sampling system.

4. FIRST RESULTS

For system evaluation, several test runs were carried out on a dynamometer test facility with a CVS dilution tunnel. As an example, Fig. 5 gives the result of three-component exhaust gas analysis of a six-cylinder engine automobile with electronically controlled fuel injection, obtained during a US-75 driving cycle. The gas concentration signals are delayed by about 7 s with respect to the speed signal due to the gas travel time from the engine to and in the dilution tunnel. The delay time within the sampling part of the TDLS analyzer system is below 0.5 s. The rise time is essentially determined by the gas exchange time in the sample cell, which is roughly 1.5 s.

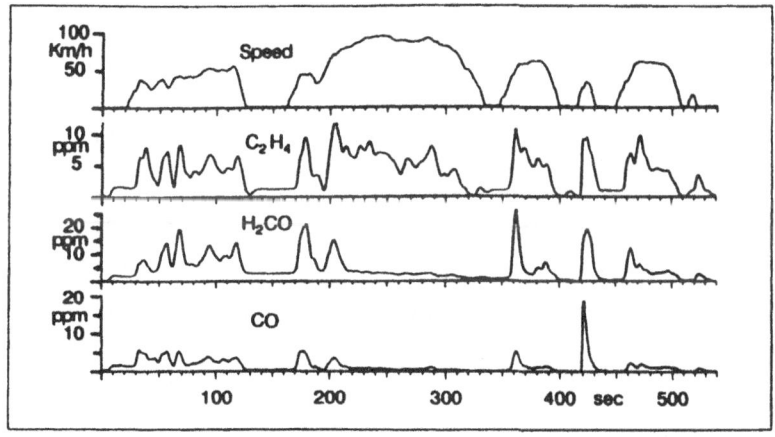

Fig. 5: Test measurements in a dynamometer CVS tunnel

So far, trace concentrations of ethene, formaldehyde and carbon monoxide in the 10 ppm range were measured without observable interference by other exhaust gas components. An example is given in Fig. 5. The observed noise levels correspond to concentration values in the 10 ppb range. Significant differences in the time profiles of the two hydrocarbons ethene and formaldehyde are observed. Information of this kind is not obtainable by flame ionisation detectors, which determine the total hydrocarbon content. Different constituents in the exhaust gas may have considerable different origins within the combustion and implications on the environment. It is not the aim of this work to discuss these and the observed profiles in more detail.

5. CONCLUSIONS

The automated multi-component gas analysis system reported here is suited for real-time analysis of trace components in automobile exhaust gas in correlation with simultaneosly measured major constituent concentrations. First experimental results obtained on a dynamometer test facility with dilution tunnel have shown detection limits to be typically in the range of some 10 ppb without any observable interference. The time resolution is better than 2 s. Apart from carbon monoxide, carbon dioxide, nitric oxide and nitric dioxide as major car exhaust gas constituents, the system was so far shown to be capable of detecting formaldehyde and ethene in trace concentrations, but should also be suited for analysis of e.g. methanol, acetaldehyde, acetylene, sulfur dioxide or ammonia.

REFERENCES

(1) STAAB, J., KLINGENBERG, H., HERGET, W.F. and RIEDEL, W.J. (1984). Progress in the Prototype Development of a new Multicomponent Exhaust Gas Sampling and Analyzing System, SAE paper No. 840470.
(2) HILL, J.C. AND MAJKOWSKI, R.F. (1980). Time-Resolved Measurement of Vehicle Sulfate and Methane Emissions With Tunable Diode Lasers", SAE paper 800510.
(3) WOLF, H., KLOCKE, U., RIEDEL, W.J. and NITSCHKE, E. (1991). Time-resolved Exhaust Gas Analysis by Infrared Diode Laser Spectroscopy, Proceedings of 24th ISATA Dedicated Conference Mechatronics, Paper 911248.
(4) SCHMIDTKE, G., KOHN, W., KLOCKE, U., KNOTHE, M., RIEDEL, W.J. and WOLF, H. (1989). Diode Laser Spectrometer for Monitoring up to Five Atmospheric Trace Gases in Unattended Operation, Applied Optics 28, 3665.
(5) SCHIFF, H.I., HARRIS, G.W. and MACKAY, G.I. (1987). Measurement of Atmospheric Gases by Tunable Diode Laser Absorption Spectrometry, in: Monitoring of Gaseous Pollutants by Tunable Diode Lasers, p. 4, Reidel, Dordrecht.
(6) RIEDEL, W.J., WISSLER, R., FICHTER, O., GREGORIUS, K., MATTAUCH, G. and SCHÖRNER, H. (1991). Device for the cooling of optoelectronic components and use of a flange joint used thereof, US Patent No. 4,985,805.
(7) RIEDEL, W.J., KNOTHE, M., KOHN, W. and GRISAR, R. (1989). An Anastigmatic White Cell for IR Diode Laser Spectroscopy, in: Monitoring of Gaseous Pollutants by Tunable Diode Lasers, Vol. 2, p. 165, Kluwer, Dordrecht.

APPLICATION OF DIODE LASER SPECTROSCOPY ON THE MEASUREMENT OF BOUNDARY LAYER - INDUCED TEMPERATURE CHANGES IN SHOCK TUBES

L.K. MOSER and F.J. HINDELANG

Universität der Bundeswehr München,
Institut für Strömungsmechanik,
8014 Neubiberg,
Germany

SUMMARY

Nitric Oxide (diluted in Ar) has been monitored in a shock tube by IR diode laser absorption spectroscopy. The effect of the boundary layer on the temperature of the shock tube core flow has been investigated experimentally by the two-line absorption method. The wavelength of an IR diode laser was rapidly modulated by a sawtooth current over two nearly coincident rotational vibrational transitions of NO originating from the vibrational ground state and the first excited vibrational level, respectively. From the recorded absorption profiles the temperature was determined by assuming a Boltzmann - distribution among the initial states of the two spectrally adjacent transitions. The measured temperatures are in good agreement with the predictions of the shock tube boundary layer model of Mirels including the transition point from laminar to turbulent boundary layer. This forms a further proof of Mirels's theory, but it also shows, that the weak boundary layer - induced temperature changes in shock tube core flows can be measured by the two - line absorption method.

1. INTRODUCTION

Due to the extremely narrow sub-Doppler linewidth and the continuous wavelength tunability of IR diode lasers, single rotational vibrational absorption line profiles can be spectrally resolved. Species concentrations can be inferred from the amplitude or the shape of one line profile. For temperature measurements two absorption profiles are required, so that the diode laser has to be modulated in wavelength across a spectral region encompassing two nearly coincident rotational vibrational transitions. From the amplitude of the two spectrally adjacent line profiles the ratio of the populations of the initial states can be calculated by using Lambert Beer's law of radiation absorption. From this the temperature can be determined by assuming a Boltzmann-distribution among the two initial states.

In this study this two-line technique has been used to measure the temperature behind incident shock waves as a function of time. According to the ideal shock tube theory the temperature remains constant between shock and contact front. In a real shock tube the effect of the wall boundary layer causes a weak temperature increase in the core flow with increasing distance from the shock front. This temperature change is very important for chemical kinetics shock tube studies, because it has a marked effect on the decomposition or formation of chemically reactive species[1].

The effect of the wall boundary layer on the shock tube core flow has been described by Mirels in several theoretical studies[2-4]. For given experimental conditions (inner diameter of the shock tube, initial test gas pressure, Mach number) corrections of the thermodynamic variables (temperature, density, pressure) and the flow velocity can be calculated from his theory including the transition point from laminar to turbulent boundary layer. However, there are still some discussions about the reliable applicability of Mirels's model, especially with respect to his prediction of the transition point[5].

This is the main reason for our experimental interest. This study serves to test the safe applicability of Mirels's theory to our shock tube conditions. An accurate knowledge of the core flow temperature profile is especially important for a reliable boundary layer correction of chemical kinetics shock tube data.

2. EXPERIMENTAL SET-UP

The shock waves were produced in a cylindrical stainless steel pressure driven shock tube (inner diameter: 100 mm). The length of the driver section (maximum pressure: 300 bar) was 2 m. As driver gas Helium was used. The length of the ultra-high vacuum driven section was 19 m. This part could be evacuated by two turbomolecular pumps to about 10^{-6} mbar before it was filled with test gas to 20 mbar. The test gas mixtures consisted of 5 % nitric oxide (NO) and Argon, which formed the heat bath. Additionally to the IR absorption the velocity of the shock was measured at the observation windows by Piezo-electric pressure transducers.

The IR diode laser spectrometer is shown schematically in Fig. 1. The KBr-lens served to transfer the divergent diode laser beam to a parallel beam. The diameter of the beam inside the shock tube was about 10 mm (spatial resolution of our experiment). Improvement of the spatial resolution was not necessary because the distance between shock and contact front was typically 1 m in our experiments. After leaving the shock tube the laser beam was focused on the entrance slit of a grating monochromator which served to select one mode. By means of the chopper the incident light intensity was determined. After leaving the exit slit only one laser mode was focused on a liquid nitrogen cooled HgCdTe-detector (rise time: 0.8 μsec). After amplification the signal was recorded by a digital oscilloscope.

The IR diode laser was rapidly modulated by a sawtooth current (5 - 10 kHz) across a spectral interval encompassing two nearly coincident rotational vibrational transitions of NO, the $v = 0 \longrightarrow v = 1$ $^2\Pi_{3/2}$ R(18.5)-transition originating from the vibrational ground state and the $v = 1 \longrightarrow v = 2$ $^2\Pi_{3/2}$ R(30.5)-transition originating from the first excited vibrational level.

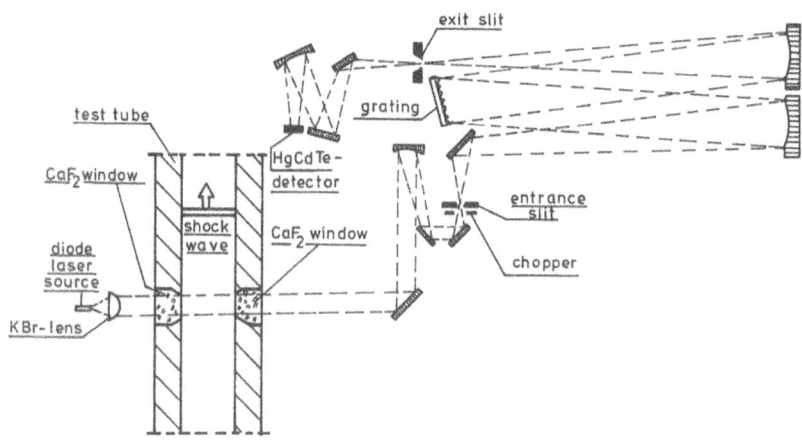

Fig. 1: Schematic drawing of the experimental set-up

3. EXPERIMENTAL RESULTS

Fig. 2 shows a typical record of the IR shock tube measurements. The lead salt diode laser, which emits around 5.2 µm, had a parabolic emission characteristic when it was modulated with a sawtooth current. Diode temperature and diode current were chosen to place the absorption profiles in the minima of the parabolas.

The arrival of the shock front at the observation windows produced a Schlieren peak in the IR signal (see Fig. 2). This peak divides the plot in a pre- and post-shock region. Before arrival of the shock front only the $v = 0 \rightarrow v = 1$ $^2\Pi_{3/2}$ R(18.5) - transition of NO can be observed. After arrival of the shock front also the spectrally adjacent $v = 1 \rightarrow v = 2$ $^2\Pi_{3/2}$ R(30.5) - transition can be detected because of the higher thermal population of the initial state in the post-shock region (temperature: ≈ 3000 K). The decrease of the post-shock absorption profiles is due to the thermal decomposition of NO.

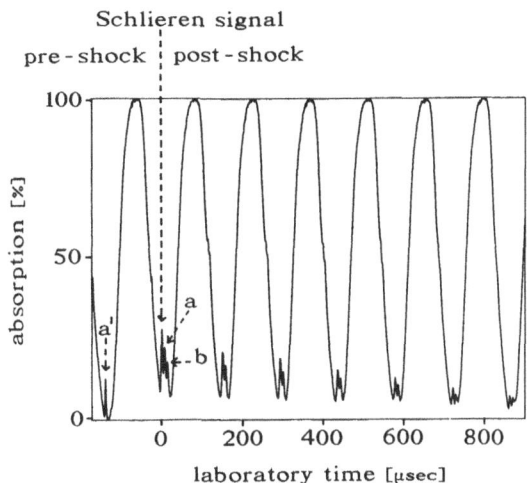

Fig. 2: IR absorption profiles of nitric oxide (diluted to 5% in Ar). Pre-shock conditions: 300 K, 22 mbar. Post-shock conditions: 3000 K, 805 mbar.
a': $v = 0 \rightarrow v = 1$ $^2\Pi_{3/2}$ R(18.5) - transition (before arrival of the shock front)
a: $v = 0 \rightarrow v = 1$ $^2\Pi_{3/2}$ R(18.5) - transition (after arrival of the shock front)
b: $v = 1 \rightarrow v = 2$ $^2\Pi_{3/2}$ R(30.5) - transition

From the two spectrally adjacent absorption profiles the temperature was determined by assuming a Boltzmann-distribution among the initial states and by assuming local equilibrium between translational, vibrational and rotational degrees of freedom. The fulfillment of the equilibrium condition between the different degrees of freedom has been verified by the authors in previous studies[6,7]. The results are listed in Table 1. These temperature values agree very well with the value of 3000 K which was determined additionally from the measured shock velocity and the ideal shock tube theory.

laboratory time [µsec] (see Fig. 2)	12	155	298	441	584	727
temperature [K]	2888	2916	2902	2989	3043	3060

Table 1: Post-shock temperatures measured by the two-line absorption method

For a comparison with the shock tube boundary layer theory of Mirels the temperature T_{2s} immediately behind the shock front has to be known. T_{2s} cannot be measured by the two-line absorption method because the IR signal would be overlapped by the Schlieren peak (see Fig. 2). Therefore we calculated T_{2s} from the first measured temperature value of the post-shock region (2888 K) and from the corresponding ratio of T_2/T_{2s} predicted by Mirels. This resulted in a value of 2864 K for T_{2s}.

With this value and the IR data from Table 1 the temperature ratio T_2/T_{2s} has been plotted as a function of the distance from the shock front. The result is shown in Fig. 3. The additional continuous line represents the shock tube boundary layer model of Mirels. The agreement between experiment and theory is very good. Even the transition point from laminar to turbulent boundary layer seems to be correctly predicted by Mirels. This result shows, that this theory can obviously be used for a reliable boundary layer correction of the shock tube core flow temperature.

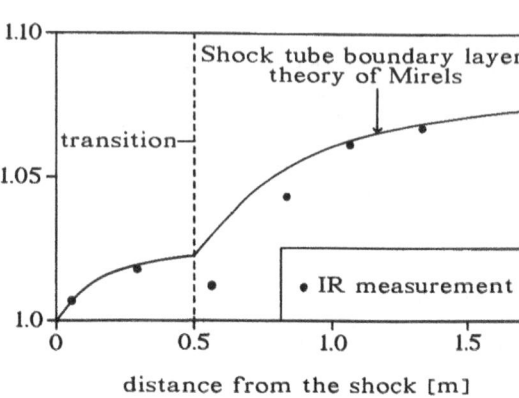

Fig. 3: Temperature ratio T_2/T_{2s} in the post shock region, where T_{2s} denotes the temperature immediately behind the schock front.

4. CONCLUSIONS

IR diode laser absorption spectroscopy has been used to monitor the gaseous pollutant nitric oxide in a shock tube. The present results show, that the theory of Mirels can be applied safely to boundary layer corrections of shock tube core flows. This is especially important for shock tube investigations of chemical reactions. This study also shows, that the two-line absorption method can be used for the measurement of boundary layer-induced temperature changes behind incident shock waves.

REFERENCES

(1) MOSER, L.K. and HINDELANG, F.J. (1992). Boundary layer effect on thermal NO decomposition behind incident shock waves. To be published in 'Shock Waves'.
(2) MIRELS, H. (1964). Shock Tube Test Time Limitation Due to Turbulent-Wall Boundary Layer. AIAA Journal, 2:84.
(3) MIRELS, H. (1966). Flow Nonuniformity in Shock Tubes Operating at Maximum Test Times. The Physics of Fluids, 9:1907.
(4) MIRELS, H. (1971). Boundary Layer Growth Effects in Shock Tubes. Proc. 8th Int. Shock Tube Symposium, London.
(5) FUJII, N., KOSHI, M., ANDO, H., ASABA, T. (1979). Evaluation of Boundary-Layer Effects in Shock-Tube Studies of Chemical Kinetics. Int. Journal of Chemical Kinetics, 11:285.
(6) MOSER, L.K. and HINDELANG, F.J. (1989). Vibrational relaxation of NO behind shock waves. Experiments in Fluids, 7:67.
(7) MOSER, L.K. and HINDELANG, F.J. (1989). Shock-Tube study of the vibrational relaxation of nitric oxide. Proc. 17th Int. Symposium on Shock Waves and Shock Tubes, Bethlehem PA, USA.

A TDL- AND FT-IR STUDY OF THE REACTION
NO$_3$ + HO$_2$ -> OH + NO$_2$ + O$_2$

J.HJORTH, F.CAPPELLANI and G.RESTELLI
Commission of the European Communities
Joint Research Centre-Environment Institute
I-21020 ISPRA (VA), Italy

SUMMARY

The kinetics of the reaction between the nitrate radical (NO$_3$) and the hydroperoxyl radical (HO$_2$) for the channel leading to the formation of the hydroxyl radical (OH), has been investigated at room temperature and atmospheric pressure. A novel technique has been applied for the determination of the OH radical concentration based on the measurement of $O^{16}C^{13}O^{18}$ build-up from the reaction of OH radicals with $C^{13}O^{18}$ added to the reaction mixture. Reactants and products have been measured by IR-FT spectroscopy except for the $O^{16}C^{13}O^{18}$ evaluated by second derivative tunable diode laser absorption spectroscopy. The results interpreted by a chemical model applied to the system appear to indicate that the rate constant of the above reaction has an upper limit about one order of magnitude lower than the only existing value in literature, which is equal to $3.6 \times 10^{-9} \, cm^{+3} molec^{-1} \, s^{-1}$.

1. INTRODUCTION

During these last years interest in the role played by chemistry driven at night by the nitrate radical (NO$_3$), generated from the reaction between NO$_x$ and O$_3$ in the troposphere, has considerably increased[1]. Recently, laboratory and field studies, performed for the condition of simultaneous presence in the air mass of organic species, NO$_x$ and O$_3$, suggest the occurrence of high nighttime peroxy radicals concentrations, possibly exceeding daytime values[2]. These conditions might be encountered in forests or in coastal areas when the air masses are affected by urban emissions or biomass burning. In this context, the reaction between the nitrate radical and the hydroperoxyl radical as a key component of a chain process capable of triggering at night OH radical chemistry, appears to have a prominent role.

A reaction between NO$_3$ and HO$_2$ was suggested for the first time in the study of the continuous photolysis of O$_3$/N$_2$/H$_2$CO mixtures in air at 700 Torr[3]; an estimation of the rate constant was given for a unique channel producing HNO$_3$

NO$_3$ + HO$_2$ --> HNO$_3$ + O$_2$ (15)

(chemical reactions are numbered according to the list of Table II).

The first direct study of the reaction between NO$_3$ and HO$_2$ was performed by using the discharge flow method, at 298 K and at pressures in the range 0.5-0.7 Torr[4]. Since OH was found in the reaction, a second channel

NO$_3$ + HO$_2$ --> OH + NO$_2$ + O$_2$ (11)

was added and the rate constants were measured for the two channels. The results indicate a total rate constant equal to

$(4.5\pm1)\times10^{-12}$ cm^{+3}molec^{-1} s^{-1} and a fractional branching for reaction (11) equal to 0.8; $k_{11}=(3.6\pm0.9)\times10^{-12}$ cm^{+3}molec^{-1} s^{-1}[4]. A modulated photolysis study[5] was performed in the same period at 1 atm pressure and in the temperature range 263-338 K. No OH radicals could be measured above the experimental detection limit of 5×10^{9} molec cm^{+3} and an upper limit of 0.60 was given for the ratio of reaction (11) to the overall rate constant. Good agreement on the value of the overall rate constant was obtained in a study using flash photolysis of HNO$_3$–H$_2$O$_2$ mixtures at 20 Torr and at the temperatures of 298 K, 323 K and 353 K[6]; no information could however be obtained on the reaction mechanism and the branching between the two channels.

The present study, specifically addressed to the kinetics of reaction (11), was performed at atmospheric pressure and room temperature, reacting NO$_3$ and HO$_2$ in a complex chemical system that was analysed using the FACSIMILE computer chemical model[7]. The concentration of OH radicals was determined by adding C^{13}O^{18} to the reaction mixture and measuring the formation of O^{16}C^{13}O^{18} by the reaction

$$O^{16}H + C^{13}O^{18} + O_2 \longrightarrow O^{16}C^{13}O^{18} + HO_2 \qquad (12).$$

Most of the reactants and products, present at ppmv mixing ratios were measured from Fourier transform infrared spectra recorded at 1 cm^{-1} or at 0.06 cm^{-1} instrumental resolution; the formation of O^{16}C^{13}O^{18}, expected to be in the low ppbv range, was measured by using second derivative tunable diode laser (TDL) absorption spectroscopy.

Highly sensitive TDL systems have been used from 1979 as a dedicated analytical tool in laboratory studies of gas-phase atmospheric chemistry[8-11], in addition to their most successful application in ambient air monitoring. The advantages offered in kinetics studies where a target species must be measured at high sensitivity, specificity and with a short time response constant are obvious when the gas system can be analysed at reduced pressure. The recent availability of TDL operated at or above 77K, permitting the use of liquid nitrogen dewars in place of the closed cycle cryogenic generators, have added an important advantage for the extension of the use of this analytical device in terms of simplicity and stability of the laser emission.

The results of this study show a definitively smaller concentration of OH radicals to be formed in the reaction between NO$_3$ and HO$_2$, than expected using the literature value for k_{11}. This conclusion is tentative because of the large uncertainty associated to the literature rate constant of one of the reactions considered in the model. However by critically analysing the data it can be preliminarly concluded that, for the conditions applied, i.e. room temperature and 750 Torr, an upper limit for k_{11} about an order of magnitude lower than the only value previously determined[4] can be estimated.

2.EXPERIMENTAL

The experiments were performed in the dark, using a 60 cm diameter reaction chamber lined with Teflon and containing a

White type mirror reflection system operated at a total beam path length of 81 m. The FT and the TDL spectrometer have both access to the multiple reflection system, as described in Ref.[10].

Purified air with a CO_2 content below 10 ppmv was used as diluent gas. N_2O_5 was prepared "in situ" by adding O_3 to excess NO_2; HO_2NO_2 was prepared from pure NO_2 and 70 percent H_2O_2 [12] and introduced into the chamber by a stream of purified air passing over the HO_2NO_2 solution while raising the temperature to room temperature. The half-life of N_2O_5 and of HO_2NO_2 in the reaction chamber was measured in excess of two hours. $C^{13}O^{18}$ 99 percent pure in C^{13} and 98 percent pure in O^{18} (ICL Isotopes) was finally introduced into the chamber and the reaction monitored by recording FT infrared spectra at 1 cm^{-1} instrumental resolution at 5 minutes time intervals. Each experimental run was allowed to continue until the concentration of the N_2O_5 had reached a level considered negligible, which occurred in 2-3 hours. At the end of each run the pressure was reduced to 50 Torr and the TDL second derivative spectrum of the $O^{16}C^{13}O^{18}$ absorption recorded.

Pressure and temperature in the cell were measured by using respectively an MKS Baratron capacitance manometer (±1%) and a thermocouple located in the cell (±1 K).

FT-IR spectra. O_3, NO_2, N_2O_5, HO_2NO_2 were measured from the integrated absorption of their bands centred respectively at 1042 cm^{-1}, 1617 cm^{-1}, 1245 cm^{-1} and 802.7 cm^{-1} .Spectral subtraction with reference spectra calibrated in this laboratory or using literature values for the absolute integrated band absorptions (O_3, HO_2NO_2), were used to invert the spectral features into molecular concentrations. An estimated accuracy of ±10% is attributed to the derived data. The concentration of $C^{13}O^{18}$ introduced into the cell was calibrated by sampling a known volume of gas with a syringe; the precision of this operation was checked by repeated introduction of the sample in the cell subsequently calibrated by recording IR spectra at 0.06 cm^{-1} instrumental resolution. The absorption lines at 2070.53 cm^{-1}, 2073.75 cm^{-1}, 2076.93 cm^{-1} and 2080.09 cm^{-1} taken from the 1990 edition of the Hitran data base[13] were used for the calculation of the molecular concentrations. The sampling procedure was found highly reproducible and a 10% uncertainty could be attributed to the $[C^{13}O^{18}]$ value.

TDL second derivative measurement of $O^{16}C^{13}O^{18}$. $O^{16}C^{13}O^{18}$ has its strongest absorption band (00011-00001) centred at 2265 cm^{-1} in a region containing transitions of other isotopic species of CO_2, of N_2O and water vapour. The problem posed by the need to identify an interference free absorption line of $O^{16}C^{13}O^{18}$ to be used to quantify this isotopic species, was solved by computer simulation of the second derivative spectrum[14], using the ATMOS[15] spectral data base. Suitable regions have been identified at 2250-2251 cm^{-1}, 2254-2255.5cm^{-1} and 2258-2259 cm^{-1}. The computer simulation of the spectrum of ambient air in the last spectral region, selected according to the characteristics of the diode laser, and for the experimental conditions applied, temperature, pressure and modulation amplitude, is shown in Fig 1.

$O^{16}C^{13}O^{18}$ is present in ambient air with a natural abundance of 0.0000444, which means that in the unperturbed troposphere the mixing ratio of this isotopic species is equal to 15.8 ppbv (355 ppmv total CO_2). Ambient air, whose CO_2 concentration was assumed constant was then used, admitted into the cell to construct a calibration plot for the $O^{16}C^{13}O^{18}$ signal. To account for variations in the transmitted laser intensity between the experimental run and the calibration, a correction factor was applied, derived by comparing the signal recorded with a closed cell containing CO_2 and N_2O, insertable in the laser optical beam path. Due to absence of important local sources of CO_2 at the air sampling site the condition of constant CO_2 concentration was considered acceptable and the complete procedure was estimated to lead to an accuracy of ±3% in the conversion of the second derivative signal into $O^{16}C^{13}O^{18}$ molecules.

Fig.2 shows the experimental spectra of ambient air, of a reference cell containing CO_2 and N_2O and of the reaction mixture. The spectral features closely follows those of the computer simulation except for an absorption feature located at 2258.28 cm^{-1}. This line which is present only after introduction of the $C^{13}O^{18}$ in the cell could not be attributed to any of the transitions of the 54 molecules tabulated in the data base; however it must be reminded that not all the bands of these molecules are included in the tabulation. The signal intensity does not correlate with that of any other signals evident in the spectral region analysed.

Because of the low $C^{13}O^{16}_2$ content of the $C^{13}O^{18}$ (0.33%) and of CO_2 in the pure air used as diluent, variations in the intensity of the spectral feature at 2258.534 cm^{-1} due to the absorption of $C^{13}O^{16}_2$ were used for correcting the $O^{16}C^{13}O^{18}$ signal for contamination by ambient air, which would increase the $O^{16}C^{13}O^{18}$ signal, during the measurement at reduced pressure. This correction was however found negligible in all the experimental runs performed.

The diode laser (Laser Photonics Analytics Div.) was operated at 80 K, using a liquid nitrogen dewar, close to its threshold current to obtain single mode emission; this condition was verified by comparison of the second derivative signals of the two N_2O lines at the extremes of the scan, found strictly proportional to their tabulated strengths. The current ramp (an 80/20 % saw-tooth) encompassing the spectral region of interest, was scanned at 0.8 Hz frequency with a superposed 2 kHz triangular modulation. The modulation amplitude was optimized to achieve a high signal-to-noise ratio while maintaining the spectral features well separated. No periodical noise was apparent in the spectrum and the wavenumber stability better than 2x10^{-3}cm^{-1} for observation times of minutes; coadding 16 scans for each record and using 4-6 measurements for a total measuring time of few minutes, a precision below 1.5 % could be obtained in the evaluation of the second derivative signal due to the $O^{16}C^{13}O^{18}$ absorption.

3.RESULTS and DISCUSSION

The most important problem encountered in the experiment was the presence in the $C^{13}O^{18}$ sample of $O^{16}C^{13}O^{18}$ as impurity; in spite of its small concentration in the sample (0.056%), this was giving an initial concentration that resulted larger than the $O^{16}C^{13}O^{18}$ build-up at the end of the reaction by a factor up to 6. This increased the uncertainty in the determination of the $O^{16}C^{13}O^{18}$ formation in the reaction; accounting for all the possible sources of error, the accuracy in the different runs, was finally estimated as ±10-20%.

The initial conditions of the 6 experimental runs are shown in Table I together with the measured yield of $O^{16}C^{13}O^{18}$, the duration of the experiment and the temperature of the gas mixture.

The chemical reaction scheme that is expected to describe the kinetics of the reaction mixture is shown in Table II. OH radicals are generated by reaction (11)

$$NO_3 + HO_2 \longrightarrow OH + NO_2 + O_2 \qquad (11)$$

but also by reaction (4)

$$NO_3 + NO_2 \longrightarrow NO + NO_2 + O_2 \qquad (4)$$

followed by reaction (13)

$$NO + HO_2 \longrightarrow OH + NO_2 \qquad (13).$$

Also reaction (7)

$$NO_3 \longrightarrow NO + O_2 \qquad (7)$$

followed by reaction (13) leads to OH production, but for the concentrations of NO_3 and NO_2 as used in the experiments, this source of OH is of minor importance:

$(k_7 [NO_3]/k_4 [NO_3] [NO_2] = 0.05)$.

The kinetics of the experimental system was simulated by the FACSIMILE software package applying the reaction scheme of Table II and the measured initial conditions for each run; the results are shown in Table III.

The reaction between OH radicals and $HO_2 NO_2$ (reaction 16) was assumed to form NO_2 H_2O and O_2, although this may not be the only reaction channel; for the overall kinetics of the system this assumption is however of little importance.

The agreement between measured and predicted build-up of $O^{16}C^{13}O^{18}$ (i.e. the concentration of OH radicals) was found to be much better when reaction (11) was left out of the chemical reaction scheme (condition indicated as "Ratio 2") than when it was included (condition indicated as "Ratio 1"). It was also observed that in the case of "Ratio 2", minor adjustments of the rate constant of reaction (4) or of the value applied for the highly temperature sensitive equilibrium constant $K_{eq}(2,3)$ were sufficient to explain the disagreement between measured and simulated concentrations of $O^{16}C^{13}O^{18}$. This was impossible in the case of "Ratio 1", with reaction (11) included in the scheme with its the rate constant previously reported in literature[4]. These results seem to indicate that the rate constant of reaction (11) as in Table II is too high.

Two approaches were followed to estimate an upper limit for k_{11}. In the first case the value of the equilibrium constant $K_{eq}(8,9)$ of reactions (8,9)

$$HO_2 NO_2 + M < \longrightarrow HO_2 + NO_2 + M \qquad (8,9)$$

and the rate constant of reaction (4) k_4, were set at their lower limits by dividing them by the uncertainty factors given in Table II and setting the measured concentrations of $O^{16}C^{13}O^{18}$ formed at their upper limits (i.e. +15% of the measured values). For these extreme assumptions the upper limit calculated for k_{11} in the different runs (see Table III) are not much lower than the literature value, but the agreement between simulated and measured data (HO_2NO_2, NO_2, N_2O_5 and $O^{16}C^{13}O^{18}$ build-up) is quite unsatisfactory.

In order to check the accuracy of the literature value of $K_{eq}(8,9)$, whose lower limit appears to represent a very conservative choice, experiments were performed following the decay of HO_2NO_2 in air in the cell. The decay rate of HO_2NO_2 in the cell is determined by its wall decay

HO_2NO_2 + wall --> products

and by the loss of HO_2 radicals by the reaction

HO_2 + HO_2 --> H_2O_2 + O_2 (10).

The wall decay of HO_2NO_2 which dominates at high NO_2 concentrations was found to be slow when compared to losses due to reaction (10) at the NO_2 levels applied in the HO_2-NO_3 experiment. The application of the literature values of $K_{eq}(8,9)$ and of k_{10} reproduced the measured data of the decay of HO_2NO_2 rather accurately. According to these results it appears that the value of $K_{eq}(8,9)$ cannot be much different from that reported in the literature and an uncertainty factor of 1.5 seems to be more appropriate than a value of 5.

Based on these considerations an estimate of the upper limit of k_{11} was obtained by taking the lower limit of k_4, the upper limit of the value of $O^{16}C^{13}O^{18}$ build-up and leaving for $K_{eq}(8,9)$ the literature value. The results shown in Table III appear to indicate that the rate constant for reaction (11) k_{11} may be at least a factor 10 lower than previously thought.

The interest in a correct evaluation of k_{11} make it necessary to improve the sensitivity of this study, while maintaining the basic procedure of deriving the OH radical concentration by the TDL measurement of $O^{16}C^{13}O^{18}$. The most important improvement will be the reduction of the NO_2 concentration in the reaction; condition could be achieved by generating the NO_3 radicals from the thermal decomposition of N_2O_5 introduced into the cell from its solid phase. A lower NO_2 concentration will increase the rate of reaction (11) and reduce the competition of reaction (14) vs reaction (12), with evident advantages. From the point of view of the accurate evaluation of the $O^{16}C^{13}O^{18}$ build-up the precision of the measurement could be improved by reducing the presence of this CO_2 isotopic species in the $C^{13}O^{18}$ sample by absorption over KOH pellets while introducing the gas into the cell. Further experimental studies in this direction are in progress.

ACKOWLEDGEMENTS

The authors gratefully acknowledge the contribution of G.Melandrone and G.Ottobrini for the experimental measurements. This study has been supported by the CNR-ENEL Project - Interactions of energy systems with human health and environment - Rome Italy.

REFERENCES

(1) WAYNE, R.P., BARNES, I., BIGGS, P., BURROWS, J.P., CANOSA-MAS, C.E., HJORTH, J., LE BRAS, G., MOORTGAT, G.K., PERNER, D., POULET, G., RESTELLI, G. and SIDEBOTTOM, H. (1991) The nitrate radical. Physics, chemistry and the atmosphere. Atmos. Environment, Special issue **25a**, Nr. 1, 1-206.

(2) PLATT, U., LE BRAS, G., POULET, G., BURROWS, J.P. and MOORTGAT, G. (1990) Peroxy radicals from nighttime reaction of NO_3 with organic compounds. Nature **348**, 147-149.

(3) CANTRELL, C.A., STOCKWELL, W.R., ANDERSON, L.G., BUSAROW, K.L., PERNER, D., SCHMELTEKOPF, A. CALVERT, J.G. and JOHNSTON, H.S.(1985) Kinetic study of the NO_3-CH_2O reaction and its possible role in nighttime tropospheric chemistry. J. Phys. Chem. **89**, 139-146.

(4) MELLOUKI, A., LE BRAS, G. and POULET, G. (1988) Kinetics of the reaction of NO_3 with OH and HO_2. J. Phys. Chem. **92**, 2229-2234.

(5) HALL, I.W., WAYNE, R.P., COX, R.A. JENKIN, M.E. and HYMAN, G.D. (1988) Kinetics of the reaction of NO_3 with HO_2. J. Phys. Chem. **92**, 5049-5054.

(6) EWIG, F., HOFFMANN, A. and ZELLNER, R. (1988) An experimental and computational study of the reaction of NO_3 radicals with Cl ,and OH. X^{th} Int. Symposium on Gas Kinetics, Swansee (uk), 24-29 July.

(7) CHANCE, E.M., CURTIS, A.R., JONES, I.P. and KIRBY, C.R. (1977) FACSIMILE , UKAEA, Harwell Report AERE-R 8775.

(8) STREIT, G.E., WELLS, J.S., FEHSENFELD, F.C. and HOWARD, C.J. (1979) A tunable diode laser study of the reaction of nitric and nitrous acid. J. Phys. Chem. **70**, 3439-3443.

(9) COX, R.A. and JENKIN, M.E. (1985) Kinetic studies of HO_2 reactions using diode laser spectroscopy. in Chemistry related to tropospheric ozone. CEC Report AP/54/86.

(10) HJORTH, J., CAPPELLANI, F., NIELSEN, C.J. and RESTELLI, G. (1989) Determination of the $NO_3 + NO_2 -> NO + O_2 + NO_2$ rate constant by IR diode laser and FT spectroscopy. J. Phys. Chem. **93**, 5458-5461.

(11) BECKER, K.H., BROKMANN, K.J. and BECHARA, J. (1990) Production of hydrogen peroxide in forest air by reaction of ozone with terpenes. Nature, **346**, 256-258.

(12) SCHWARZ, R. (1948) Uber die peroxy-salpetersaure. Zeitschrift fur anorganische chemie, **256**, 3-9.

(13) ROTHMAN, L.S. et al.(1991) The HITRAN molecular absorption data base, 1991 Edition. Appl. Optics

(14) HAURIE, Y., PAGNY, C., CAPPELLANI, F. and RESTELLI, G. (1989) Formulae, algorithms, procedures and program for application of infrared spectroscopic measurements. EUR Report 12527 EN.

(15) BROWN, L.R., FARMER, C.B., RINSLAND, C.P. and TOTH, R.A. (1987) Molecular line parameters for the atmospheric trace molecule spectroscopy experiment. Applied Optics, **26**, 5154-5182.

(16) DE MORE, W.B. ET AL. (1990) Chemical kinetics and photochemical data for use in stratospheric modeling, evaluation nr. 9. JPL Publication 90-1.

TABLE I.

Run No.	$[HO_2NO_2]$ ppm	$[N_2O_5]$ ppm	$[NO_2]$ ppm	$[C^{13}O^{18}]$ ppm	$\Delta[O^{16}C^{13}O^{18}]$ ppb	Temp. ^0C	Time min.
1	7.4	4,4	7.4	10.4	1.2	22.5	111
2	8.9	4.6	7.5	10.4	2.3	22.5	112
3	9.0	5.5	5.6	5.2	0.70	20.5	151
4	8.2	3.6	7.5	5.2	0.94	25.0	152
5	7.6	3.9	7.5	5.2	1.08	25.0	163
6	7.3	4.1	6.2	10.4	1.65	22.5	197

Initial conditions, duration and $O^{16}C^{13}O^{18}$ build-up measured in the 6 experimental runs.

TABLE III.

Run No.	Ratio 1	Ratio 2	$k_4 \times 10^{16}$	$k_{11}a \times 10^{12}$ $cm^3molecule^{-1}s^{-1}$	$k_{11}b \times 10^{12}$
1	4.5	1.8	3.6	3.4	0.3
2	2.3	0.9	8.1	8.6	2.1
3	5.5	2.1	3.3	2.0	0.2
4	3.9	1.7	4.2	2.7	0.6
5	3.6	1.6	4.2	3.0	0.4
6	4.3	1.7	3.5	2.5	0.3

Results of model-calculations "Ratio 1" is the ratio between the average OH concentration predicted by a model including reaction (11) with its literature value for the rate constant k_{11} and the average OH concentration derived from the measurement. "Ratio 2" is obtained by excluding reaction (11) in the model. "k_4" is the value of the rate constant of reaction (4) that gives the best fit to the experimental data, when reaction (11) is excluded from the reaction scheme. "$k_{11}a$" is the upper limit of k_{11} obtained by a best fit to the experimental data, applying the upper limit of the build-up of $O^{16}C^{13}O^{18}$ and the lower limits of k_4 and k_9. "$k_{11}b$" is the upper limit for k_{11} obtained by taking the upper limit for the build-up of $O^{16}C^{13}O^{18}$, the lower limit for k_4 and the literature value for k_9.

TABLE II.

Reaction	Rate constant (298 K)	Uncertainty
(1) $NO+NO_3 \rightarrow 2NO_2$	$3.0 \times 10^{-11} cm^3 molecule^{-1} s^{-1}$	1.3
(2) $NO_2+NO_3+M \rightarrow N_2O_5+M$	$1.41 \times 10^{-12} cm^3 molecule^{-1} s^{-1}$	1.5
(3) $N_2O_5+M \rightarrow NO_2+NO_3+M$	$3.6 \times 10^{-2} s^{-1}$	K_{eq}: 1.5
(4) $NO_2+NO_3 \rightarrow NO+NO_2+O_2$	$6.0 \times 10^{-16} cm^3 molecule^{-1} s^{-1}*$	(1.5)
(5) $2NO_3 \rightarrow 2NO_2+O_2$	$3.3 \times 10^{-16} cm^3 molecule^{-1} s^{-1}*$	(1.5)
(6) $N_2O_5+wall \rightarrow products$	variable ($\sim 10^{-4} s^{-1}$)	
(7) $NO_3 \rightarrow NO+O_2$	$2.9 \times 10^{-3} s^{-1}*$	(2)
(8) $HO_2+NO_2+M \rightarrow HO_2NO_2+M$	$1.4 \times 10^{-12} cm^3 molecule^{-1} s^{-1}$	1.2
(9) $HO_2NO_2+M \rightarrow HO_2+NO_2+M$	$0.09 s^{-1}$	K_{eq}: 5
(10) $2HO_2 \rightarrow H_2O_2+O_2$	$2.9 \times 10^{-12} cm^3 molecule^{-1} s^{-1}$	1.3
(11) $HO_2+NO_3 \rightarrow HO+NO_2+O_2$	$3.6 \times 10^{-12} cm^3 molecule^{-1} s^{-1}*$	(1.3)
(12) $HO+CO+O_2 \rightarrow CO_2+HO_2$	$2.4 \times 10^{-13} cm^3 molecule^{-1} s^{-1}$	1.3
(13) $HO_2+NO \rightarrow HO+NO_2$	$8.3 \times 10^{-12} cm^3 molecule^{-1} s^{-1}$	1.2
(14) $HO+NO_2+M \rightarrow HNO_3+M$	$1.4 \times 10^{-11} cm^3 molecule^{-1} s^{-1}$	1.5
(15) $HO_2+NO_3 \rightarrow HNO_3+O_2$	$9.2 \times 10^{-13} cm^3 molecule^{-1} s^{-1}$	2.0
(16)[a] $HO+HO_2NO_2 \rightarrow H_2O+O_2+NO_2$	$4.6 \times 10^{-12} cm^3 molecule^{-1} s^{-1}$	1.5
(17) $HO+NO_3 \rightarrow HO_2+NO_2$	$2.3 \times 10^{-11} cm^3 molecule^{-1} s^{-1}$	(1.3)
(18) $HO+HO_2 \rightarrow H_2O+O_2$	$1.1 \times 10^{-10} cm^3 molecule^{-1} s^{-1}$	1.3

Reaction scheme, rate constant and uncertainty factors at 298 K applied in the computer simulation of the experiment. All rate constants are adopted from Ref.[16], except for those labeled * which are taken from Ref.[1]. Uncertainty factors for first order reactions and equilibrium constants are the estimates given in Ref.[16], those for second order reactions are estimated by the authors from standard deviations given in Ref.[16]. Factors in parenthesis are estimated by the authors according to the discussion in Ref.[1]. [a]) The products of this reaction are discussed in the text.

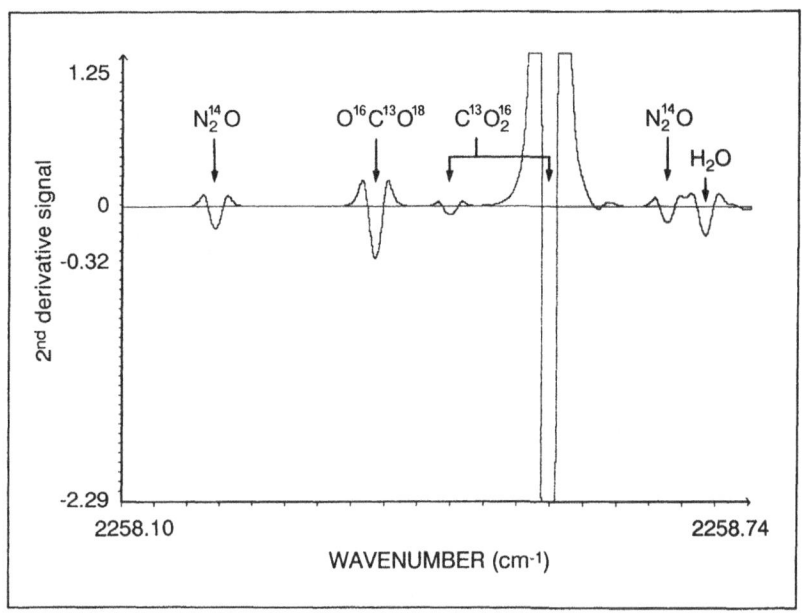

Fig.1

Computer generated second derivative absorption spectrum of ambient air in the spectral region from 2258.1 cm^{-1} to 2258.74 cm^{-1}. Conditions are those applied in the study: 298 K, 50 Torr, modulation amplitude 0.012 cm^{-1} (this value is only an estimate of that corresponding to the current modulation amplitude as used in the experiment). CO_2: 355 ppm, N_2O: 310 ppb, H_2O: 4000 ppm.

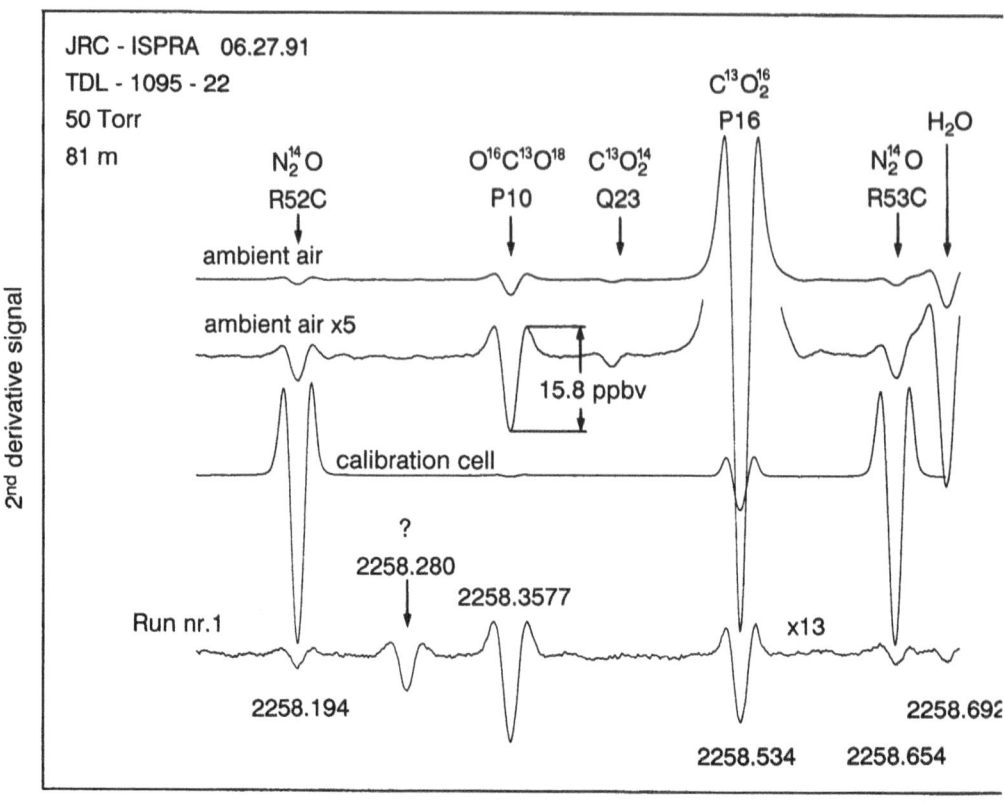

Fig.2

Experimental second derivative absorption spectra, from top: ambient air; reference cell containing CO_2 and N_2O; gas reaction mixture at the end of run 1. The "?" indicates the spectral feature for which no attribution has been found. Line positions and notations are taken from Ref.[15].

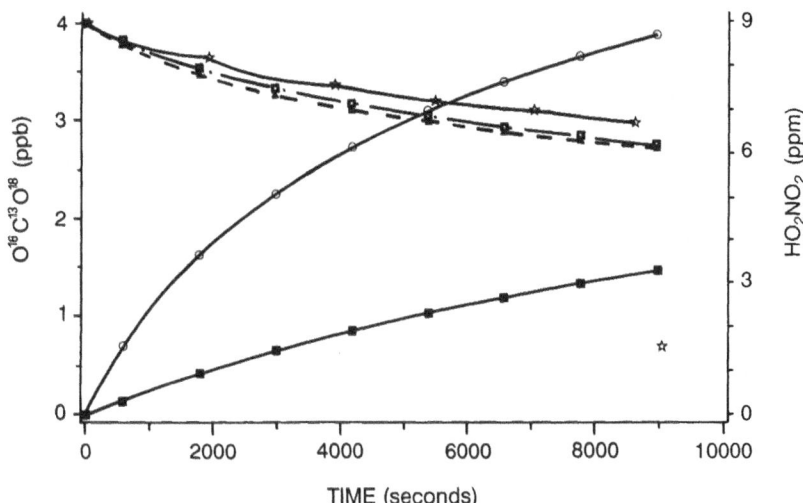

Fig.3
Simulated vs. measured concentrations of $O^{16}C^{13}O^{18}$ (ppbv) and of HO_2NO_2(ppmv) in run 3 (see Table I). Measured values:star; values simulated by the model not including reaction (11): squares; values simulated by the model including reaction (11) with its literature rate constant: dots.

MEASUREMENTS OF THE $^{13}C/^{12}C$ RATIO IN METHANE USING A TUNABLE DIODE LASER ABSORPTION SPECTROMETER

M.Schupp, P. Bergamaschi, G. W. Harris
Max Planck Institute for Chemistry
Air Chemistry Department
P.O. Box 3060
W-6500 Mainz
Federal Republic of Germany

SUMMARY

A tunable diode laser absorption spectrometer (TDLAS) for measuring the $^{13}C/^{12}C$ isotope ratio of methane has been developed and tested. In an optical triple path arrangement the spectra of a CH_4 sample, a $^{13}CH_4/^{12}CH_4$ isotope standard and pure $^{13}CH_4$ are simultaneously recorded and compared to determine the $^{13}CH_4/^{12}CH_4$ ratio of the sample, using a $^{13}CH_4$ and $^{12}CH_4$ absorption line pair near 3007 cm^{-1}. Systematic effects in the different channels arising from variations in temperature, pressure, and optical density were investigated for this pair of rovibronic absorptions. Optical interference was effectively suppressed by linearly polarizing the laser beam and using Brewster windows for gas cells and detectors. The overall $\delta^{13}C$ accuracy vs. the PDB scale is about ± 1 ‰ for a sample size of 10 μmoles CH_4 (0.22 STP cm^3). The measurements requires only a few minutes and does not necessitate chemical conversion of the methane to CO_2.

1. INTRODUCTION

In recent years extensive efforts have been invested in the evaluation of the global budget of atmospheric methane [1-4], because of its contribution to the global warming and its important role in atmospheric chemistry. In addition to direct measurements of CH_4 flux densities from various sources and extrapolation to the global scale, parallel CH_4 isotope measurements ($^{13}CH_4$, CH_3D, $^{14}CH_4$) provide important constraints on the relative contributions of the CH_4 sources since their isotopic signatures differ [5,6,7], in particular for methane from the different production pathways (biogenic, thermogenic [8] and incomplete combustion of biomass [6,9]). Usually the stable isotope ratios are expressed in the δ-notation:

$$\delta^{13}C = \left[\frac{R_{sample}}{R_{PDB}} - 1\right] \times 1000 \, [^o/_{oo}] \quad \textbf{(1)}$$

where R_{sample}, R_{PDB} are the $^{13}C/^{12}C$ ratios of the sample and the Pee Dee Belimnite (PDB) carbonate isotope standard, to which in general $^{13}C/^{12}C$ ratio measurements are referred to [10].

Up till now $^{13}CH_4/^{12}CH_4$ measurements have been performed by mass spectrometry, which requires oxidizing the CH_4 to CO_2 and measuring the $^{13}CO_2/^{12}CO_2$ ratio[1]. In addition a time consuming separation of the methane sample from CO_2, CO, H_2O and higher hydrocarbons is necessary [9,12]. Typically the complete preparation and measurement of a single sample requires several hours. Thus, to allow larger numbers of samples, an alternative approach to measure the $^{13}CH_4/^{12}CH_4$ ratio by tunable diode laser spectroscopy has been developed. The main advantage is the short measurement time of 10-15 min including the calibration procedure and the direct measurement on the CH_4 molecule.

We know of only a few efforts to apply TDLAS to atmospheric isotopic measurements: for $^{13}CO_2/^{12}CO_2$ [13,14], $C^{18}O/C^{17}O/C^{16}O$ [15] and $N_2^{18}O/N_2^{16}O$ [16]. In this paper we report the first measurements of the $^{13}CH_4/^{12}CH_4$ ratio by means of TDLAS and discuss several effects which in general have limited the precision previously attainable for TDLAS isotope ratio measurements.

2. EXPERIMENTAL

The experimental apparatus is illustrated in Fig. 1. The Pb-salt diode laser (output power: 50-150 μW in the main mode; Fraunhofer Institut für physikalische Meßtechnik, Freiburg, FRG) is mounted in a liquid nitrogen Dewar and is temperature controlled in the range from 77 to 100 K. The laser output is transmitted through a slanted CaF_2 window and collimated to a parallel beam by an off-axis-parabolic mirror mounted outside the Dewar. A monochromator can be switched into the beam to measure the wavelength or to separate different longitudinal laser modes if necessary. A set of 12 CaF_2 Brewster plates linearly polarize the laser beam before it is split into three parts using a wedged window as a beamsplitter. An absorption cell is located in each path and the laser power is focused onto three InSb detectors. The three cells contain respectively 99.7% enriched $^{13}CH_4$, a methane isotope standard and the unknown sample, each in N_2 bath gas (total gas pressure: 30 mbar). The temperature of the gas is controlled by circulating water through a jacket around the cells. The water temperature is stabilized to ± 0.1 K by a cryostat. The bath gas is allowed into the cells via valves and the CH_4 sample introduced through septa using gas syringes. The gas contents are circulated by a pump to assure thorough mixing of the methane with the bath gas. The cell windows as well as the detector windows are mounted at the Brewster angle. For alignment purposes a HeNe laser beam can be co-aligned with the infrared beam using a beam combiner.

The laser current is slowly ramped (computer generated with

---

[1] This conversion is necessary because of the mass identity of $^{13}CH_4$ and CH_3D, whereas the mass interference between $^{13}C^{16}O_2$ and $^{12}C^{17}O^{16}O$ can be corrected for by parallel measurement of the $^{12}C^{18}O^{16}O$ content [11].

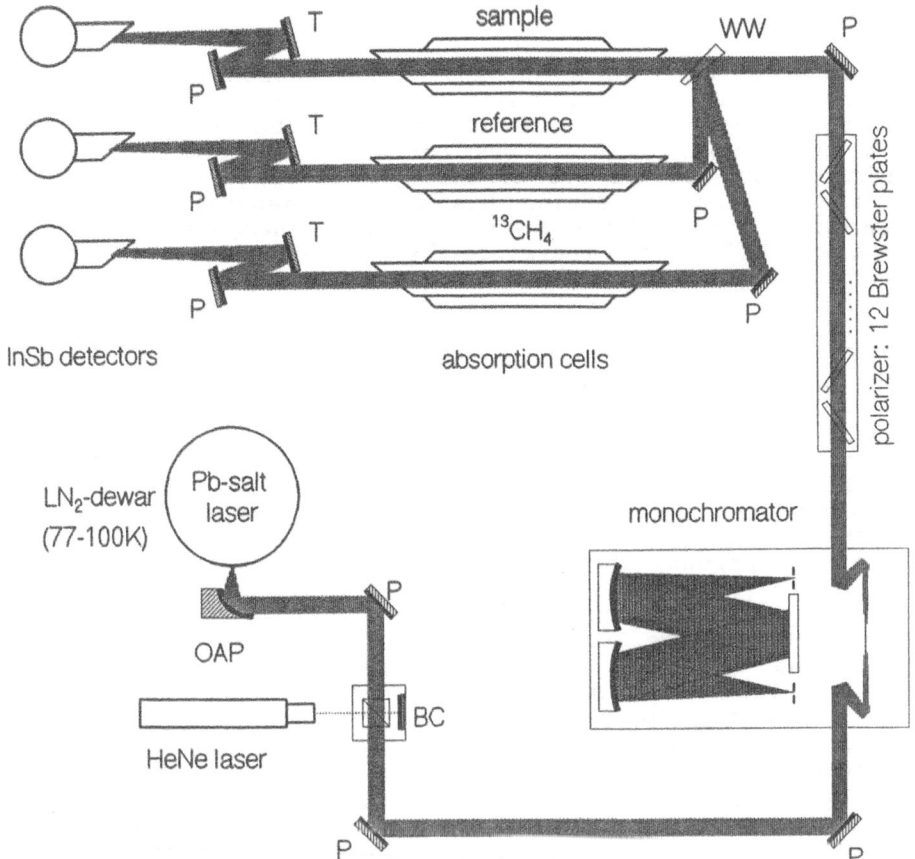

Fig. 1: Optical setup. P...plane mirror, T...toroidal mirror, WW... wedged window, BC...beam combiner, OAP...off-axis-parabolic mirror

a scan time of 2 s) to tune the laser frequency over the absorption lines of interest. A 40 kHz sinusoidal modulation is added to the laser current. Each detector output is amplified and then demodulated at the second harmonic of the modulation frequency by a Lock-in Amplifier [LiA]. The LiA outputs (time constant 10 ms) are digitized by 16 bit ADCs. All spectra are recorded synchronously with 1024 points per scan, and the analysis performed by a digital signal processor mounted inside an IBM compatible AT 80286 computer.

For calibration of the TDLAS system the isotopic composition of standard gases were measured by mass spectrometry. The sample preparation was as described by Bösinger [17].

3. MEASUREMENTS

In order to cover a $^{13}CH_4$ and a $^{12}CH_4$ absorption line in one continuous scan a pair of adjacent lines was sought. Since the natural isotopic abundance is about 1:100, the line pair should have an approximately inverse ratio of line strengths (per molecule) for optimal use of the dynamic range of the amplifiers. Line pairs satisfying this criterion generally arise from transitions having different ground state energies. The ground state populations, and therefore the line intensities of the two absorptions, usually reveal strongly differing temperature dependencies. Among several candidate pairs we chose a $^{13}CH_4$ line at 3007.15 cm^{-1} (Q(0) branch of the ν_3 vibrational band, (7 F1,1 → 7 F2,1) [18]), and a neighbouring $^{12}CH_4$ line at 3007.08 cm^{-1} (Q(0) branch of ν_3, (16 F1,1 → 16 F2,1) and (16 F2,0 → 16 F1,0) [19]). It is important to note that the magnitude of the systematic errors discussed below, as well as the reported accuracy, refer strictly to the line pair selected.

Fig. 2 shows a set of three quasi-simultaneously recorded 2f spectra for the three measurement cells. To compensate for drifts of the laser temperature and current, the position of the $^{13}CH_4$ peak is determined after each scan and the start voltage for the next ramp adjusted accordingly (line locking via laser current). The recorded spectra are treated by two multiple linear regressions. The first regression gives the proportionality constant between the $^{13}CH_4$ absorption signals in the reference cell and the $^{13}CH_4$ cell. This regression is carried out over the spectral region of the $^{13}CH_4$-line only:

$$spec_{13CH_4}(i) = a_0 spec_{ref}(i) + \quad (2)$$
$$+polynomial\ terms$$

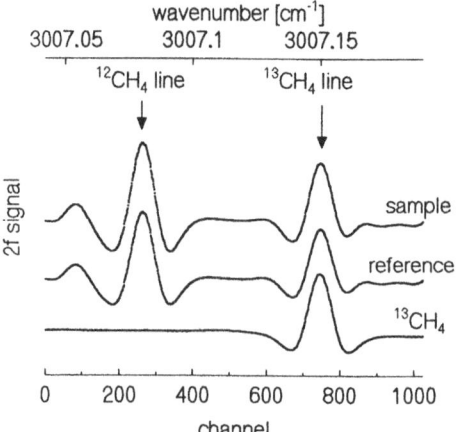

Fig. 2: Recorded 2f spectra of a CH_4 sample, a $^{13}CH_4/^{12}CH_4$ standard ($\delta^{13}C = -27.79 \pm 0.04$ ‰) and 99.7% enriched $^{13}CH_4$

where spec(i) is the ith point of the indicated spectrum and a_0 is the regression coefficient. A second linear regression fits a linear combination of the reference spectrum, the $^{13}CH_4$ spectrum and a polynomial to the spectrum of the unknown sample over both lines:

$$spec_{sample}(i) = b_0 spec_{ref}(i) + b_1 spec_{13CH_4}(i) + \quad (3)$$
$$+polynomial\ terms$$

The $\delta^{13}C$ value is obtained from the regression coefficients a_0,

b_0 and b_1:

$$\delta^{13}C_{sample} = [1000 + \delta^{13}C_{ref}] \times [1 + \alpha \times \frac{a_o \times b_1}{b_0}] - 1000 \qquad (4)$$

where α is a calibration factor which has to be determined experimentally by the measurement of two gases of different $\delta^{13}C$ content determined by mass spectrometry. This ensures the correct coupling to the PDB scale.

A typical measurement consists of about 40 scans. Each scan is analyzed separately (on line) and the $\delta^{13}C$ values subsequently averaged.

4. RESULTS AND DISCUSSION

Potential Sources of Systematic Error

Constraints are put on the accuracy of the method by several systematic effects. In the pressure region near 30 mbar the resulting absorption lineshape is a Voigt profile (i.e. a convolution of a Lorentzian and a Gaussian profile) with $\gamma_L = 0.2-0.3 \gamma_D$. The lineshape depends on the temperature and pressure. In addition, as mentioned above, the intensity ratio of the two lines depends strongly on the temperature, since their ground state energies are different. Further, at optical densities of several percent the linear approximation of the Beer Lambert law does not hold with sufficient accuracy (10% optical density: deviation of ~5%), thus the lineshape of the recorded 2f spectra depends on the optical density as well.

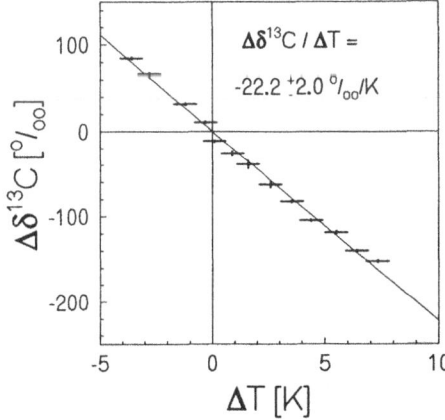
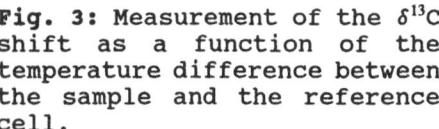
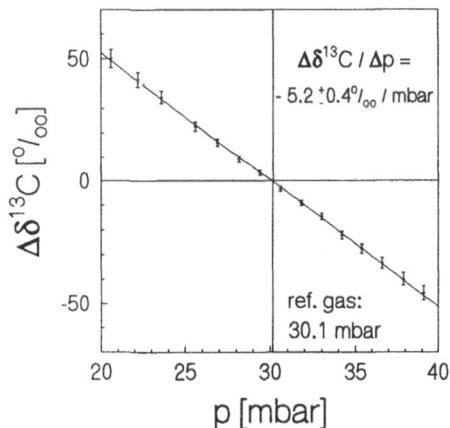

Fig. 3: Measurement of the $\delta^{13}C$ shift as a function of the temperature difference between the sample and the reference cell.

Fig. 4: Measurement of the $\delta^{13}C$ shift as a function of the pressure difference between the sample and the reference cell.

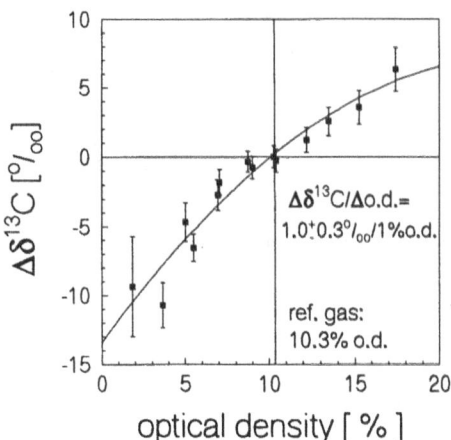

optical density [%]

Fig. 5: Measurement of the $\delta^{13}C$ shift as a function of the difference in the optical densities between the sample and reference cell.

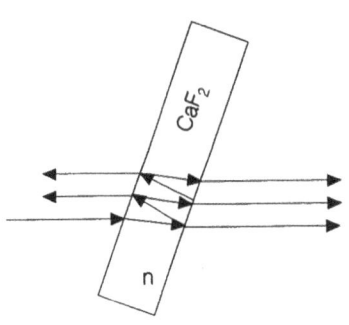

Fig. 6: multiple beam interferences (Fabry Perot Etalon Fringes)

These three parameters (temperature, pressure, and optical density) must therefore be controlled in all the three cells. Measurements are most sensitive to these parameters for the sample and the reference cell, while their adjustment is of lesser importance for the $^{13}CH_4$ cell. Systematic effects arising from deviations of these parameters among the different cells are investigated by varying **one** of these parameters in the sample cell and measuring the apparent $\delta^{13}C$ shift using the reference gas as sample. The measurements yielded a temperature coefficient of 22.2 ± 2 ‰/K, a pressure coefficient of 5.2 ± 0.4 ‰/mbar and an optical density coefficient of 1.0 ± 0.3 ‰/(1% optical density) around optical densities of ~10% (cf. Fig. 3-5). These measurements show that, given a desired maximum systematic error of 0.5 ‰, it is necessary to match the sample and reference cell temperatures to ≤ 0.023 K, the pressures to ≤ 0.1 mbar and the optical densities to ≤ 0.5%. Whereas the latter two conditions are rather easy to achieve, the temperature is the most critical parameter.

Further systematic errors due to multiple beam interference (Fabry Perot Etalon fringes) can occur. In our case, the most important effect arises from multiple reflection within optical elements with two parallel surfaces, such as the windows used for gas cells and detectors, producing fringes with a free spectral range of about 2 cm^{-1} (cf. Fig. 6)[2]. This interference causes the transmitted laser power to vary

---

[2] The free spectral range is given by v_{FSR} = 1/2nl (n=refractive index, l=distance of the parallel surfaces). For a given reflectivity R the transmitted power varies F=dI$_{max}$/I= 4R/(1-R)^2. For normal incidence CaF$_2$ (n=1.41) has a reflectivity of 2.9%, v_{FSR}=1.8cm^{-1} and F=0.12 (l=2mm). With a line separation of 0.08cm^{-1} the maximum power difference between the line centers is 11 ‰ due to the fringes of only one window.

significantly with wavelength, differently in the different optical paths. These power variations strongly depend on laser beam alignment and its change with time due to vibrations and thermal effects (beam wander). These interferences were very effectivly suppressed by linearly polarizing the laser beam and mounting the windows for cells and detectors at the Brewster angle. Interference within the beamsplitter is prevented by its wedged shape.

In the application of TDLAS to methane field samples, the occurrence of neighbouring absorption lines from other gases is of great importance. A recorded spectrum of ambient air using a 200m path White cell revealed a weak H_2O absorption line between the two methane lines. However, drying samples is easy to achieve. Further interfering absorption lines were not observed. Nevertheless attention must be paid to possible spectroscopic interferences particularly for samples from sources such as landfills or fuel combustion which may contain a wide variety of compounds.

Accuracy vs. the PDB Scale

The accuracy vs. the PDB scale has been tested by mixing the reference gas ($\delta^{13}C$ = -27.79 ‰) and a second gas with $\delta^{13}C$ = 19.1 ‰ in various ratios at a constant optical density of 10.0%. The mixing proportions and thus the isotopic composition were determined directly by the magnitude of the $^{12}CH_4$ absorption signals after injecting the first gas alone and after adding the second gas. The error of this isotopic composition evaluation is about $\Delta\delta^{13}C$ = 0.4 ‰ due to the error in the mixing ratio determination. Afterwards the $\delta^{13}C$ values for the various mixtures were measured by the TDL system and compared to the evaluated values. The results are shown in Fig. 7. The standard deviation, σ, between the measured and the evaluated $\delta^{13}C$ values in this set of 8 samples is 0.8 ‰. These measurements also

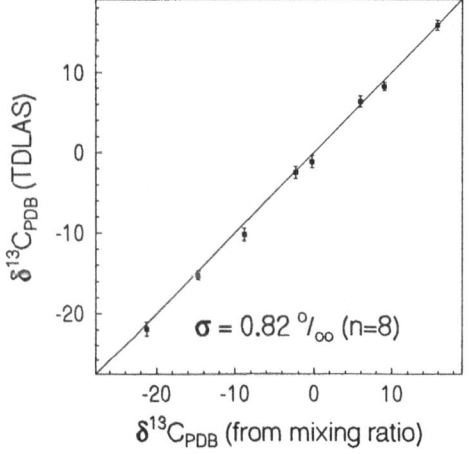

Fig. 7: $\delta^{13}C$ accuracy determined by measuring various mixtures from two gases of know isotopic composition ($\delta^{13}C$ = -27.79 ‰, 19.1 ‰).

demonstrate the linearity of the $\delta^{13}C$ scaling against the reference. For field sample measurements further standard gases will be used to cover the typical $\delta^{13}C$ range of methane samples (-20 ‰ $\leq \delta^{13}C$(atmospheric sources) $\leq$ -80 ‰).

The TDLAS isotope measurements reported by other authors cited above either employ a similar approach using adjacent absorption lines of comparable intensity [13,14,16], or a dual path cell with a long and a short path length in a length ratio approximately inverse to the ratio of the isotopes observed [15]. The reported precision for the $^{13}CO_2/^{12}CO_2$ measurements by Wong is 2.5 ‰, for the $N_2^{18}O/N_2^{16}O$ measurements by Wahlen 0.4 ‰. In neither case, however, is there a discussion about systematic errors, in particular concerning the temperature dependence of measurements, although all the studies cited used line pairs with differing ground state energies.

5. CONCLUSIONS

We have demonstrated the feasibility of $^{13}CH_4/^{12}CH_4$ ratio measurement by means of TDLAS. At present, in order to achieve a $\delta^{13}C$ accuracy of $\Delta\delta^{13}C \leq 1$ ‰, optical densities of greater than 5% are necessary corresponding to concentrations of 2.5 vol% CH_4 for the 36 cm long absorption cells used. This accuracy is poorer than for mass spectrometer measurements (0.04-0.3 ‰) [7,9,17,20,21], but nevertheless should be sufficient for most investigations of methane **sources** since their $\delta^{13}C$ variabilities (temporal, spatial, etc.) are typically much greater. Up to the present detailed methane source measurements have been limited by the substantial effort of sample preparation for mass spectrometer measurements. In contrast, the TDLAS measurements need 10-15 min for one sample, including the calibration procedure. The planned application of a multipath reflection cell (White cell) with an absorption length of approximately 200 m should decrease the current concentration limit of 2.5 vol% (using our present laser) by a factor of 20000/36 corresponding to ~50 ppm. This should be sufficient for most methane source measurements without sample pre-enrichment in the laboratory.

ACKNOWLEDGEMENTS

We are grateful to I. Levin for the permission to use her CH_4 sample preparation apparatus and to C. Junghans for her careful work performing the mass spectrometer analyses.

REFERENCES

[1] Aselmann, I., and P.J. Crutzen, Global distribution of natural freshwater wetlands and rice paddies, their net primary productivity, seasonality and possible methane emissions, *Journal of Atmospheric Chemistry*, 8, 307-358, 1989.

[2] Cicerone, R. J., and R. S. Oremland, Biogeochemical aspects of atmospheric methane, *Global Biogeochemical Cycles*, Vol. 2, No. 2, 299-327, 1988.

[3] Crutzen, P.J., Methane's sinks and sources, *Nature*, 350, 380-381, 1991.

[4] Ehhalt, D.J., How has the atmospheric concentration of CH_4 changed?, Report on the Dahlem Workshop on The Changing Atmosphere, 1988.

[5] Levin, I., D. Trapp, and P. Bergamaschi, Stable isotopic signature of methane from different sources in Western Europe, submitted to *Chemosphere*, 1991.

[6] Stevens, C.M., and A. Engelkemeir, Stable carbon isotopic composition of methane from some natural and anthropogenic sources, *J. Geophys. Res.*, 93 (D1), 725-733, 1988.

[7] Quay, P.D., S.L. King, J. Stutsman, D.O. Wilbur, L.P. Steele, I. Fung, R.H. Gammon, T.A. Brown, G.W. Farwell, P.M. Grootes, and F.H. Schmidt, Carbon isotopic composition of atmospheric CH_4: fossil and biomass burning source strenghts, *Global Biogeochemical Cycles*, Vol.5, No.1, 25-47, 1991.

[8] Whiticar, M.J., E. Faber, and M. Schoell, Biogenic methane formation in marine and freshwater environments: CO_2 reduction vs. actetate fermentation - isotope evidence, *Geochimica et Cosmochimica Acta*, 50, 693-709, 1986.

[9] Wahlen, M., N. Tanaka, R. Henry. B. Deck, J. Zeglen, J.S. Vogel, J. Southon, A. Shemesh, R. Fairbanks, and W. Broecker, Carbon-14 in methane sources and in atmospheric methane: The contribution from fossil carbon, *Science*, 245, 286-290, 1989.

[10] Craig, H., The geochemistry of the stable carbon isotopes, *Geochimica et Cosmochimica Acta*, 3, 53-92, 1953.

[11] Craig, H., Isotopic standards for carbon and oxygen and correction factors for mass-spectrometric analysis of carbon dioxide, *Geochimica et Cosmochimica Acta*, 12, 133-149, 1957.

[12] Lowe, D.C., C.A.M. Brenninkmeijer, S.C. Tyler, and E.J. Dlugkenoky, Determination of the isotopic composition of atmospheric methane and its application in the antarctic, *J. Geophys. Res.*, 96 (D8), 15455-15467, 1991.

[13] Wall, D.L., R.S. Eng, and A.W. Mantz, SP5100 Isotope ratio measurement system, Development of a tunable diode laser isotope ratio measurement system, Spectra-Physics laser Analytics Divisions, Bedford, MA 01730, Internal Report.

[14] Wong, W.W., Comparison of infrared and mass-spectrometric measurements of carbon-13/carbon-12 ratios, *Int. J. Appl. Radiat. Isot.*, Vol. 36, No. 12, 997-999, 1985.

[15] Lee, P.S., and R.F. Majkowski, High resolution infrared diode laser spectroscopy for isotope analyis - measurement of isotopic carbon monoxide, *Appl. Phys. Lett.*, 48(10), 619-621, 1986.

[16] Wahlen, M., and T. Yoshinari, Oxygen isotope ratios in N_2O from different environments, *Nature*, Vol. 313, No. 6005, 780-782, 1985.

[17] Bösinger, R., Isotope measurements of methane in the at-

mosphere and near sources (in German), Doctoral Thesis, University of Heidelberg, 1990.

[18] Pinkley, L.W., N. Rao, M. Dang-Nhu, G. Tarrago, and G. Poussigue, Forbidden lines of the ν_3 band of $^{13}CH_4$: Ground-state constants, *Journal of Molecular Spectroscopy*, 63, 402-444, 1976.

[19] Gray, D.L., A.G. Robiette, and A.S. Pine, Extended measurement and analysis of the ν_3 infrared band of methane, *Journal of Molecular Spectroscopy*, 77, 440-456, 1979.

[20] Stevens, C.M., and F.E. Rust, The carbon isotopic composition of atmospheric methane, *J. Geophysic. Res.*, 87 (C7), 4879-4882, 1982.

[21] Tyler, S.C., Stable carbon isotope ratios in atmospheric methane and some of its sources, *J. Geophysic. Res.*, 91 (D12), 13232-13238, 1986.

APPLICATION OF TUNABLE DIODE LASERS FOR HUMAN EXPIRATION DIAGNOSTICS

E.V.STEPANOV, I.I.ZASAVITSKII,
K.L.MOSKALENKO, A.I.NADEZHDINSKII
Institute of General Physics of the Academy
of Science of the USSR,

Summary

Results on the application of tunable diode lasers gas analysis to determining the trace components of human breath are presented. Schemes of the analyzers specially developed for measurements of both carbon oxides in expiration are described. A few results illuminating possible applications of TDL in medicine high sensitive diagnostics have been obtained. For nonsmoker persons, expired concentration of CO are slightly higher than that in inhaled air. Specific surplus value depends on the age of person. The surplus CO content proved to be increased by more than order of magnitude just after intensive physical exercises, e.g. jogging. For smokers farmacokinetical curve of abundant CO removal from organism could be investigated. The smoking status of tested person becomes easy available. Breath-hold simultaneous measurements of CO and CO_2 have shown the difference in the dependencies of their concentrations on breath-holding time. The possibility to investigate phenomena like molecular pulmonary diffusion of alveolar-capillary membrane and organism's compensation reactions to oxygen shortage seems to become real. Perspective leads for development and application of diode laser spectroscopy methods to analysis of gaseous microimpurities in medicine are discussed.

1. INTRODUCTION

Medicine diagnostic seems to be a prosperous field for future applications of tunable diode laser spectroscopy (TDLS). Due to several outstanding fiatures available when TDL are used for gas mixture analysis, common medicine diagnostic techniques in some cases may be forced out by more powerful, sensitive and reliable ones based on diode lasers application. As well as new advanced methods of diseases diagnostic could be developed on this basis in nearest future.

These attractive analytical advantages of TDLS are as

follows:

Diode-laser spectroscopy are extremely sensitive to small gas concentrations being measured and possess a wide dynamic detection range. Their sensitivity to many gases can be made as high as 0.1-1 ppb. This sensitivity is attained at lower power (less then 1 mW) of laser IR radiation and with the use of either open optical paths, of about 100 m, or multipass cells. In the latter case , the gas sample volume necessary for the analysis is sufficiently small (0.5-3 liters). At fixed measurement parameters, the dynamic range reaches 10^3 of magnitude and can be increased by choice of the optical path length and the analytical line.

The method is highly selective in measuring the gas mixture components specified in advance. It also allows simultaneous detection by one laser of different molecules or their isotopic modifications, whose absorption spectra lie in the same region. These TDLS prosperities owe their existence to the broad band piecewise-continuous frequency tuning in combination with high spectral resolution inherent in diode lasers. In TDL gas analysis, high concentration measurement accuracy and fast detection of gases are provided since the absorption line profile of the substance being measured is detected virtually undistorted during the time of about 1 microsecond. In routine measurements, the gas content is determined to better than 10 percent accuracy. This value can be made as good as tenths of a percent in special cases when needed. Most problems of gas analysis can be solved in real time (the operation time realized is down to 10 ms).

Finally among the advantages of the method are the versatility of the gas detection principle, i.e. its independence of the measured object (the main criterion of the method applicability is the presence of a sufficiently contrast ro-vibrational structure in the IR absorption spectra) and the possibility of almost complete automatization of measurement procedure owing to electronic control of TDL radiation parameters. Frequency scanning and adjustment are performed by varying the pump current parameters and the laser temperature.

Unfortunately there are some features hindering wider usage of the method. For one, middle-IR (4-46 microns) Pb-salt lasers operate at cryogenic temperatures of < 100K created by liquid cryogens (He, N_2) or by powerful cooling units and their achievable power in one lasing mode is low, <1 mW. The latter circumstance makes it necessary to cool photodetectors too to improve their sensitivity. Near-IR lasers (0.7-2.8 microns) are more powerful and can operate at room temperature and lower temperature maintained with the help of Peltier-elements. This spectral region, however, contains only overtones of ro-vibrational absorption bands with the intensity by the several orders of magnitude lower than the fundamentals which decreases the detection sensitivity. Another drawback of the method, as far as its applicability to gas analysis is concerned, is its relatively high cost and intricacy. Control of and data storage in TDL based systems should be computerized and attended by experienced operators.

Therefore, when designing and developing a special purpose TDL-based gas analyzer the result to be obtained should be compared with the expected cost of the project. Medicine, in our opinion, would offer some TDLS applications, where the results to be obtained might pay for the themselves. One of medical applications could be the trace components diagnosis of human breath.

The air breathed out by a man is a complex gas mixture whose composition and components contents may indicate the health status of the entire organism and of its separate systems. As is known, human expiration mainly contains N_2, O_2, CO_2, and H_2O. Besides, microquantities of NH_3, CO, H_2S, CH_4, NO_2, and other ingredients may also be present. Most of these gases, except for N_2, can be TDLS-detected in human breath in the amounts comparable with their concentrations in breathed air to an accuracy of several percents and at the operation time less than 0.1s. It's mainly due to the fact that TDLS methods for high sensitive detection of above mentioned molecules have been already developed in frames of atmosphere monitoring projects in different countries (1,2).

This paper dwells on our first result of the studies carried out in this field. The TDLS-based methods of gas analysis have been applied for detection of both carbon oxides, CO and CO_2, in human breath.

Carbon monoxide is known as quiet popular gas in conventional pulmanory tests. Besides, the information about CO removal processes are very important for medical treatment of CO-poisioning (3-6). As is known, the CO content in atmospheric air is about 10^{-5} percent or 0.1 ppm. It may reach several units of ppm in cites and industrial premises. Thus, the CO content in the air breathed out by a man living under real conditions can be determined with the help of an analyzer which would detect the CO content of about 1 ppm with dynamic range 2-3 orders of magnitude. In order to observe the breathing process dynamics, the analyzer short range resolution should be less than 0.1s.

Carbon dioxide is an active component of human methabolism and an important product of human life. So its concentration varies essentialy in breath. CO_2 concentration in the atmosphere air is on the level of 330 ppm (or about 0.04%).and reachs several percent (3%) in exhaled air. Carbon dioxide in breathed air caracterizes activity of organizm and playes an important role in compensation reacton regulation of body.

The first our data on CO trace concentrations measurements in exhaled air (presented in chapters 4 and 5) was obtained by very simple TDL gas analyser. These results have demonstrated us that valuable diagnostic information like pharmacokinetical curves for chemical substances removal from organism, smoking status are available with used method. Consequences of intensive lung ventilation and physical exercises may be also investigated by this technique. The main our aim on the next stage was to improve this TDL gas analyzer and to fit it specially to breath investigation experiments. Important attention was paid to the possibility of systematic investigations. Special fiatures of analyzer functioning and parameter achieved before and after these modifications are

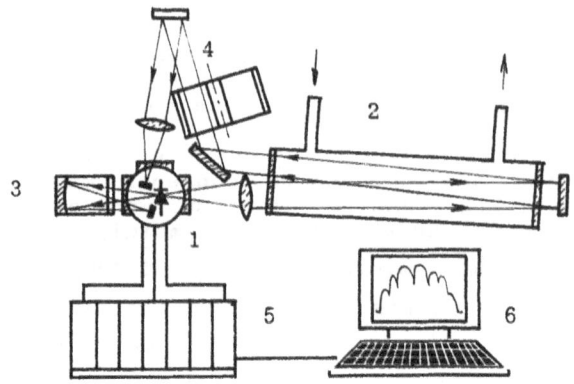

Fig.1. Diagram of TDL gas analyzer for CO recording in
expiration. 1, Nitrogen cryostat with diode laser and
photodetectors; 2, measuring cell; 3, TDL-radiation
stabilization cell; 4, calibration cell; 5, gas analyzer
electronic system; 6 IBM XT computer.

discussed below. Possibilities of the analyzer are demonstrated
by a set of results on CO and CO_2 concentration measurements in
exhalation of several volunteers.

2.APPARATUS
The experimental setup possessing the above-mentioned
properties and used by us in measuring the CO content in human
breath is in effect, an extremely sensitive tunable diode laser
spectrometer (Fig.1) based on the automated CO analyzer which
has been developed and applied earlier to atmospheric pollution
studies (7,8).
This first version of TDL medical analyzer was extremely
simple and was intended to demonstrate possibilityi to measure
trace components of expired air. The PbSSe diode laser which
operate at a wavelength around fundamental ro-vibrational CO
absorption band at 4.7 microns was placed in a nitrogen
cryostat and cooled to 77K. Two InSb photodetectors for laser
radiation detection in the analytical channel and in the
stabilization channel of TDL frequency sweep cycles were also
placed on the cryostat heat sink. In the analytical channel,
laser radiation escaping the cryostat was twice passed through
a 2-m long, 35mm aperture, 2-1 volume cell. The cell was
pre-evacuated and then filled with the analyzed gas sample to
the atmospheric pressure. The radiation past the cell was
directed back into the cryostat to one of the photodetectors.
In the stabilization channel, the radiation was passed through
a short cell containing the reference gas mixture and then also
returned to the cryostast's other photodetector. The cells for
the CO content magnitude calibration were inserted as needed
into the analytical channel.

Repetitive rectangular current pulses applied to the laser heated the laser crystal thus producing laser radiation with the frequency changing piecewise-continuously with time. The pulse repetition rate was about 100 Hz and pulse duration up to 100 ms. Absorption lines of the gas being analyzed were detected oweing to laser frequency sweep during the pulse. Recorded absorption in CO lines at used optical path and measured gas concentrations (from 1 ppm to 50 ppm) was quite weak, of about 0.01-1%. So the setup sensitivity to absorption was improved by filtering the photodetector electric signal, for example, with the help of a differentiating and integrating RC-networks cutting low and high frequencies. The CO content was measured by the resonanse peaks magnitude of differenciated absorption line profile. This value was then compared with ones obtained with calibration cells (the so called reference calibration procedure).

At the optical length used (4m), the sensitivity without the additional signal accumulation, i.e. at a speed of 10 ms, was about 0.2 ppm. It was only limited by the signal modulation due to radiation interference at the spectrometer optical components.

The state-of-the-art gas analyzer is controlled by the IBM XT personal computer. The recording and control system is constructed in the Euro-standard (9) and is a set of modules connected through an 8-bit data exchange unibus. The modules used are as follows:

-the interface module and the adapter circuit board for the IBM-compatible computer which ensures its random access to the data exchange unibus;

-the TDL temperature control module (control is in error by 0.003K);

-the laser pumping module for complete control of current pulse parameters;

-the programmable broad-band amplifier for signal analog preprocessing;

-the 8-bit, 50 ns analog-digital converter with buffer storage of 16 kB;

-the fast-acting 32-bit digital accumulator with storage of 16k*32 bit.

Algorithmic and programming support of the method ensures visualization of spectral and service information, clamp-on of the data management and record system, setup automatic calibration cells, and its automatic adjustment on the stabilization channel absorption line.

3.THE GAS ANALYZER IMPROVEMENT

The setup described above has been modified to improve its accuracy, reliability, and dynamic range in carrying out methodic experiments on the CO-contents analysis in human breath and on simultaneous measurements of bouth carbon oxides (CO and CO_2). For example, in the analytical channel, the simple cell was replaced with its multipass counterpart and the gas content refference calibration was replaced with the absolute one. The latter becomes possible using exact preliminary assignment of the analytical lines. This changes makes it possible to reach better setup parameters while

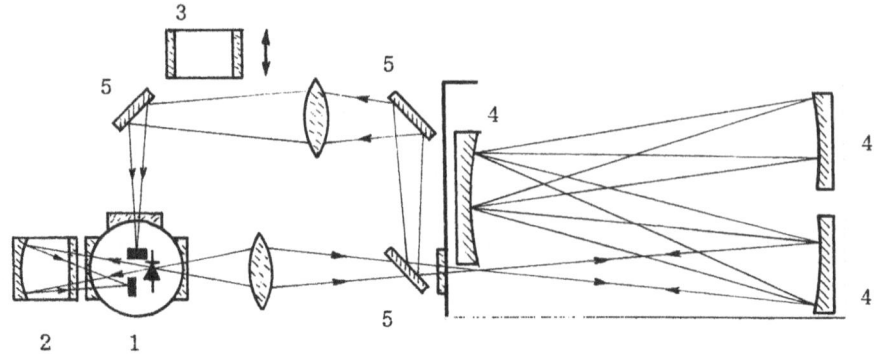

Fig.2. Optical lay-out of a gas analyzer with multipass cell.
1 - nitrogen cryostat for TDL and photodetectors;
2 - TDL-radiation stabilization cell; 3 - calibration
cell; 4 - cofocal spheric mirrors; 5, mirror plates.

retaining small air sample volume (not more than 400-500 cm^3
breathed out in one expiration act) and minimizing analyzer
size. This also simplifies the calibration procedure and
ensures its reliability. In what follows we will dwell on the
operation features and parameters of the thus modified
gas-analyzing spectrometer.

3.1.IMPROVED OPTIC SETUP
The modified optical scheme is shown on Fig.2. TDL
radiation leaving out liquid nitrogen cryostat is focused on an
inlet of the multipass cell aligned by Barskaya scheme.(10).
Then laser beam past through the cell and reflected by an
outlet mirror is focused on an IR detector plased in the
cryostate. If it was necessary additional reference gas cell
was put on laser beam to identify absorption spectra.
Optimization of numbers of laser beam passes through the
cell is necessary to achieve higher accuracy and sensitivity of
absorption line parameters when multipass cell is used. This
optimum is caused by two main reasons. The first one is a
raise of resonance absorption with passes number increase and
the opposite one is a loss of radiation intensity due to
dispersion of light on cell mirrors. Besides, interference
disturbances could affect analytical accuracy due to laser
beams overlapping when TDL is used (11).
An optimum number of passes follows from the Beer-Lambert
law. When only reflection losses are in consideration and
disturbances limiting sensitivity of the analyzer are
independent on light intensity, the optimum number of passes
(N) is given by:

$$N=(\varkappa pL_o)^{-1}\ln(1-\varkappa pL_o/\ln\ q) \qquad (1)$$

where: $\varkappa$ is line absorption coefficient, p is partial pressure of analyzed absorbent, L_o is base length of multipass cell, q is mirror reflection coefficient.

At this N value a ratio of absorption signal to noise reaches maximum. The relation can be used when relatively high gas concentrations are measured with high accuracy and optical density ($\varkappa pL_oN$) is close to unit. At microconcentration measurements when sensitivity is more important the relation (1) is modified to:

$$N=-1/\ln\ q \qquad (2)$$

If accuracy and sensitivity of absorption line detection depends on radiation intensity or passes number (for examples, due to interference) the optimum number may differ from (2).

In our experiment an electronic noise in detector circuit was the main reason determining analyzer sensitivity. At reflection coefficient of multipass cell mirrors of about 0.975 the optimum passes number for microconcentration detection (on the level of 0.1 ppm) is as much as 40. If the strongest lines of carbon monoxide with intensities of about 10 $cm^{-2}atm^{-1}$ are used and minimum detectable level of optical density is of the order of 10^{-4}, this optimum path allows to detect concentrations at ppb level with cell base length L=35 cm and gas volume of 500 cm. This sensitivity is higher than the value necessary for detection of some other gases like methane, ammonium et.al. which may also be present in human exhalation at ppb level. Besides, choice of convenient number of laser radiation passes through the cell permits to increase dynamic range of measurements up to 1.5 orders.

3.2.ABSOLUTE CALIBRATION OF CONCENTRATION

At this stage of measurements the determination of analyzed gas concentration was carried out directly by the Beer-Lambert law. A priory data on strength and broadening coefficients of detected absorption lines were involved to this calibration procedure. This procedure of observed data handling is considered to be an absolute calibration on the contrary to the refference calibration mentioned above. The absolute calibration eliminates influences of additional interference disturbances of signal occurring when reference cell are put in optical path as well as errors connected with degradation of reference mixtures of gases due to storage or external causes. To realize this way of calibration an exact assignment of analytical lines and its spectral parameters data are necessary. So the precision of gas concentration defenition depends only on errors of a priory data obtained usually from preliminary experiments or modeling calculations.

The assignment of analytical lines was carried out in our experiment using different isotopic modifications of carbon monoxide such as $^{12}C^{16}O$, $^{13}C^{16}O$, $^{12}C^{18}O$ and hot band absorption

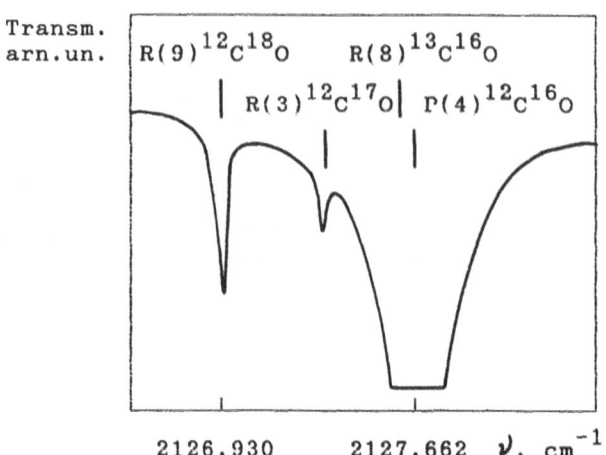

Transm.
arn.un.

R(9)^{12}C^{18}O   R(8)^{13}C^{16}O

| R(3)^{12}C^{17}O| P(4)^{12}C^{16}O

2126.930   2127.662 ν, cm^{-1}

Fig.3. Absorption spectrum of nature CO isotopic mixture at a
pressure of 30 Torr and cell length of 85 cm, which was
use for analytic line identification.

lines of CO_2. Quite simple and reliable assignment of
analytical line is possible in this case due to an isotope
shift of ro-vibrational band centers and differences in
rotational constants for various isotope modifications (12). As
consequence, a view of detected spectra of carbon monoxide
natural mixture in the region of 4.7 microns is very specific
which allows to determine spectrum region by reciprocal
positions and relative intensities of observed lines. For
initial frequency scaling of used spectral region with accuracy
of about 20 % the width of absorption lines broadened by air
could be used. Finally the accuracy of absolute frequency
scaling of identified spectral region is less than 0.001 cm^{-1}.

Figure 3 demonstrates a piece of the absorption spectrum
of carbon monoxide natural isotope mixture at 2127 cm^{-1} region
detected for analyzer calibration. The assignment of observed
lines is also presented. The spectrum has been recorded at
partial pressure of about 30 Torr and optical path length of 85
cm. Lines assignment may be carried out in analyzer
stabilization channel, if necessary, to release the analytical
channel from additional disturbances.

After spectrum assignment measured concentration may be
calculated directly using the meanings of analytical line
absorption coefficient $\varkappa$ from spectral data bases. Its value at
the center of line broadened by buffer gases was taken as
usualy in the form:

$$\varkappa = S/\ \pi\gamma \qquad (3)$$

here: S is a strength of analytical line;
γ is the width of the line determined by $\gamma = \odot\gamma_i p_i$, where
p_i- is partial pressure of gases in investigated mixture

and i- is line broadening coefficient for i-th buffer gas.

In our investigation the strength of analytical line was determined from model calculations of fundamental band 1-0 of CO. A band strength value of 283 cm^{-2}atm^{-1} and a calculation model were taken from Ref.13. Collisional broadening coefficients of CO absorption lines for main atmospheric constituents N_2, O_2, CO_2, H_2O published in Ref.13-15 were used.

Special attention was paid to several other sources affected accuracy and reliability of obtained data. Temperature variations of gas mixture in the cell were measured and taken into account when absorption coefficient was calculated. Influence of expired water vapor on final data have been investigated carefully. Due to high selectivity of TDLS the presented technique proved to be practically insensitive to the presence of water vapor in measured gas mixture . However it is preferable, for the sake of fast operation, the breathed air should pass through the drying system in order to decrease the time of multipass cell evacuation.

The influence of correct optical zero determination on accuracy of observed data was investigated. It's necessary to eliminate errors due to TDL superluminescence when greating is not used for spectral selection. Usually, full absorption of IR laser radiation in narrow spectral region (e.g. saturated absorption line) allows us to determine the zero level of optical signal. Usage of longer optical paths instead of higher partial pressures of gases involved was found to be a more correct procedure in this case.

4.THE SMOKER'S BREATH ANALYSIS

The data presented in this chapter, as well as in the next one, have been obtained on the first simple version of breath diagnostic TDL analyzer. We began our study with the CO-content analysis in the air breathed out by a smoker. It could be easily predicted that in this case the CO-content exceeds that in a nonsmoker's breath. Figure 4 illustrates the experimental results.

The person being tested, who smoked regularly about one cigarette per hour, breathed out air into the buffer volume and then the sample gas was transfered into the measurement cell. The first experimental point in Fig.4 was obtained prior to his smoking the cigarette. It corresponds to the CO content of about 20 ppm which is approximately two-fold daily average safety limit for in the outdoor atmosphere air (dash-dot straight line) and approximately 20-fold the CO content in the laboratory premises (lower dashed straight line). The lower dashed straight line indicates the CO content in the nonsmoker's expiration virtually equal to that in the air inhaled by him.

The next measurement was performed during the process immediately after inhalation. The CO content increased to its peak value of about 50 ppm. The rest points were obtained after the cigarette had been smoked and the smoker abstained from smoking another one. As is seen, the moment the smoking process terminates, the CO content abruptly falls off down to 30 ppm due to lungs ventilation in breathing. Then the CO content

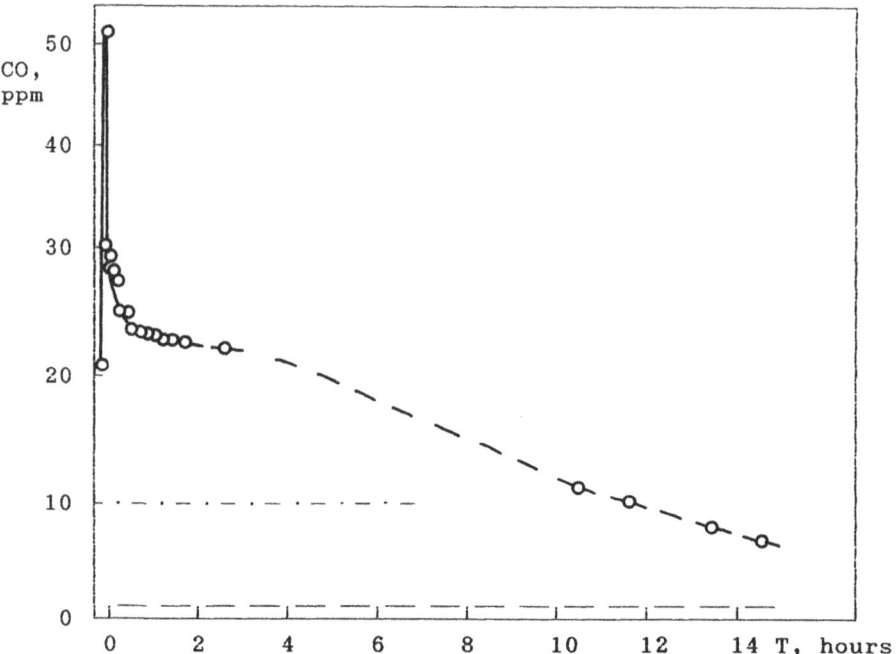

Fig.4. CO content time variation in smoker's expiration.
— — — CO content in atmosphere; — · — · — average daily
safety limit of CO content in atmosphere.

exhibits slow exponential decrease with the time constant $t_1 = 40$
min. In about two hours, the expiration CO content approaches
the constant of about 20 ppm close approximating its magnitude
before smoking where upon the smoker did not smoke and the
excess CO content in his expiration continued to fall off with
a greater time constant $t = 700$ min. Figure 4 implies that only
in two days of abstentions the excess CO content reduces to
zero.

The result obtained may be interpreted as follows. Besides
very fast process (with the time constant of about 1s) of gas
exchange in the lungs due to ventilation, there are two added
processes of CO molecular exchange between human organism and
the atmosphere. The first process is likely to be responsible
for CO molecule removal from blood circulating in blood vessels
and is connected with the substitution of O_2 for the CO
molecules in erythrocytes. The second, slower process consists
in CO removal from the organism's tissues (fats, muscles)
supplied with CO enriched blood due to regular smoking. In this
case, erythrocytes which contain haemoglobin transmit CO from
tissues to blood where it is replaced with oxygen. This process
is reversible, therefore, the blood CO enrichment in the
smoker's lungs leads to its transfer to his tissues. The
CO-content dynamic equilibrium is thus maintained in the
regular smoker's body and can be characterized by its
expiration value at the instant the process occurring with the

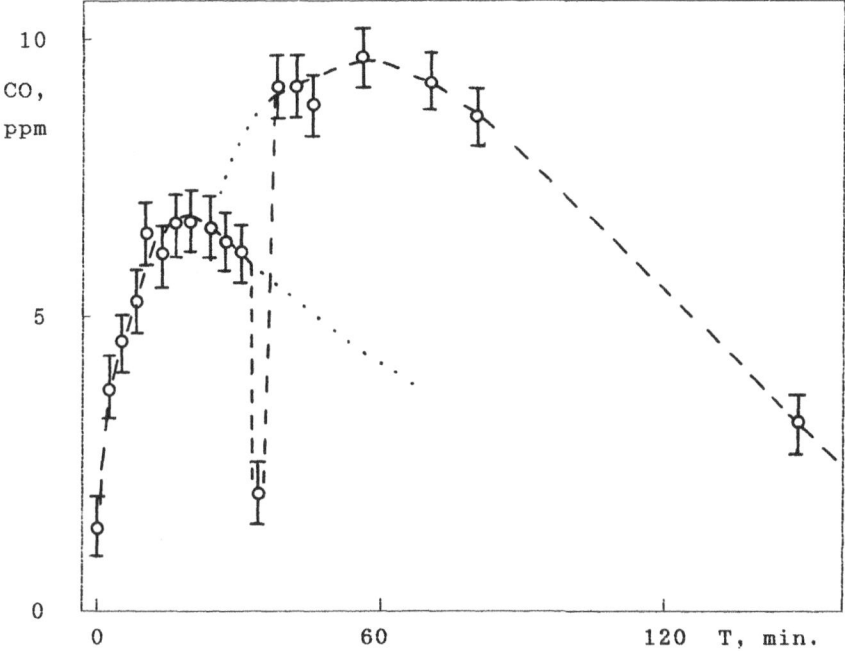

Fig.5. CO content time variation in sporsman's expiration.

time constant t_1 goes over to the process with time constant t_2. This level could indicate a smoking-status of patient. In effect, the dependence obtained is a so named pharmocokinetic curve characterizing CO removal from the organism.

5.SPORTSMAN'S BREATH ANALYSIS

The second experiment was concerned with the CO-content analysis in a sportsman's breath after intense physical load (Fig.5). Initially, the person being tested ran for 5 min at a considerable speed till his beats was at 180. The CO content in his expiration prior to and immediately after running did not differ within the experiment accuracy and was approximately equal to the CO content in the atmospheric air or about 1 ppm (the first two points in the curve). At the running termination, the CO content in the sportsman's expiration increased sharply and, in 15 min after measurement initiation, maximally exceeded its atmospheric content by 6 ppm which evidences active CO removal from the sportsman's body at rest. At the 20-th minute of the experiment, the sportsman started running again, when he stopped, the CO content in his breath was equal to that in the atmosphere. This can be attributed to active lungs ventilation in running. It termination again caused the breathed CO content to grow sharply. By the 50-th minute the CO content had reached its second maximum 10 ppm (the safety limit level). Then it fell off slowly and reached its atmospheric level in about two hours. It should be noted that at the instants corresponding to the CO-content maxima the sportsman felt some giddiness and muscle tremor which is typical not only of physical tiredness but of CO-poisoning as well.

Thus, we observed that the human organism under physical load is a peculiar CO generator. This phenomenon may be explained as follows. In running there occurs intensive energy consumption by the organism provided by oxidation of amino acids, fats, proteins, and carbohydrates. Carbon dioxide known to be a product of these reactions is removed from the organism via the blood circulation system and respiration. As the physical load is removed, man's breathing recovers and becomes less intense, the molecule decomposition persists, however, for some time. The organism's are then insufficiently supplied with oxygen (due to a decrease in the breathing rate) and carbon monoxide molecules may appear which are carried away by means of erythrocytes from tissues to lungs where they are finally removed.

6. STATISTICS FOR NON SMOKERS

The investigations presented in the next two chapters were carried out as a continuation of previous measurements but with use of modifyed analyzer, Fig.2. The analyzer described in the chapter 3 have been used to provide series of regular and long-time measurements of CO concentrations in expiration of several volunteers and to investigate the dependence of CO and CO_2 concentrations in exhalation on breath holding time.

Aims of the former experiment have consisted in accumulation of statistic data, in ascertaining of general dependences and representativity of obtained data. Small group of TDLS laboratory stuffs participated in the measurements have consisted of 11 persons aged from 21 to 40 years old. This group involved smokers and nonsmokers, as well as men and women.

At first, air from one exhalation was collected on buffer volume of ~1 liter and then was put on the preliminary evacuated multipass cell volume. As a rule, delay time between inhalation and exhalation did not exceed one second. Side by side with breath investigation, content of carbon monoxide in laboratory air was detected.

Obtained data show that only investigations of nonsmoking persons provide the most representative results for diagnostic purpose. Wide dispersion of CO concentrations in exhalation of smokers is due to its strong dependence on quality of cigarettes, frequency of smoking and time after the last smoking. The similar results have been presented early (17).

CO content in nonsmoker exhalation is more determined. Fig.6a shows variations of this value on time for three persons. Variation of CO concentration in atmosphere at the same time is also presented here (lower curve). CO concentration in exhalation correlates strongly with atmosphere one, Fig.6b, and exceeds atmosphere level on 0.4-0.7 ppm for all men. The specific surplus CO content seems to be dependent on the age of tested person and in any cases could caracterize the red blood cell formation rate (haemopoiesis) in organism. These additional values may be connected with abundant carbon

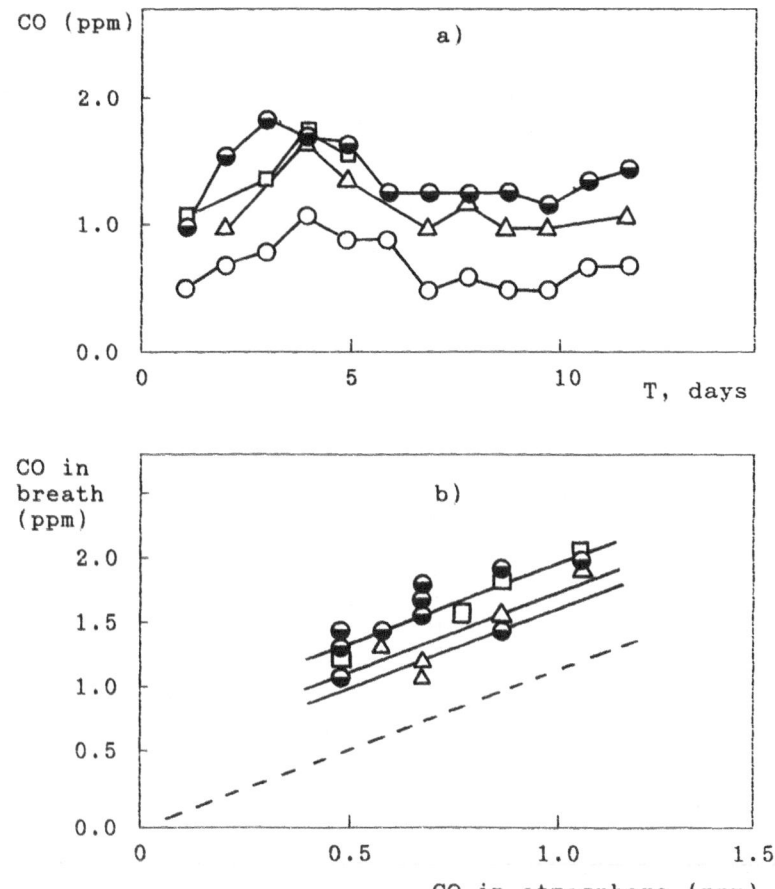

Fig.6. Days variations of CO concentration in expiration of tested persons and in the atmosphere, a), and the correlation of these values, b)

monoxide removal from organism.

7.SIMULTANEOUS DETECTION OF CO AND CO_2 IN EXHALED AIR.

Identification of the spectral regions used shows that absorption lines of CO_2 and H_2O were sometimes detected simultaneously with lines of CO. For example, Fig.7 shows a piece of absorption spectrum at 2115 cm^{-1} measured in breath experiments and assignment of the detected lines. The spectrum has been obtained at 16 meters optical path length and atmosphere pressure of investigated gas mixture. As seen from Fig.7, CO_2 absorption lines belonging to hot combination ro-vibrational band 20^00-01^10 with center at 2129.756 cm^{-1} have been detected simultaneously with CO absorption lines from P-branch of 1-0 band. The most strong lines of this CO_2 band

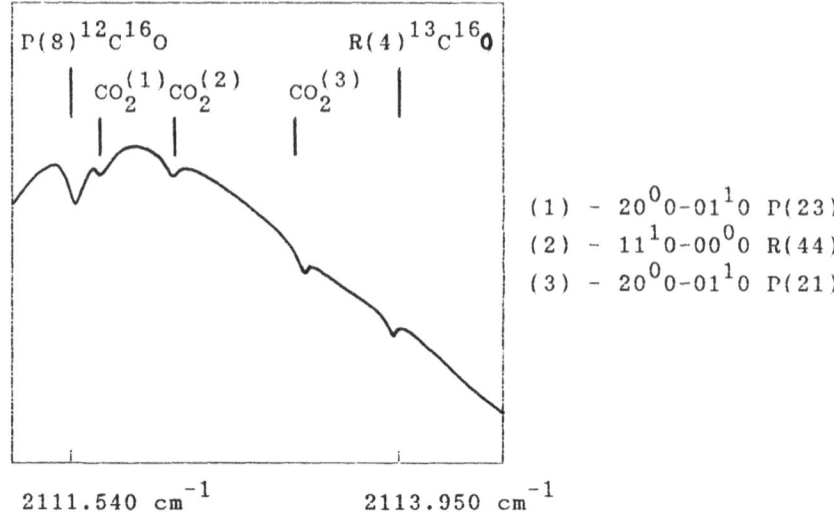

$$P(8)^{12}C^{16}O \qquad R(4)^{13}C^{16}O$$

$$CO_2^{(1)} CO_2^{(2)} \qquad CO_2^{(3)}$$

$(1) - 20^00-01^10 \; P(23)$
$(2) - 11^10-00^00 \; R(44)$
$(3) - 20^00-01^10 \; P(21)$

2111.540 cm^{-1} 2113.950 cm^{-1}

Fig.7. The analytical absorption spectrum at atmospheric
pressure and optical path length through the multipass
cell of 16 meters which was used for simultaneous
registration of caron oxides in exhaled air

are located in the vicinity of the $P(5)-P(11)$ transitions of
1-0 CO band. The $P(13)-P(21)$ absorption lines of CO overlap
with another CO_2 hot combination band (12^20-01^10) with the
center at 2043.345 cm^{-1}. This spectral region is also very
convenient for simultaneous detection of CO and CO_2. The
strengths of hot combination bands of CO_2 are approximately of
4 orders less than the strength of fundamental ro-vibrational
band 1-0 of CO. The distinction in line strength could
compensate an essential difference between concentration levels
of CO and CO_2 in breathed air which also reaches about 4
orders. Thus concentration measurements of both gases in
exhaled air could be provided using the same dynamic range of
the analyzer. On contrast to CO detection the reference
calibration of concentration was realized for CO_2 measurements
in our experiments.
     Simultaneous detection of carbon monoxide and carbon
dioxide was carried out when dependence of exhalation content
on breath holding time was investigated. In the experiment
nonsmoking persons varied the time delay between inhalation and
exhalation from 0 to 60 seconds. Every next exhalation test was
provided in 3-5 minutes after the last one to recover breath
and beats rate. One of the most typical result obtained for 33
years old man is presented in Fig.8. This figure shows the
difference in dependences of CO and CO_2 concentrations on
breath holding time. CO_2 concentration increases from 3% at
zero holding time till to about 8% at 60 seconds delay

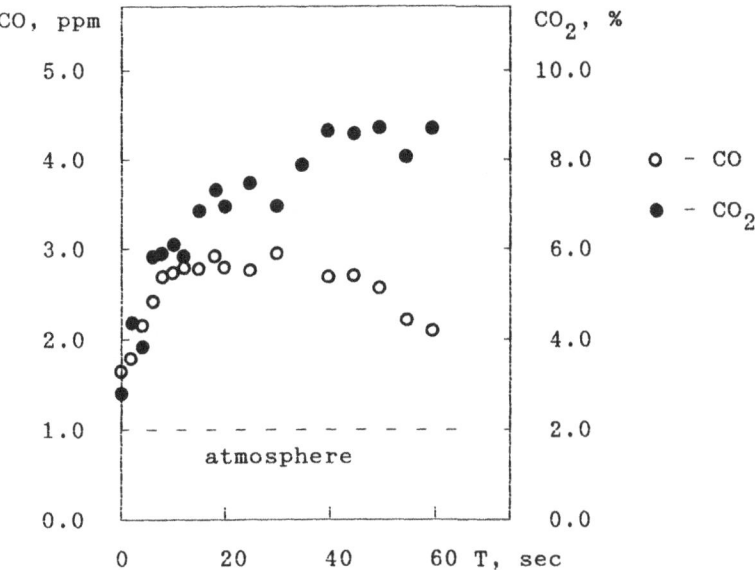

Fig.8. Behaviour of the CO and CO_2 concentration in exhaled air as functions of breath-holding time obtained in simultaneous measurements by TDL analyzer.

monotonously. The concentration of CO demonstrates more complicated behavior. At times from 0 to 10 second, nearly linear increase is observed. For this time CO concentration in exhalation increases twice as compared with initial level. Then the concentration holds at the same level of ~2.8 ppm till to 30-35 seconds of breath holding and decreases down to 3/4 of maximum level after 35-th second. The same dependences were also observed for several other involved persons. The curves obtained for different people vary in slopes and extreme values of the measured concentrations.

The presented dependences show the difference in increase rate of CO and CO_2 concentrations in exhalation with holding time. Since the CO and CO_2 removal is due to their blood partial pressures excess on the partial pressures in lung air, the distinction of the rate seems to be caused by different pulmonary diffusing capacity of lung alveolar-capillary membrane for the gases. Special attention should be paid to decrease of CO content in exhalation at longer breath-holding times. This decrease shows reverse motion of CO from lung air to blood. The reverse motion is caused by a partial pressure drop for carbon monoxide solved in blood. In turn the change in blood may be caused by erythrocytes content increase which is known to be one of the most fast compensation reactions starting in organism at hypoxia (acute oxygen shortage). This oxygen shortage is provided by three times increase of exhaled CO_2 concentrations detected at simultaneous measurements of CO and CO_2, Fig.8. Additional erythrocytes appearing in blood and aiming to increase an efficiency of oxygen transport, absorb a part of carbon monoxide solved in blood and reduce its partial pressure which originates revers motion of CO from lung to blood. Finally this causes the observed dependence of CO content on breath-holding time.

8.CONCLUSION.
The above presented results demonstrate possibilities and advantages arising when TDLS methods are used for medical diagnostics. They can successfully be applied to investigation of metabolic processes occurring in human organism, gas exchange and respiration peculiarities, especially in oxygen debt. The method sensitivity and speed make it possible to study fine short-run processes which cannot be observed by conventional methods of gas analysis now applied in medicine. The TDLS may appear to be an alternative method in diagnostic of various diseases.
The results presented above show that farmacokinetical dependence for CO and other similar gases and smoking-status of smokers may be obtained easily. Consequences of physical exercises (jogging, etc.) on exhaled gas content also may be tested. Data presented in this article gives information on correlation between CO concentrations in exhaled air and in atmosphere. This correlation indicates an excess of CO concentration in organism as compared with clean atmosphere. Also the possibility of simultaneous measurements of CO and CO_2 content in expiration is demonstrated. The diagnostic provides data on pulmonary diffusing capacity as well as on hypoxia compensation reactions of organism. Now conventional breath tests use for inhalation gas mixtures with excess CO concentration. TDLS diagnostics free from this disadvantage and so gives really noninvasive method.
The technique may also be very useful in conventional human breath diagnostics developing now in the following directions: 1) screening breath tests; 2) detection of CO poisoning sources and their compensation methods, smoking-status analysis; 3) determination of pulmonary diffusing capacity; 4) CO exhaled concentration research in connection with carboxihaemoglobine content in blood. Some conventional methods of breath diagnostic like breath-holding may be very useful in combination with TDLS analysis.
Future development of TDLS based medical diagnostic is connected with realization of gas analysis for other actual molecules such as NH_3, CH_4 et.al. and of multicomponent detection. Fiber-optical design of TDL sensors is also very hopeful for medical applications.

REFERENCES

(1) "Monitoring of gaseous pollutants by tunable diode laser" Proc. of the Int. symposium, held in Freiburg, FRG, 11-13 November 1986, Edited by GRISSAR, R., PREIER, H., SHMIDTKE G., RESTELLI, G. 176 p. D.Reided Publishing Company,

Dordrecht, Holland, (1987).

(2) "Monitoring of gaseous pollutants by tunable diode laser" Proc. of the Int. symposium, held in Freiburg, FRG, 17-18 October 1986, Edited by GRISSAR, R., SHMIDTKE G., TACKE, M., RESTELLI, G. 305 p., Kluwer Academic Publishers, Dordrecht, Holland, (1989).

(3) WANG,J.N., CAO,S.R., LI,Z., ZHANG,Y., LI,S.M. (1988) "Human exposure to carbon monoxide and inhalable particulate in Beijing, China", Biomed-Environ-Sci., vol.1, no.1, pp.5-12.

(4) KLESGES,R.C., BROWN,K., PASCALE,R.W., MURPHY,M., WILLIAMS,E., CIGRANG,J.A. (1988) "Factors associated with participation, attrition, and outcome in a smoking cessation program at the workplace", Health-Psychol., vol.7, no.6, pp.575-589.

(5) BELL,M., TATE,L., FOWLER,D.L.(1989) "Narcolepsy mimicking suicidal carbon monoxide poisoning", Am-J-Forensic-Med-Pathol., vol.10, no.3, pp.226-228.

(6) HALPERN-JS,J.S. (1989) "Chronic occult carbon monoxide poisoning", JEN, vol.15, no.2(Pt 1), pp.107-111.

(7) ANGELOVA,M., DONCHEV,A., ZASAVITSKII,I.I., KOSICHKIN,YU.V., KRYSTEV,T., KUZNETSOV,A.I., PEROV,A.N., PENCHEV,S.P., STEPANOV,E.V., FILIPPOV,A.N., TSANEV,V., SHOTOV,A.P. (1986) An automated open path gas analyzer, based on tunable diode laser and its application in the monitoring of atmosphere pollutions. Sov.Physical-Lebedev Institute Reports No10:54-58

(8) BLOKH,M.A., KNJAZEV,V.N., KOSICHKIN,YU.V., KVASHNIN,A.A., KUZNETSOV,A.I., STEPANOV,E.V., SHVETS,E.V. (1988) Automated diode laser based gas analyzer. Proc. 20-th Soviet Congress on Spectroscopy 416 (in Russian)

(9) KUZNETSOV,A.I., NADEZHDINSKII, A.I., STEPANOV,E.V. (1990) "Computerized infrared fiber-optic system for gas analysis based on diode lasers", Proc. SPIE "Infrared Fiber Optics II", Vol.1228, pp.262-265.

(10) BARSKAYA,E.G. (1968) "Multipass optical cell", Patent of the USSR, #206857, Published in 1968, Moscow, USSR

(11) OGRAM,G.L., NORTHRUP,F.J., EDWARDS,G.C., (1988) Journal of Atmospoheric and Oceanic Technology, Vol.5, pp.521-527.

(12) GUELACHVILI,G., (1979) "Absolute wavenumbers and molecular constants of the fundamental bands $^{12}C^{16}O$, $^{12}C^{17}O$, $^{12}C^{18}O$, $^{3}C^{16}O$, $^{13}C^{17}O$, $^{13}C^{18}O$ and of the 2-1 bands of $^{12}C^{16}O$ and $^{13}C^{16}O$, around 5mm, by Fourier Spectroscopy under Vacuum", JMS, vol.75, pp.251-269.

(13) VARGHESE,P.L., HANSON,R.K. (1980) "Tunable infrared diode laser measurements of line strengths and collision width of $^{12}C^{16}O$ at room temperature", JQSRT, vol.24, pp.479-489.

(14) VARGHESE,P.L., HANSON,R.K. (1981) "Room temperature measurements of collision width of CO lines broadened by H_2O", JMS, vol.88, pp.234-235.

(15) HARTMANN,J.M., ROSENMANN,L., PERRET,M.Y., TAINE,J. (1988) "Accurate calculated tabulations of CO line broadening by H_2O, N_2, O_2, and CO_2 in the 200-3000-K temperature",

      Appl.Opt., vol.27, no.15, pp.3063-3065.
(16) ROTMAN,L.S., YOUNG,L.D.G. (1981) "Infrared energy levels of carbon dioxide,II", JQSRT, vol.25, pp.505-524.
(17) ZACNY,J.P., STITZER,M.L. (1988) "Cigarette brand-switching: effects on smoke exposure and smoking behavior", J-Pharmacol-Exp-Ther., vol.246, no.2, pp.619-627.

AUTHOR INDEX

Agne, M. 93
Anselm, N. 231
Armerding, W. 283

Baranov, A.N. 79
Bauer, G. 150
Bauerecker, S. 291
Becker, K.H. 13
Bergamaschi, P. 353
Berger, J. 241
Bessey, E. 319
Blunier, S. 147
Bomse, D.S. 31
Böttner, H. 63, 69, 139, 151
Brockmann, K.J. 13
Buckhard, H. 183
Buhleier, R. 140
Burrows, J.P. 183

Cammenga, H.K. 291
Cantow, H.-I. 151
Cappellani, F. 329
Comes, F.J. 283
Crété, D.G. 149

Daddato, R. 85
De Andelis, M. 257
Di Lonardo, G. 275
Dietrich, S. 41
Dornisch, D. 303
Drummond, J.R. 03

Eberius, H. 311
Elsaesser, T. 140
Ernst, K. 275

Fach, A. 63, 69
Feit, Z. 105
Fink, T. 217
Fox, J. 03
Fried, A. 03

Giesen, Th. 231
Grisar, R. 303, 319

Halford, B. 69
Harris, G.W. 183, 353
Harter, M. 231

Heiner, Y. 249
Henry, B. 03
Herbert, A. 283
Hermann, K.H. 139
Herrmann, F. 41
Herrmann, Ka. 241, 249
Hindelang, F.J. 325
Hjorth, J. 329
Hoshino, T. 147
Hovde, D.C. 31
Hüdepohl, R. 13

Imenkov, A.N. 79
Inguscio, M. 257, 275
Iwaschkina, D.A. 146

Johnson, T.J. 183
Just, Th. 311

Karecki, D.R. 21
Kästle, R. 303
Klann, R. 140
Klocke, U. 319
Knothe, M. 319
Koga, R. 51
Kostyk, D. 105
Kowalczyk, L. 145
Kronfeldt, H.-D. 241
Kühnemann, F. 41, 241, 249
Kurbel, R. 69
Kurtenbach, R. 13
Kuznetsov, A.I. 203

Lambrecht, A. 69, 85, 93, 139, 140
Leidig, K. 141
Lemechov, N.V. 249
Lemke, F. 111
Leute, V. 115

Mackay, G.I. 21
Magonov, S.N. 151
Maissen, C. 147, 148
Mak, P. 105
Marin, F. 257, 275
Masek, J. 147
Mathet, V. 149
Menge, D. 115
Michaelis, W. 291

Mikhailova, M.P. 79
Mocker, M. 111
Möllmann, K.-P. 139
Moser, L.K. 325
Moskalenko, K.L. 203, 353
Mücke, R., 169
Mülberg, M. 144
Mürtz, M. 191

Nadezhdinskii, A.I. 155, 203, 353
Nadler, S.D. 21
Nguyen-van-Dau, F. 149

Oh, D.B. 31
Orphal, J. 41, 241
Overhamm, M. 311

Pavone, F.S. 257, 275
Pelz, N. 319
Petri, M. 217
Pisano, J.T. 21
Post, R. 144
Pustogov, V.V. 41, 241

Restelli, G. 329
Riedel, W.J. 319
Rooth, R.A. 265

Sams, R. 03
Schaefer, M. 191
Schelb, S. 151
Schieder, R. 231
Schiessl, U. 85, 93, 191
Schiff, H.I. 21
Schneider, M. 191
Scholz, C. 303
Schottka, P. 319
Schupp, M. 353
Shotov, A.P. 125
Siche, D. 146
Silver, J.A. 31
Slemr, F., 169
Spiekermann, M. 283
Spilker, G. 85
Springholz, G. 150
Stanton, A.C. 31
Stepanov, E.V. 203, 249, 353
Stocker, W. 151
Story, T. 143
Sultan, A. 148
Sumpf, B. 41, 241, 249

Szczerbakow, A. 142

Tacke, M. 63, 69, 85, 93, 139, 151, 191, 303
Taucher, F. 291
Teodoropol, S. 148
Tino, G.M. 257
Tomm, J.W. 139

Urban, W. 191, 217

Walter, J. 283
Weitkamp, C. 291
Wells, J.S. 191
Werle, P., 169
Wienhold, F.G. 183
Wiesen, P. 13
Winnewisser, G. 231
Wolf, H. 319
Woods, R.J. 105

Yakovlev, Yu.P. 79

Zasavitskii, I.I. 353
Zogg, H. 147, 148

The manufacturer's authorised representative in the EU is Springer
Nature Customer Service Centre GmbH, Europaplatz 3, 69115 Heidelberg,
Germany. If you have any concerns regarding our products, please
contact ProductSafety@springernature.com

Printed and bound by CPI Group (UK) Ltd, Croydon, CR0 4YY

23/04/2026

02095629-0005